Helen R. Irving • Christoph Gehring
Editors

Plant Signaling Peptides

Springer

Editors
Helen R. Irving
Monash Institute of Pharmaceutical Sciences
Monash University
Parkville
Australia

Christoph Gehring
Division of Chemistry
Life Science and Engineering
4700 King Abdullah University of Science and Technology
Thuwal
Kingdom of Saudi Arabia

ISSN 1867-9048 ISSN 1867-9056 (electronic)
ISBN 978-3-642-27602-6 ISBN 978-3-642-27603-3 (eBook)
DOI 10.1007/978-3-642-27603-3
Springer Heidelberg New York Dordrecht London

Library of Congress Control Number: 2012940197

Printed on acid-free paper

Springer is part of Springer Science+Business Media (www.springer.com)

Preface

Proteins and peptides form integral parts of all living cells where their function is essential for the survival of the cell and organism. Proteins are dynamic biomolecules that function in maintaining structure, metabolism, and also cellular communication. Peptides are loosely defined as small proteins containing 50 amino acids or less. In plants, as in animals and other organisms, peptides have diverse roles and participate in communication between cells. The focus of this volume is on the diverse roles that peptides and small proteins have in intercellular and intracellular communication in plants. In part because of their immobile nature, plants have evolved a complex array of signaling molecules to facilitate their growth and development and their interactions with the environment. A vast number of different peptide molecules make an important, but until relatively recently overlooked, component among these signaling molecules. As discussed in chapter "Plant Peptide Signaling: An Evolutionary Adaptation," plant signaling peptides have evolved in several independent events with distinct and separate phylogenies to create a diverse repertoire of signaling molecules.

This volume focuses on the roles of various peptide signaling molecules in plant growth, development, defense, and homeostasis. The roles of plant peptides in growth and development are discussed in chapters "Peptides Regulating Apical Meristem Development," "Peptides Regulating Root Growth," "Peptides Regulating Plant Vascular Development," and "The S-LOCUS CYSTEINE RICH PROTEIN (SCR): A Small Peptide with a High Impact on the Evolution of Flowering Plants." Chapter "Peptides Regulating Apical Meristem Development" reviews the well-understood role of peptide signaling in the shoot apical meristem of the model plant *Arabidopsis* that in turn has led to the discovery of related peptides in other plants. In fact, the major peptide protagonist in *Arabidopsis* CLAVATA 3 (CLV3) and its maize homolog EMBRYO SURROUNDING REGION (ESR) contributed to the naming of one of the largest signaling peptide groups, the CLE peptides. CLE peptides are involved in regulating organogenesis and have roles in the root growth and development which is discussed in chapter "Peptides Regulating Root

Growth." Members of the CLE peptides are also involved in regulating the development of the vascular cambium which is reviewed in chapter "Peptides Regulating Plant Vascular Development." In addition, members of the CLE family are co-opted by legumes during the symbiosis between legumes and rhizobia bacteria, as described in chapter "The Role of Plant Peptides in Symbiotic Interactions." However, the story of peptide signaling is not restricted to one family of peptides and certainly in development is integrated with signals from other plant growth regulators. Other peptides that contribute to organogenesis and the maintenance of stem cells include phytosulfokines (PSKs), ENOD40, rapid alkalinization factors (RALFs), and the recently discovered root growth factor (RGF). Specific and novel peptides are involved in various developmental processes. The family of S-LOCUS CYSTEINE RICH PROTEINs (SCRs) has an important role as a determinant of self-incompatibility in members of the *Brassicaceae* which is discussed in chapter "The S-LOCUS CYSTEINE RICH PROTEIN (SCR): A Small Peptide with a High Impact on the Evolution of Flowering Plants." While another recently characterized small family of peptides called EPIDERMAL PATTERNING FACTORs (EPFs) are also cysteine-rich peptides that have a role in regulating stomatal development, as reviewed in chapter "Peptides Modulating Development of Specialized Cells."

Signaling peptides also function in a wide range of plant defense responses. In fact, the first signaling peptide to be discovered and characterized was systemin which induces synthesis of proteinase inhibitors in leaves as a wound response. Since then, a myriad of plant defense proteins with diverse structures have been identified. Many of these are antimicrobial proteins and include defensins, thionins, and knottin-like peptides, as described in chapter "Plant Antimicrobial Peptides." Other signaling peptides function as endogenous amplification signals of plant innate immune responses as part of the pattern and/or microbe-associated molecular pattern (PAMP/MAMP) response which is discussed in chapter "Peptides as Danger Signals: MAMPs and DAMPs." The signal exchange initiated by rhizobia employs and/or co-opts several plant signaling peptides in the host legume, as reviewed in chapter "The Role of Plant Peptides in Symbiotic Interactions." Plant signaling peptides also have roles in maintaining overall plant homeostasis in addition to organogenesis and development, and in chapter "Peptides and the Regulation of Plant Homeostasis," the role of the small protein plant natriuretic peptide is described.

As it is highly likely that to date, only a few of the signaling peptides are known, and further plant peptide signaling molecules remain to be discovered, the last section of this volume takes a practical look at methods to identify new peptides and characterize their function. Signaling peptides usually contain an N-terminal signal motif and are secreted into the extracellular matrix (apoplast) where, in some cases, they are proteolytically cleaved. Peptides such as PSK and RGF are also sulfated on tyrosine residues, and some CLE peptides are hydroxylated on proline residues before secretion. The processing of peptides is described in chapter "Processing of Peptides" along with strategies for investigating these processes. In chapter "Methods to Isolate and Identify New Plant Signaling Peptides," the principles and methods for peptide purification are discussed. For signaling peptides to be successful as communicators in the plant, specific partners are required, and genetic and biochemical approaches to identify these partners are described in chapter "Methods to Identify New Partners of Plant Signaling Peptides." Finally, a computational approach is outlined in chapter "Computational-Based Analysis to Associate the Function of Plant Signaling Peptides with Distinct Biological Processes" where proteins co-expressed with PROPEP2 are identified.

May 2012
Parkville, Australia — Helen R. Irving
Thuwal, Kingdom of Saudi Arabia — Chris Gehring

Contents

Plant Peptide Signaling: An Evolutionary Adaptation

Janet I. Wheeler and Helen R. Irving

Abstract Peptide signaling molecules are well characterized in animal systems, but it is only over the last three decades that they have been recognized in plants. In this chapter, we compare some of the major features of animal peptide signaling molecules with the new classes that have been identified in plants. We introduce the concept of modular signaling and discuss how this adaptable feature can be evolutionarily advantageous to multicellular organisms. Most signaling peptides have been identified in angiosperms (both monocot and dicot) although representative signaling peptides occur in moss and green algae. Some classes contain peptides with highly diverse sequences (within and across species) while other peptide signaling classes are small or represented by a single peptide or only found in a single family of plants. The different classes of plant signaling peptides are not phylogenetically related indicating that they have been independently selected to enable modular or "mix and match" signaling.

1 Introduction

Plants, due to their sessile nature, need to respond rapidly to fluctuations in their environment which can range from changes in humidity and temperature to predatory attacks by herbivores or pathogens. Plants successfully withstand these challenges as they have evolved complex and highly interconnected signaling networks that operate both intra- and extracellularly to relay cellular responses. The signaling processes are mediated by ligands that include gases, small organic molecules, and peptides and small proteins. Specific receptors for these molecules recognize the signal and activate signaling cascades that relay the message within

J.I. Wheeler • H.R. Irving (✉)
Monash Institute of Pharmaceutical Sciences, Monash University, 381 Royal Parade, Parkville, VIC 3052, Australia
e-mail: helen.irving@monash.edu

H.R. Irving and C. Gehring (eds.), *Plant Signaling Peptides*,
Signaling and Communication in Plants 16, DOI 10.1007/978-3-642-27603-3_1,

specific cells to stimulate the appropriate physiological responses. Peptide and small protein molecules contribute as signaling ligands in a wide variety of plant functions ranging from plant cell differentiation to host defense responses. In addition, plants produce an arsenal of plant defense peptides and proteins with antimicrobial activity and many proteins that act as protease inhibitors. In this book, the contributions of various peptides and small proteins to plant growth and development are investigated. The roles of signaling peptides in modulating plant growth and development are examined in chapters "Peptides Regulating Apical Meristem Development, Peptides Regulating Root Growth, Peptides Regulating Plant Vascular Development, The S-LOCUS CYSTEINE RICH PROTEIN (SCR): A Small Peptide with a High Impact on the Evolution of Flowering Plants, and Peptides Modulating Development of Specialized Cells," and the signaling peptides that participate in defense and homeostasis are examined in chapters "Plant Antimicrobial Peptides, The Role of Plant Peptides in Symbiotic Interactions, Peptides as Danger Signals: MAMPs and DAMPs, and Peptides and the Regulation of Plant Homeostasis." Various strategies for working with plant signaling peptides are explored in chapters "Processing of Peptides, Methods to Isolate and Identify New Plant Signaling Peptides, Methods to Identify New Partners of Plant Signaling Peptides, and Computational-Based Analysis to Associate the Function of Plant Signaling Peptides with Distinct Biological Processes."

In this chapter, we introduce the secreted peptide signaling molecules and briefly review the role of the various classes of the molecules in plants and their phylogenetic relationships. In the process, we discuss several generalized themes that have emerged following recent developments in biochemical, genetic, and molecular biology studies focusing on the evolutionary adaptations that the peptides offer as a preface to the detailed discussions in the chapters following.

2 Animal Peptide Signaling Systems

The isolation of insulin revealed that proteins could act as hormones in mammals (Banting et al. 1922a, b), and since then, a plethora of signaling peptides have been discovered that act as hormones or as paracrine and/or autocrine molecules. Hormones are secreted from an exocrine gland and distributed via the bloodstream to distant regions of the body, whereas paracrine molecules are secreted and act on nearby cells in the tissue and autocrine signals act on the secreting cell. Paracrine signaling is a very ancient form of cell communication also evident in unicellular organisms as well as multicellular organisms. Unicellular organisms evolved mechanisms to respond to the presence of environmental constraints and other cells. Quorum sensing is used by bacteria to detect the presence of other bacteria and, when a certain level of bacterial population is reached (the quorum), signals the generation of various compounds including antibiotics directed at other bacteria. Gram-positive bacteria use peptide ligands to undertake quorum sensing (Antunes et al. 2010; Hibbing et al. 2010; Thoendel and Horswill 2010). Specific peptide ligands are also secreted by haploid cells of *Saccharomyces cerevisiae* as pheromones that are recognized by the opposite haploid cell mating type (Bardwell 2004). In both cases, peptide ligands form an important part of the signaling network. This trait has been preserved through evolution as peptides form major groups of paracrine signaling molecules in animals. An example is the cytokine family of secreted small protein signals that regulate immune and hematopoietic cell development and includes interleukins, interferons, and erythropoietin. The cytokines and their receptors all have similar tertiary structural homology indicating that all the ligands are derived from a common ancestral protein as are all the receptors (Ozaki and Leonard 2002). So it is not surprising that the cytokine receptors all activate similar signaling pathways involving kinase enzymes [Janus kinase (JAK)] that in turn phosphorylate and activate transcription factors [signal transduction and activation of transcription (STAT) proteins] forming the JAK/STAT pathway (Yamaoka et al. 2004). The cellular response to a particular cytokine is dictated by whether the receptor is present and the particular group(s) of transcription factors present in that cell. Thus, a type of mix and match of ligand and receptor class modules has coevolved with transcription factors to modulate gene transcription to regulate cell development.

3 Discovery of Peptide Signaling Systems in Plants

Over the last two decades, the number of known peptide signaling molecules has increased dramatically from the first molecule identified in tomato to now over 15 diverse families that influence plant growth and development. A list of representatives of peptide signaling families found in *Arabidopsis thaliana* and a brief description of their known function(s) are presented in Table 1. Systemin is an 18-amino-acid peptide which was the first peptide signaling molecule identified as the factor that induced synthesis of proteinase inhibitors in wounded tomato leaves (*Solanum lycopersicum*) (Pearce et al. 1991). Systemin obtained its name as it was associated with a systemic response where proteinase inhibitors were also produced in tomato leaves above the wounded leaves. Prosystemin is the precursor peptide containing the systemin peptide within its sequence (McGurl et al. 1992). Overexpression of prosystemin in tomato induces the systemic response involving protease inhibitors (McGurl et al. 1994). The systemic response involving systemin is entwined with the production of the lipid-derived hormone jasmonic acid which is also associated with the wound response (Farmer et al. 1992; Farmer and Ryan 1992). Grafting experiments revealed that synthesis of jasmonic acid and systemin in the lower leaves at the site of wounding is necessary for strong upregulation of proteinase inhibitors in the upper leaves following perception of jasmonic acid (for a review, see Schilmiller and Howe 2005). Systemin is restricted to members of the Solanaceae and hence is not listed in Table 1.

Small peptides secreted into plant cell culture media were also identified about this period. The first identified was phytosulfokine (PSK) which is a pentapeptide

Table 1 Summary of plant peptide signaling molecules present in *Arabidopsis thaliana*

Propeptide	Gene family	Propeptide size (kDa)	Processed peptide, size	Function	Site of action	Receptor	References
CEP	6	8.5–11.5	CEP1, 14 AA	Inhibits root growth	Lateral root primordia	Unknown	Ohyama et al. (2008)
CLE (and CLV3)	32	7.8–14.5	CLE or mCLV3, 12–14 AA	Stimulates organogenesis and inhibits meristematic growth; can stimulate vascular development	Floral, shoot, and root meristems; vascular	CLV1, BAM1, CLV2, RPK2	Clark et al. (1995, 1997), Cock and McCormack (2001), Fiers et al. (2006), Kondo et al. (2006), DeYoung and Clark (2008), and Kinoshita et al. (2010)
DVL/RTFL	24	4.6–16.2	Not processed	Polarity dependence, cell proliferation, leaf development	Stem, rosette leaves, pedicles, siliques	Unknown	Narita et al. (2004), Wen et al. (2004) and Ikeuchi et al. (2011)
EPF	11	11.5–14.3	Unknown	Promotes epidermal cell division leading to guard cell (stomatal) formation	Epidermis and meristemoid mother cells	TMM, ER, ERL1, ERL2	Hara et al. (2007, 2009) and Hunt and Gray (2009)
IDA and IDL	6	8.4–13	EPIP	Inhibits floral abscission	Abscission zone	HAS, HSL	Butenko et al. (2003) and Stenvik et al. (2006)
PROPEP	7	9.3–12.3	Pep1, 23 AA	Promotes innate immune responses (a danger signal)	Widespread, leaves	Pep1R	Huffaker et al. (2006), Yamaguchi et al. (2006), Pearce et al. (2008), and Qi et al. (2010)
PNP	2	13–14	Unknown	Extracellular, cell expansion, water/ion movement, stomatal opening, inhibits ABA-induced stomatal closure	Leaves, mesophyll and guard cells, root stele, stem	Unknown	Gehring et al. (1996), Maryani et al. (2001), Ludidi et al. (2002), Rafudeen et al. (2003), Wang et al. (2007), Gottig et al. (2008), Ruzvidzo et al. (2011), and Wang et al. (2011)

POLARIS	1	4.6	36 AA	Required for root elongation, lateral root formation, leaf vascular patterning	Embryonic root and seedling	Unknown	Casson et al. (2002) and Chilley et al. (2006)
PSK	6	8.7–9.7	PSK-α, 5 AA	Promotes cell proliferation and longevity, root elongation	Widespread, mesophyll cells, roots	PSKR1	Matsubayashi and Sakagami (1996), Lorbiecke and Sauter (2002), Matsubayashi et al. (2002, 2006), and Kutschmar et al. (2009)
PSY	3	7.9	PSY1, 18 AA	Promotes cellular expansion and proliferation, upregulated by wounding	Mesophyll cells, roots	PSYR1	Amano et al. (2007)
RALF and RALFL	39	7–14	RALF, 25–30 AA	Associated with danger signals, affects growth—inhibits root growth	Widespread in plants	Unknown	Pearce et al. (2001), Silverstein et al. (2007), and Wu et al. (2007)
RGF1	9	13	13 AA	Maintain root stem cell niche in inner layer of columella cells	Root meristem	Unknown	Matsuzaki et al. (2010)
SCRL	27	9.2–11.5	Not processed	Prevents self-fertilization (but not in *A. thaliana*)	Pollen	SRK	Schopfer et al. (1999) and Vanoosthuyse et al. (2001)
TPD	2	19.5	TPD1	Anther development promoting tapetum formation	Anthers	EMS1	Yang et al. (2003) and Jia et al. (2008)

sulfated on its two tyrosine residues [Y(SO_3H)IY(SO_3H)TQ] (Matsubayashi and Sakagami 1996). PSK was first discovered as an essential cell proliferation agent necessary to maintain low-density cell cultures in several species including asparagus, rice, and carrot (Matsubayashi and Sakagami 1996; Matsubayashi et al. 1997; Kobayashi et al. 1999; also see chapters "Peptides Regulating Root Growth, Peptides Regulating Plant Vascular Development, Peptides and the Regulation of Plant Homeostasis, and Processing of Peptides"). It was later found that PSK occurs across the taxa, and now, it is used to promote taxol production in cell cultures (Kim et al. 2006).

This was followed by the identification of a peptide factor from tobacco leaf extracts that induced a rapid alkalinization response in tobacco suspension cells (Pearce et al. 2001). This peptide is a member of a large family known as the Rapid ALkalinization Factors (RALF) or RALF-like (RALFL) found in many plant families with a common highly conserved cysteine-rich motif (CX{4,14}CX{22,51}CX{6,12}CX{5,14}CX{5,6}C) in a relatively divergent polar peptide (Pearce et al. 2001; Silverstein et al. 2007). RALFL peptides are likely to have a ubiquitous role in plants as their sequences have also been found in many plant families (Pearce et al. 2001; Olsen et al. 2002; Haruta and Constabel 2003; Germain et al. 2005; Silverstein et al. 2007; also see chapters "The Role of Plant Peptides in Symbiotic Interactions, Peptides and the Regulation of Plant Homeostasis, and Processing of Peptides"). Since then, several other secreted peptides have been identified from extracellular extracts including peptide 1 (Pep1) (Huffaker et al. 2006), C-terminally encoded peptide 1 (CEP1) (Ohyama et al. 2008), and plant peptide containing sulfated tyrosine (PSY1) (Amano et al. 2007).

Various genetic screens have also resulted in the identification of peptide signals (also see chapter "Methods to Identify New Partners of Plant Signaling Peptides"). These screens are used to reveal components necessary for development, and knockout mutants are selected on the basis of abnormal phenotypic growth patterns. The first peptide to be identified using these screens was CLAVATA 3 (CLV3) where the *clv3* mutant has abnormally large shoot and floral meristems due to excess stem cells (Clark et al. 1995; Fletcher et al. 1999). Mutant studies have the advantage that identifying mutants with similar phenotypes may in turn reveal other components of the signaling pathway that can then be experimentally confirmed. The mutant *clv1* also has abnormally large shoot and floral meristems, and CLV1 encodes a full-length leucine-rich repeat receptor-like kinase (LRR-RLK) (Clark et al. 1997). More recently, it has been shown biochemically that the processed CLV3 peptide directly interacts with the external leucine-rich domain of the CLV1 receptor (Ogawa et al. 2008). The CLE family is named for CLV3 (Clark et al. 1995; Fletcher et al. 1999) from *Arabidopsis* and *EMBRYO SURROUNDING REGION* (*ESR*) from maize (Opshal-Ferstad et al. 1997) and forms one of the largest families of plant peptide signaling molecules that is present throughout the plant kingdom (Cock and McCormack 2001; Oelkers et al. 2008; Sect. 5; chapters "Peptides Regulating Apical Meristem Development, Peptides Regulating Root Growth, and Peptides Regulating Plant Vascular Development," containing 32 annotated genes in *Arabidopsis*. CLE family members contain the CLE domain

(14 amino acids) near the C-terminal (Cock and McCormack 2001), and CLV3 is the best characterized member. Mutant screens that detected abnormal stomatal patterns identified two EPIDERMAL PATTERNING FACTOR (EPF) proteins, EPF1 and EPF2, which are involved in determining epidermal cell division events leading to stomatal formation (Hara et al. 2007, 2009; Hunt and Gray 2009; see chapter "Peptides Modulating Development of Specialized Cells"). More recently, combinations of biochemical and genetic approaches have been used to identify additional peptides regulating plant growth and development such as the root meristem growth factor (RGF) (Matsuzaki et al. 2010, Sect. 9). Another interesting example is the pollen tube attractants (LURE) that have been identified in *Torenia fournieri* (Okuda et al. 2009; Higashiyama 2010), but to date, homologues of these proteins in *Arabidopsis* have not been found, and they are excluded from the phylogenetic analysis in Sect. 5.

An alternative approach based on peptide analogy to animal systems was used to identify and purify a small protein now known as plant natriuretic peptide (PNP) which is immunoreactive to antisera specific for the animal peptide factor, atrial natriuretic peptide (Vesely and Giordano 1991; Gehring et al. 1996; Maryani et al. 2001; Ludidi et al. 2002; see chapter "Peptides and the Regulation of Plant Homeostasis"). The 33-amino-acid active region of PNP-A is the only region with similarity to atrial natriuretic peptide (Wang et al. 2007), and phylogenetic analysis suggests that the similarities between the animal and plant form most likely are the result of convergent evolution (Gehring and Irving 2003).

4 Processing of Peptide Ligands

Once peptide ligands were discovered, it then became evident that they were formed from precursor proteins that had been secreted into the apoplast where the proteins were further processed (Fig. 1). That is, the precursor proteins include an N-terminal signal sequence targeting the proteins for export (preproprotein), and this is cleaved during protein translation to/from the proprotein that is exported by the default secretory pathway (Denecke et al. 1990). Following export, the actual active peptide ligand is proteolytically cleaved from the proprotein. The processes underlying these events are discussed in more detail in chapter "Processing of Peptides." Systemin is the active peptide released from the C-terminal end of the 200-amino-acid precursor protein prosystemin by proteolytic processing systems in the apoplast (McGurl et al. 1992). CLE, PSK, and RALF are also formed from precursor proteins where the final processing into the active peptide occurs in the apoplast, and this is the case for several other peptide molecules as listed in Table 1. Enzymes such as tyrosylprotein sulfotransferase catalyze the sulfation of the tyrosine residues in proPSK as the protein is processed through the Golgi network before being secreted (Hanai et al. 2000). Alternative processing events involve hydroxylation of proline residues which occurs in the CLE motif of the

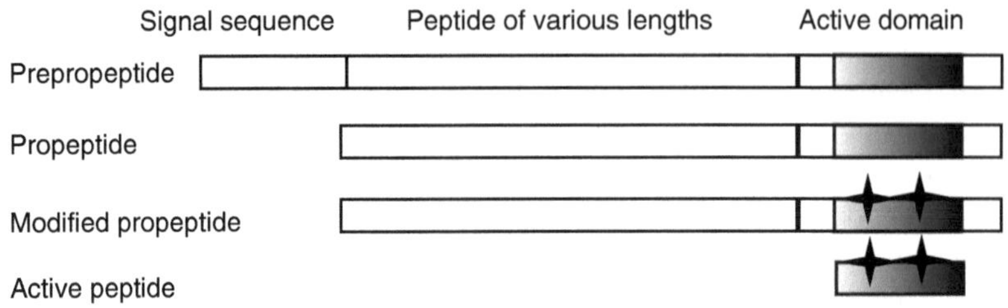

Fig. 1 Processing of peptide signaling molecules. After transcription, the prepropeptide has its secretory signal sequence cleaved in the endoplasmic reticulum forming the propeptide. As the propeptide is processed through the Golgi apparatus, amino acid modifications are made (indicated by *dagger line*) before the propeptide is secreted into the apoplast where the active peptide is released by further proteolytic processing as described in the text. The active domain is generally located in the C-terminal region of the peptide and is indicated by the *darker rectangle* with a dibasic region 5–20-amino-acid residues upstream (*dark line*) that is a proteolytic cleavage point

active mature hydroxyproline CLV3 (Fiers et al. 2006; Kondo et al. 2006) and also CEP1 (Ohyama et al. 2008). Once secreted, further processing occurs in the apoplast, and it is known that specific subtilisin serine proteases cleave dibasic residues upstream from the PSK sequence in proPSK to release an 8-amino-acid fragment (Srivastava et al. 2008). The enzymes necessary to form the final active product PSK are not yet known. However, it is evident that several enzymes and ProPSK need to come together in the apoplast for these processing events to be successful. Presumably for this to occur, the processing enzymes and proproteins are secreted from the same or adjacent cells indicating a large degree of coordinated protein regulation. Equally well-programmed events where the proteolytic enzymes are secreted (or present) in addition to the proprotein are necessary in the processing of RALFL (Srivastava et al. 2009). ProCLV3 is secreted into the meristematic apoplast (Rojo et al. 2002) where it is processed into the active short peptide (Kondo et al. 2006; Ni and Clark 2006). The active small peptides are generally found at the C-terminal of the propeptide molecule, and this region has homology (or is even identical) with other peptides of the same class but not between classes (Sect. 5; Fig. 2).

However, not all proteins appear to be processed into short active peptides. The Low-molecular-weight Cysteine-Rich (LCR or pollen coat protein) and S-locus Cysteine-Rich (SCR) and SCR-like (SCRL) proteins are characterized by eight conserved cysteine residues throughout the relatively small (ranging from 4 to 11 kDa) secreted protein (Vanoosthuyse et al. 2001). The cysteine residues are found throughout the secreted protein and thus contribute to the folding of the SCR or SCRL (see chapter "The S-LOCUS CYSTEINE RICH PROTEIN (SCR): A Small Peptide with a High Impact on the Evolution of Flowering Plants"). The solution structure of an SCR has been resolved, and it folds into an α/β sandwich structure that resembles that of plant defensins with a unique loop bulging out from the body of the protein containing the hypervariable region (Mishima et al. 2003; see chapter "Plant

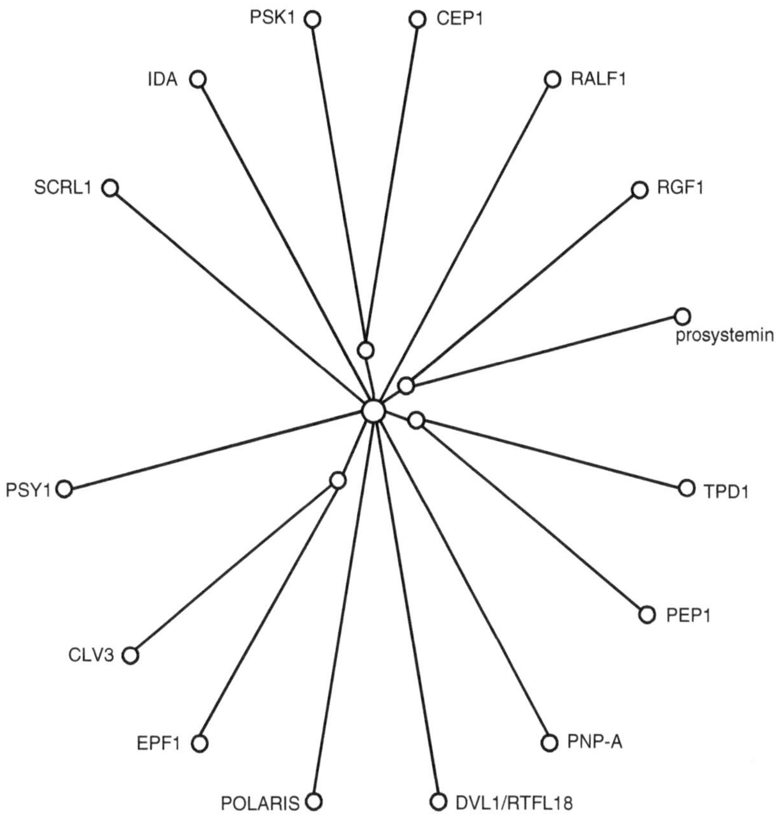

Fig. 2 Phylogenic diagram demonstrating that the different peptide classes are not related. This Radial Cobalt Tree was produced using the National Center for Biotechnology Information (NCBI) COBALT multiple alignment tool. All 175 available *Arabidopsis* amino acid sequences of the different peptide classes as well as the tomato prosystemin segregated into the different peptide classes. In the Radial Cobalt Tree shown, a single-protein sequence was selected to represent each group: CEP (6 sequences), CLE/TDIF (32), DVL/RTFL (24), EPF (11), IDA (6), PROPEP (7), PNP (2), POLARIS (1), PSK (6), PSY (3), RALF (39), RGF (9), SCRL (27), and TPD1 (2). Refer to Table 1 for abbreviations and a description of the known function for each class

Antimicrobial Peptides"). The LURE peptides identified as additional components of the pollen pistil interaction also contain conserved cysteine residues throughout and, like LCRs and SCRs, are defensin-like proteins (Higashiyama 2010; chapter "Plant Antimicrobial Peptides"). The protein PNP-A is secreted from mesophyll cells (Wang et al. 2011), but it is still unknown if PNP is further processed despite some initial speculation (Pharmawati et al. 1998). PNP-A contains six cysteine residues throughout the secreted protein, and the region that contains specific activity occurs toward the N terminus where modeling predicts it to form part of an exposed loop from the β barrel structure (Morse et al. 2004; Wang et al. 2007).

5 Phylogenic Relationships Between Classes of Peptide Ligands

Of the small secreted peptides identified to date, most have representatives in agronomically important monocot and dicot lineages including rice (*Oryza sativa* L.), maize (*Zea mays* L.), sorghum [*Sorghum bicolor* (L.) Moench], soybean [*Glycine max* (L.) Merr.], castor oil bean (*Ricinus communis* L.), wine grape (*Vitis vinifera* L.), and the black cotton wood tree [*Populus balsamifera* L. Ssp. *trichocarpa* (Torr. and A. Gray ex hook) Brayshaw]. Some peptide groups such as EPF and CLE can be found across the plant kingdom including moss (*Physcomitrella patens*) and lycophytes (*Selaginella moellendorffii*) (Miwa et al. 2009; Floyd and Bowman 2010; Rychel et al. 2010). Examples of the CLE family have also been found in green algae (*Chlamydomonas reinhardtii*) (Oelkers et al. 2008) and the nematode *Heterodera glycines* (Olsen and Skriver 2003) where it is believed to be an example of horizontal gene transfer. The CLE peptide secreted by the nematode is thought to mimic the endogenous CLE in soybean roots (Jun et al. 2008; see chapter "The Role of Plant Peptides in Symbiotic Interactions"). PNP (sometimes annotated as expansin-like) also occurs widely across the plant kingdom (Ludidi et al. 2002). A unique analogue of PNP has been reported in the bacterial pathogen *Xanthomonas axonopodis* pv *citri* that can alter host plant homeostasis mechanisms, and it is thought that this is another example of horizontal gene transfer (Nembaware et al. 2004; Gottig et al. 2008; see chapter "Peptides and the Regulation of Plant Homeostasis"). Most other peptide groups listed in Table 1 such as CEP, DEVIL/ROTUNDIFOLIA4 (DVL/RTFL originally called DVL1/ROT4), INFLORESCENCE DEFICIENT IN ABSCISSION (IDA), Pep1, PSK, RALF, and TAPETUM DETERMINANT (TPD1) are represented in angiosperms including both monocots and dicots (Matsubayashi and Sakagami 1996; Matsubayashi et al. 1997; Pearce et al. 2001; Butenko et al. 2003; Yang et al. 2003; Narita et al. 2004; Wen et al. 2004; Huffaker et al. 2006; Silverstein et al. 2007; Combier et al. 2008; Ohyama et al. 2008; Zhao et al. 2008). A smaller number are represented only in dicots including two sulfated peptide families, PSY and the recently identified RGFs (Amano et al. 2007; Matsuzaki et al. 2010). Finally, some peptide groups are further limited to a single family such as systemin which is restricted to Solanaceae (Ryan and Pearce 1998) and SCRL to the Brassicaceae (Schopfer et al. 1999; Vanoosthuyse et al. 2001) while POLARIS is found only in *Arabidopsis* (Casson et al. 2002).

Of the 15 peptide groups examined, all but systemin have representatives in the model plant *A. thaliana*. Comparison of all 175 available *Arabidopsis* peptide sequences, along with tomato systemin, shows that peptides of the same group are more similar to each other than any peptide from another group (with the exception of RGF5 and RGF8 which are most similar to CLE46 and CLE45, respectively). This is despite the low level of homology between peptides of the same group outside the defining peptide domain. A single-protein sequence was selected to represent each peptide group, and a radial COBALT tree shows the peptides in relation to one another (Fig. 2) where, for example, CLV3 was chosen to represent the CLE family. The number of members of the peptide groups we examined varied from 1 or 2 (POLARIS and TPD) to 39 (RALFL), and groups with many members (CLE 32, SCRL 27, and DVL/RTFL 24) did not necessarily correlate with ancient lineages. An EPF sequence homologue has been found in moss while the prolific SCRL family has 27 members in *Arabidopsis* but is limited to the Brassicaceae (Schopfer et al. 1999; Vanoosthuyse et al. 2001).

Within each peptide group, the members were often spread across the chromosomes, but in some cases, the peptide-encoding genes were clustered. For example, six genes of proPep are clustered in two groups on chromosome 5 and another gene (proPep6) occurs on chromosome 2, but it is more similar to members of one of the clusters than the other (Wheeler and Irving 2010). Similarly, the more expanded signaling peptide groups such as RALFL, SCRL, and CLE are also clustered. In the case of RALFL, several encoding gene groups are clustered similarly on both the *Arabidopsis* and rice chromosomes (Silverstein et al. 2007) and examples include those most similar in sequence such as RALFL2 and 3, RALF8 and 9, RALF10–13, and RALF25 and 26. However, it should be noted that close proximity of encoding genes does not necessarily mean there is a high level of sequence similarity. Interestingly, in the IDA group of peptides, the two most closely related genes IDL2 (At5g64667) and IDL3 (At5g09805) are found in duplicated regions of chromosome 5, covering genes At5g63600–65640 and At5g08570–10570 although neither IDL2 nor IDL3 is sufficient to complement IDA, implying that these gene products are not redundant (Stenvik et al. 2008).

In fact, in some peptide families such as EPF, the functional differences among family members appear to be defined by their distinct spatiotemporal expression patterns and this is the case for EPF1 and EPF2. These genes are expressed at late and early stages of stomatal lineage progression respectively and display loss of function phenotypes consistent with their time of expression (Abrash and Bergmann 2010; Abrash and Lampard 2010; Peterson et al. 2010; Rychel et al. 2010; chapter "Peptides Modulating Development of Specialized Cells"). Differential expression of genes from the same peptide group encoding the same processed signal peptide can occur as is the case with the PSK family (Yang et al. 2001; Matsubayashi et al. 2006). It has also been shown that in rice, *Medicago*, wheat (*Triticum aestivum*), and *Selaginella,* a single CLE gene encodes multiple CLE domains (Cock and McCormack 2001; Kinoshita et al. 2007; Miwa et al. 2009). In rice, CLE75 encodes 6 CLE domains, and in *Selaginella,* SmCLE15 has 8 CLE domains encoding two different classes of CLE peptides. It is suggested that multiple peptides produced from one precursor will contribute to a rapid response and that these events may be specific to rice, *Medicago*, wheat, and *Selaginella* as they are not found in *Arabidopsis* (Kinoshita et al. 2007; Miwa et al. 2009).

Interestingly, BLAST analysis of all the *Arabidopsis* peptides from the RGF peptide group shows that each peptide is more similar to orthologs found in other species than to the other *Arabidopsis* RGF peptides. For instance, RGF6 (At4g16515) has higher homology with *Arabidopsis lyrata* (XP_002868128), castor oil bean (XP_002525906), black cotton wood tree (XP_002329915, XP_002329923), and table grape (CAN75750, CBI31912.3). Comparison of more abundantly spread peptide sequences over a range of species using PROPEP and PSK has shown that sequence similarity separates along monocot and dicot lineages (Lorbiecke et al. 2005; Huffaker et al. 2006).

6 Paracrine and Autocrine Effects

A theme that will become evident in this book is that many plant signaling peptides are expressed in specific and often restricted regions of the plant where they are secreted, further processed, and act upon nearby cells (Fig. 3). This action is similar to growth factors and cytokines regulating development in animal cells where paracrine and autocrine effects are paramount (Ozaki and Leonard 2002). This type of signaling is very ancient and evident throughout multicellular organisms and is probably based on signaling mechanisms developed in single-cell organisms such as quorum sensing in bacteria (e.g., Hibbing et al. 2010). It has the major advantage of allowing a gradient of molecules to form from the secretory to recipient cells which creates a tailored signaling microenvironment that can stimulate different responses in the nearby cells depending on the strength of the signal. The ligands involved in this type of gradient signaling are sometimes referred to as morphogens. Most of the peptides listed in Table 1 exhibit paracrine and/or autocrine effects acting within specific regions such as floral or shoot meristems as shown in Fig. 3. This is particularly marked with peptides directly affecting development such as CLE, EFP, IDA, and RGF which act at specific localized regions within the plant (see chapters "Peptides Regulating Apical Meristem Development, Peptides Regulating Root Growth, Peptides Regulating Plant Vascular Development, Peptides Modulating Development of Specialized Cells, and The Role of Plant Peptides in Symbiotic Interactions").

7 Peptide Ligand Receptors

For cells to respond to the peptide ligands, receptors have to be present, and it is thought that receptors and their ligands have evolved in parallel (Fryxell 1996). Receptors for the peptide ligands that have been identified are generally members of receptor-like protein kinase families such as the leucine-rich repeat receptor-like kinases (LRR-RLKs), and several recent reviews have discussed the different types

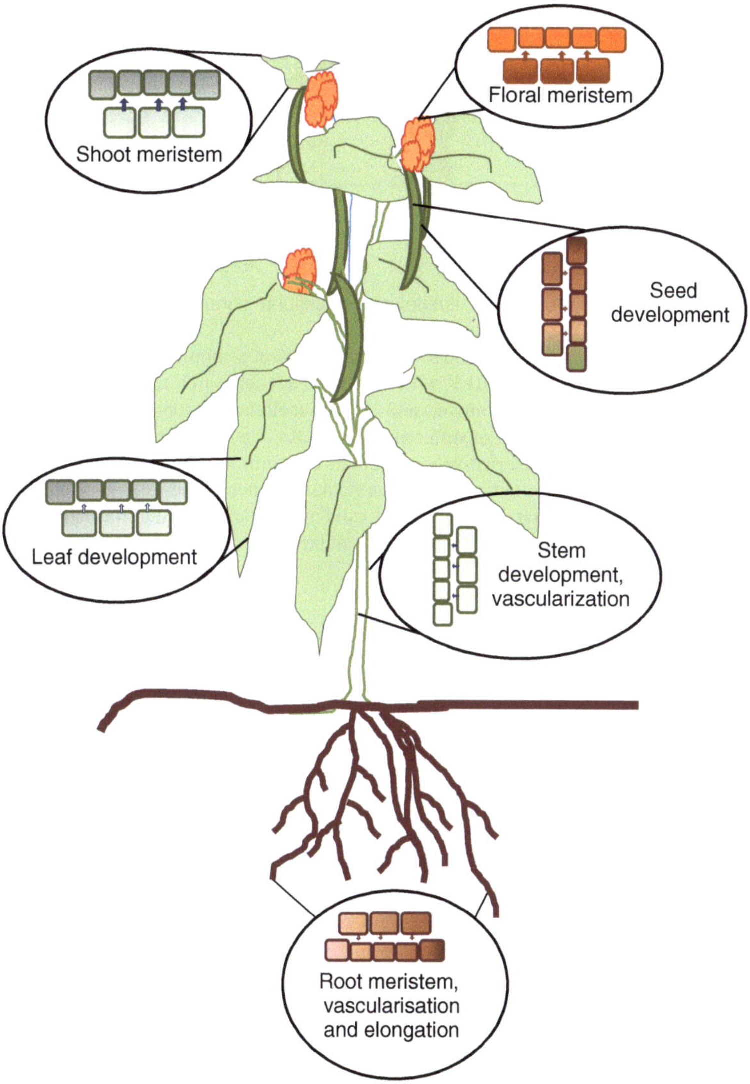

Fig. 3 Paracrine and restricted regions for peptide signaling. In many cases, peptide signaling molecules are released from a localized group of cells into the apoplast where they form a concentration gradient that most strongly acts on nearby cells containing receptors. Peptides acting locally (paracrine) are known to affect cell development in the meristems (floral, shoot, and root), seeds, vasculature, and formation of guard cells in leaves as shown

of receptors and the signaling mechanisms activated following binding of peptides or other ligands (see, e.g., Atzal et al. 2008; Boller and Felix 2009; Tör et al. 2009; Clouse 2011). The receptors for peptides can form oligomers and often exist as either homomeric or heteromeric dimers (Fig. 4). The extracellular regions of the receptors recognize particular surface patterns of the active peptide ligands that are

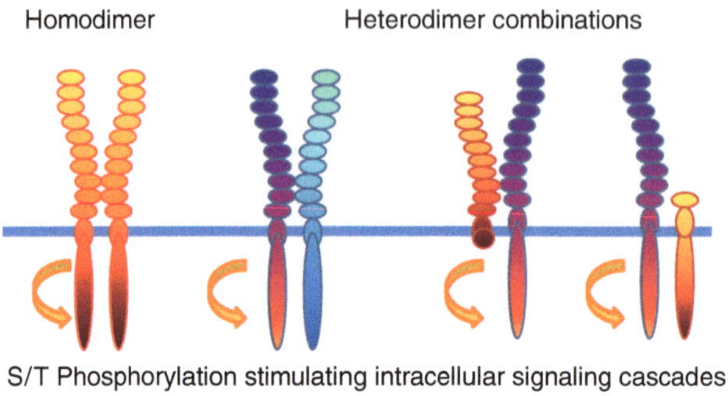

Fig. 4 Models of functional receptor oligomers that can form as homo- or heterodimers. In each case, a component will be an LRR-RLK which contains a leucine-rich repeat external domain, a single transmembrane-spanning domain, and an intracellular (cytoplasmic) kinase domain. Oligomers form between the same or different LRR-RLKs where both receptor proteins contain all of the functional domains. Alternatively, oligomers can form with one LRR-RLK and either an LRR receptor-like protein that contains no kinase domain or membrane-associated kinase protein containing a limited extracellular domain. Upon ligand binding and oligomer formation, the kinase domains autophosphorylate serine and threonine residues and initiate an intracellular phosphorylation cascade

present in the extracellular matrix. In some cases, a receptor may recognize more than one peptide perhaps due to the common surface pattern formed by the peptide. This is likely to be the case with the CLE peptides. There are 32 known CLE peptides in *Arabidopsis* which all contain a common CLE motif but are derived from different precursor proteins with specific regional expression patterns. Hence, the effects of the peptide ligands are limited to areas where the peptides are actually secreted and processed, and this has been shown with CLE ligands where expression of the receptors involved in organogenesis is restricted to localized areas (Müller et al. 2008). When CLE peptides are applied exogenously, it results in a gradient of responses, and different members can mimic each other (Whitford et al. 2008), which is indicative of overlapping redundancy in the receptor specificity for small but highly similar active peptide fragments. CLV3 was the first of these ligands discovered and it directly interacts with the extracellular domain of CLV1 which is a full-length LRR-RLK (Ogawa et al. 2008), and CLV1 has been shown to form heterodimers with BARELY ANY MERISTEM 1 (BAM1) which is a related LRR-RLK (DeYoung and Clark 2008). More recently, another LRR-RLK, homomers of receptor-like PROTEIN KINASE 2 [RPK2, also known as TOADSTOOL 2 (TOAD2)], has been shown to interact with CLV3 (Kinoshita et al. 2010).

The interaction between ligand and receptor has not been as well characterized for the other signaling peptides. Binding and cross-linking affinity studies were used to identify receptors for PSK, PSY1, and Pep1, and these receptors are all members of the LRR-RLK family (Matsubayashi et al. 2002, 2006; Yamaguchi et al. 2006; Amano et al. 2007; Jia et al. 2008, chapter "Methods to Identify New Partners of Plant Signaling Peptides"). LRR-RLK proteins contain a large extracellular leucine-rich repeat domain that has a specific peptide-recognizing region within it, a single transmembrane-spanning region, and an intracellular kinase domain thought to mediate a phosphorylation signaling cascade. Several of the peptide receptors [PSKR1, PepR1, CLV1, and ERECTA (ER)] also contain a putative guanylate cyclase catalytic center within the intracellular general kinase domain region (Kwezi et al. 2007, 2011; Qi et al. 2010), and this has led to speculation that production of cGMP may form part of the signaling pathway in addition to phosphorylation cascades. Indeed, AtPSKR1 does demonstrate some guanylate cyclase activity in vitro while PSK-α stimulates cGMP production in protoplasts (Kwezi et al. 2011).

8 Cost–Benefits of Peptide Ligands

Peptide ligands are relatively nitrogen rich (amino acids contain one to four N atoms) and, in some cases, also either sulfated due to cysteine (or methionine residues) or posttranslationally sulfated [e.g., phytosulfokines (Hanai et al. 2000)]. Both nitrogen and sulfur are nutrients that limit plant growth (Hawkesford and De Kok 2006; Elser et al. 2007), so it would appear counterintuitive for plants to invest in nitrogen-rich signaling molecules. However, it has recently been shown that ecological nitrogen limitations have influenced both the transcribed RNA and proteome nitrogen content such that crop and nitrogen-fixing plants contain higher nitrogen levels in their transcribed RNA and more amino acids with nitrogen-rich side chains than undomesticated plants (Acquisti et al. 2009). We hypothesized that the relative importance of the plant peptides to the plants may be reflected in the nitrogen level of the precursor proteins (Wheeler and Irving 2010). We observed that prepropeptide molecules collectively had higher levels of nitrogen-containing amino acids than the average across all proteins in *Arabidopsis*. This finding led us to argue that plants could efficiently use the limited nitrogen and sulfur resources by restricting expression of these peptides to areas such as the meristem (Wheeler and Irving 2010). An advantage in using peptides as the signaling molecules also lies in the fact that they can be easily upregulated in specific cells by gene transcription in response to environmental triggers. Whether this is a more efficient process than generating the classical hormones is open to question. However, a gradient signal will be produced that can act in a local (paracrine) environment over a few hours reflecting the constraints of gene transcription. More rapid responses can also occur if partially processed peptides are secreted from intracellular stores and/or further processed in the apoplast (extracellular matrix) by various proteolytic enzymes to generate the signaling peptide (Srivastava et al. 2008, 2009; chapter "Processing of Peptides"). The propeptide is relatively inactive compared with the cleaved mature active peptides which are active at nanomolar or lower concentrations (Matsubayashi and Sakagami 1996; Ito et al. 2006; Pearce et al. 2008). Thus, the presence of the additional processing enzymes in the apoplast provides a further level of control in determining the level of activity of peptide signaling molecules.

9 Mix and Match Signaling

Since the receptors for peptides form as oligomers composed of one or two (or more) separate subunits, a degree of selection is introduced to perception of the peptide ligand by receptors in various regions within the plant. In addition, different peptides can act to counteract the effect of another peptide and so bring in a degree of control at the level of perception (also see chapters "Peptides Regulating Plant Vascular Development and Peptides Modulating Development of Specialized Cells"). A further degree of adaptability is introduced by the fact that, in many cases, multiple proteins are precursors for very similar peptide ligands and these are expressed in quite particular and restricted regions of the plant. Even though this results in considerable redundancy in the effects of the peptide class, the restricted spatiotemporal patterns of expression act to counterbalance this redundancy. Here, we will use two examples to explore this concept using the CLE and RGF peptide families.

Most members of the CLE family are known as CLE A, and these act to repress cell division in the meristematic regions while the CLE B (specifically CLE41–44) influence development of cells forming the vascular system (Whitford et al. 2008). The CLE B family members are homologues of the *Zinnia elegans* tracheary element differentiation factor that suppresses xylem cell differentiation in cultured mesophyll cells (Ito et al. 2006). In addition, PSK-α appears to act in a cooperative manner with CLE B peptides. PSK-α promotes tracheary element differentiation in *Zinnia* mesophyll cell cultures in the presence of auxin and cytokinin (Matsubayashi et al. 1999; Motose et al. 2009), whereas the CLE B peptides inhibit this process (Ito et al. 2006; for a discussion, see Fukuda et al. 2007; and chapter "Peptides Regulating Plant Vascular Development"). PSK-α has a general proliferative effect and was discovered as a cell proliferation agent essential for low-density cell cultures (Matsubayashi and Sakagami 1996; Matsubayashi et al. 2006). Investigations into the CLE A family have revealed that the different members can have overlapping effects that are not normally seen due to spatial or temporal separation of their expression. For instance, CLE19 is normally found in roots, and ectopic application of synthetic peptides corresponding to the overlapping conserved CLE motifs of CLV3, CLE19, and CLE40 caused the termination of the root meristem, resulting in a similar phenotype to overexpressed CLE19 mutants (Fiers et al. 2006). These results indicate that the receptor recognizes the overlapping CLE motif but not specific CLE peptides. The receptor for CLV3 is likely to be an oligomer of the LRR-RLKs CLV1 and BAM (DeYoung and Clark 2008; Ogawa et al. 2008) or homomers of RPK2 (Kinoshita et al. 2010), but whether these receptor combinations are expressed in roots and responsive to CLE19 remains to be determined (also see chapter "Peptides Regulating Root Growth"). In addition, applications of combinations of CLE A and B peptides result in favoring proliferation of vascular development indicating that the CLE B peptides are dominating the development in this instance (Whitford et al. 2008). This suggests that a reciprocal gradient will form CLE A and B type peptides in the meristem, and this, in turn, will regulate organogenesis and vascular development. This effect is likely to be due to expression of the same or similar classes of receptors recognizing different combinations of CLE ligands—a mix and match combination of ligand and receptor. Such events are probably not surprising as the CLE ligand is relatively conserved (Cock and McCormack 2001), and it is likely that multiple combinations of CLE receptors are expressed in the developing vascular and meristematic regions (also see Fukuda et al. 2007; Jun et al. 2008; chapters "Peptides Regulating Plant Vascular Development and The Role of Plant Peptides in Symbiotic Interactions").

Another example of mix and match signaling involves the recently identified tyrosine-sulfated peptide family RGF which contains nine members in *Arabidopsis*. The mature active peptides are 13 amino acids long with a sulfated tyrosine residue and are necessary to maintain the root meristem (Matsuzaki et al. 2010). The RGF peptides were discovered by database searches after it was observed that the tyrosylprotein sulfotransferase (TPST) knockout mutant *tpst* had abnormally small root meristems that could not be rescued by application of PSK or PSY (the other two known sulfated peptides). Application of synthetic RGF1–6 clearly rescues the *tpst* phenotype, whereas RGF7 and 9 have weak activity and RGF8 has no activity. RGF acts posttranscriptionally to define the expression levels and patterns of the AP2 domain transcription factor plethora2 (PLT2), creating a gradient in the stem cell niche that has been shown using ectopic application with RGF attached to dextran beads (Matsuzaki et al. 2010). Such findings highlight the importance of both spatial and temporal differentiation in the expression patterns of the CLE and RGF (and other) peptides to prevent developmental errors.

10 Conclusions

Peptide signaling molecules represent an ancient evolutionary adaptation that is used by all organisms as part of the rapid and modular responsive system to environmental challenges and to regulate growth and development. Although peptide signals have only been identified in plants in the last two decades or so, it is now apparent that plants contain a diverse group of peptide signaling molecules that have independent lineages. Some of these peptide families contain numerous members that participate in mix and match modular signaling with receptor combinations, and this will provide the plant with flexibility in regulating responses to the peptides. In many cases, paracrine signaling appears to be an important aspect where the degree of response is modulated by a gradient which ensures that particular regions such as meristems respond to the signal. A further advantage that may be associated with peptide signaling is that relatively rapid and controlled release can be achieved by not only secreting the prepropeptide but also processing enzymes that ensure that the mature peptide is released. From a nitrogen and energy use perspective at least, it may be relatively cheap for plants to use peptide signals as often there are only a few cells making this demand on nitrogen and energy resources. It is likely that even more peptide signaling molecules will be discovered as many small peptides are not annotated in the databases (Silverstein et al. 2007), and this is attested by the recent discoveries of RGFs (Matsuzaki et al. 2010) and LUREs (Higashiyama 2010). At this stage, only a few receptor–ligand pairs are known, and further receptors are likely to be identified for the peptide ligands (see Butenko et al. 2009; chapter "Methods to Identify New Partners of Plant Signaling Peptides"). In the final section of this book, methods to study and discover peptide signaling molecules are examined in detail (chapters "Methods to Isolate and Identify New Plant Signaling Peptides, Methods to Identify New Partners of Plant Signaling Peptides, and Computational-Based Analysis to Associate the Function of Plant Signaling Peptides with Distinct Biological Processes").

Acknowledgments This work was supported by the Australian Research Council's Discovery project funding scheme (DP0557561, DP0878194).

References

Abrash EB, Bergmann DC (2010) Regional specification of stomatal production by the putative ligand CHALLAH. Development 137:447–455

Abrash EB, Lampard GR (2010) A view from the top: new ligands controlling stomatal development in Arabidopsis. New Phytol 186:561–564

Acquisti C, Elser JJ, Kumar S (2009) Ecological nitrogen limitation shapes the DNA composition of plant genomes. Mol Biol Evol 26:953–956

Amano Y, Tsubouchi H, Shinohara H, Ogawa M, Matsubayashi Y (2007) Tyrosine-sulfated glycopeptide involved in cellular proliferation and expansion in *Arabidopsis*. Proc Natl Acad Sci U S A 104:18333–18338

Antunes LCM, Ferreira RBR, Buckner MMC, Finlay BB (2010) Quorum sensing in bacterial virulence. Microbiology 156:2271–2282

Atzal AJ, Wood AJ, Lightfoot DA (2008) Plant receptor-like serine threonine kinases: roles in signalling and plant defense. Mol Plant Microbe Interact 21:507–517

Banting FG, Best CH, Collip JB, Macleod JJR, Noble EC (1922a) The effects of insulin on experimental hyperglycaemia in rabbits. Am J Physiol 62:559–580

Banting FG, Best CH, Collip JB, Macleod JJR, Noble EC (1922b) The effect of pancretic extract (insulin) on normal rabbits. Am J Physiol 62:162–176

Bardwell L (2004) A walk-through of the yeast mating pheromone response pathway. Peptides 25:1465–1476

Boller T, Felix G (2009) A renaissance of elicitors: perception of microbe-associated molecular patterns and danger signals by pattern-recognition receptors. Annu Rev Plant Biol 60:379–406

Butenko MA, Patterson SE, Grini PE, Stenvik G-E, Amundsen SS, Mandal A, Aalen RB (2003) *Inflorescence deficient in abscission* controls floral organ abscission in Arabidopsis and identifies a novel family of putative ligands in plants. Plant Cell 15:2296–2307

Butenko MA, Vie AK, Brembu T, Aalen RB, Bones AM (2009) Plant peptides in signalling: looking for new partners. Trends Plant Sci 14:255–263

Casson SA, Chilley PM, Topping JF, Evans M, Souter MA, Lindsey K (2002) The *POLARIS* gene of Arabidopsis encodes a predicted peptide required for correct root growth and leaf vascular patterning. Plant Cell 14:1705–1721

Chilley PM, Casson SA, Tarkowski P, Hawkins N, Wang KL-C, Hussey PJ, Beale M, Ecker JR, Sandberg GK, Lindsey K (2006) The POLARIS peptide of *Arabidopsis* regulates auxin transport and root growth via effects on ethylene signaling. Plant Cell 18:3058–3072

Clark SE, Running MP, Meyerowitz EM (1995) *CLAVATA3* is a specific regulator of shoot and floral meristem development affecting the same processes as *CLAVATA1*. Development 121:2057–2067

Clark SE, Williams RW, Meyerowitz EM (1997) The CLAVATA1 gene encodes a putative receptor kinase that controls shoot and floral meristem size in Arabidopsis. Cell 89:575–585

Clouse SD (2011) Brassinosteroid signal transduction: from receptor kinase activation to transcriptional networks regulating plant development. Plant Cell 23:1219–1230

Cock JM, McCormack S (2001) A large family of genes that share homology with *CLAVATA3*. Plant Physiol 126:939–942

Combier J-P, Küster H, Journet E-P, Natalija Hohnjec H, Gamas P, Niebel A (2008) Evidence for the involvement in nodulation of the two small putative regulatory peptide-encoding genes *MtRALFL1* and *MtDVL1*. Mol Plant Microbe Interact 21:1118–1127

Denecke J, Botterman J, Deblaere R (1990) Protein secretion in plant cells can occur via a default pathway. Plant Cell 2:51–59

DeYoung BJ, Clark SE (2008) BAM receptors regulate stem cell specification and organ development through complex interactions with CLAVATA signaling. Genetics 180:895–904

Elser JJ, Bracken MES, Cleland EE, Gruner DS, Harpole WS, Hillebrand H, Ngai JT, Seabloom EW, Shurlin JB, Smith JE (2007) Global analysis of nitrogen and phosphorous limitation of primary producers in freshwater, marine and terrestrial ecosystems. Ecol Lett 10:1135–1142

Farmer EE, Ryan CA (1992) Octadecanoid precursors of jasmonic acid activate the synthesis of wound-inducible proteinase inhibitors. Plant Cell 4:129–134

Farmer EE, Johnson RR, Ryan CA (1992) Regulation of expression of proteinase inhibitor genes by methyl jasmonates and jasmonic acid. Plant Physiol 98:995–1002

Fiers M, Golemiec E, van der Schors R, van der Geest L, Li KW, Stiekema WJ, Liu C-M (2006) The CLAVATA3/ESR motif of CLAVAT3 is functionally independent from the nonconserved flanking regions. Plant Physiol 141:1284–1292

Fletcher JC, Brand U, Running MP, Simon R, Meyerowitz EM (1999) Signaling of cell fate decisions by *CLAVATA3* in Arabidopsis shoot meristems. Science 283:1911–1914

Floyd S, Bowman JL (2010) Gene expression patterns in seed plant shoot meristems and leaves: homoplasy or homology? J Plant Res 123:43–55

Fryxell KJ (1996) The coevolution of gene family trees. Trends Genet 12:364–369

Fukuda H, Hirakawa Y, Sawa S (2007) Peptide signaling in vascular development. Curr Opin Plant Biol 10:477–482

Gehring CA, Irving HR (2003) Natriuretic peptides – a class of heterologous molecules in plants. Int J Biochem Cell Biol 35:1318–1322

Gehring CA, Md Khalid K, Toop T, Donald JA (1996) Rat natriuretic peptide binds specifically to plant membranes and induces stomatal opening. Biochem Biophys Res Commun 228:739–744

Germain H, Chevalier E, Caron S, Matton DP (2005) Characterization of five *RALF-like* genes from *Solanum chacoense* provides support for a developmental role in plants. Planta 220:447–454

Gottig N, Garavaglia BS, Daurelio LD, Valentine A, Gehring C, Orellano EG, Ottado J (2008) *Xanthomonas axonopodis* pv. *citri* uses a plant natriuretic peptide-like protein to modify host homeostasis. Proc Natl Acad Sci U S A 105:18631–18636

Hanai H, Nakayama D, Yang H, Matsubayashi Y, Hirota Y, Sakagami Y (2000) Existence of a plant tyrosylprotein sulfotransferase: novel plant enzyme catalyzing tyrosine *O*-sulfation of preprophytosulfokine variants in vitro. FEBS Lett 470:97–101

Hara K, Kajita R, Torii KU, Bergmann DC, Kakimoto T (2007) The secretory peptide gene *EPF1* enforces the stomatal one-cell-spacing rule. Genes Dev 21:1720–1725

Hara K, Yokoo T, Kajita R, Onishi T, Yahata S, Peterson KM, Torii KU, Kakimoto T (2009) Epidermal cell density is autoregulated via a secretory peptide, EPIDERMAL PATTERNING FACTOR 2 in Arabidopsis leaves. Plant Cell Physiol 50:1019–1031

Haruta M, Constabel CP (2003) Rapid alkalinization factors in poplar cell cultures. Peptide isolation, cDNA cloning and differential expression in leaves and methyl jasmonate-treated cells. Plant Physiol 131:814–823

Hawkesford MJ, De Kok LJ (2006) Managing sulphur metabolism in plants. Plant Cell Environ 29:382–395

Hibbing ME, Fuqua C, Parsek MR, Peterson SB (2010) Bacterial competition: surviving and thriving in the microbial jungle. Nat Rev Microbiol 8:15–25

Higashiyama T (2010) Peptide signaling in pollen-pistil interactions. Plant Cell Physiol 51:177–189

Huffaker A, Pearce G, Ryan CA (2006) An endogenous peptide signal in Arabidopsis activates components of the innate immune response. Proc Natl Acad Sci U S A 103:10098–10103

Hunt L, Gray JE (2009) The signaling peptide EPF2 controls asymmetric cell divisions during stomatal development. Curr Biol 19:864–869

Ikeuchi M, Yamaguchi Y, Kazama T, Ito T, Horiguchi G, Tsukaya H (2011) ROTUNDIFOLIA4 regulates cell proliferation along the body axis in Arabidopsis shoot. Plant Cell Physiol 52:59–69

Ito Y, Nakanoyo I, Motose H, Iwamoto K, Sawa S, Dohmae N, Fukuda H (2006) Dodeca-CLE peptides as suppressors of plant stem cell differentiation. Science 313:842–845

Jia G, liu X, Owen HA, Zhao D (2008) Signaling of cell fate determination by the TPD1 small protein and EMS1 receptor kinase. Proc Natl Acad Sci U S A 105:2220–2225

Jun JH, Fiume E, Fletcher JC (2008) The CLE family of plant polypeptide signaling molecules. Cell Mol Life Sci 65:743–755

Kim BJ, Gibson DM, Shuler ML (2006) Effect of the plant peptide regulator, phytosulfokine-a, on the growth and taxol production from *Taxus* sp. suspension cultures. Biotechnol Bioeng 95:8–14

Kinoshita A, Nakamura Y, Sasaki E, Kyozuka J, Fukuda H, Sawa S (2007) Gain-of-function phenotypes of chemically synthetic CLAVATA3/ESR-related (CLE) peptides in *Arabidopsis thaliana* and *Oryza sativa*. Plant Cell Physiol 48:1821–1825

Kinoshita A, Betsuyaku S, Osakabe Y, Mizuno S, Nagawa S, Stahl Y, Simon R, Yamaguchi-Shinozaki K, Fukuda H, Sawa S (2010) RPK2 is an essential receptor-like kinase that transmits the CLV3 signal in Arabidopsis. Development 137:3911–3920

Kobayashi T, Eun C-H, Hanai H, Matsubayashi Y, Sakagami Y, Kamada H (1999) Phytosulfokine-a, a peptidyl plant growth factor, stimulates somatic embryogenesis in carrot. J Exp Bot 50:1123–1128

Kondo T, Sawa S, Kinoshita A, Mizuno S, Kakimoto T, Fukuda H, Sakagami Y (2006) A plant peptide encoded by *CLV3* identified by in situ MALDI-TOF MS analysis. Science 313:845–848

Kutschmar A, Rzewuski G, Stührwohldt N, Bemmster GTS, Inzé D, Sauter M (2009) PSK-a promotes root growth in Arabidopsis. New Phytol 181:820–831

Kwezi L, Meier S, Mungur L, Ruzvidzo O, Irving H, Gehring C (2007) The *Arabidopsis thaliana* brassinosteroid receptor (AtBRI1) contains a domain that functions as a guanylyl cyclase in vitro. PLoS One 2:e449. doi:410.1371/journal.pone.0000449

Kwezi L, Ruzvidzo O, Wheeler JI, Govender K, Iacuone S, Thompson PE, Gehring C, Irving HR (2011) The phytosulfokine (PSK) receptor is capable of guanylate cyclase activity and enabling cyclic GMP-dependant signalling in plants. J Biol Chem 286:22580–22588

Lorbiecke R, Sauter M (2002) Comparative analysis of PSK peptide growth factor precursor homologs. Plant Sci 163:321–332

Lorbiecke R, Steffens M, Tomm JM, Scholten S, von Wiegen P, Kranz E, Wienand U, Sauter M (2005) Phytosulphokine gene regulation during maize (*Zea mays* L.) reproduction. J Exp Bot 56:1805–1819

Ludidi NN, Heazlewood JL, Seoighe C, Irving HR, Gehring CA (2002) Expansin-like molecules: novel functions derived from common domains. J Mol Evol 54:587–594

Maryani MM, Bradley G, Cahill DM, Gehring CA (2001) Natriuretic peptides and immunoreactants modify osmoticum-dependent volume changes in *Solanum tuberosum* L. mesophyll cell protoplasts. Plant Sci 161:443–452

Matsubayashi Y, Sakagami Y (1996) Phytosulfokine, sulfated peptides that induce the proliferation of single mesophyll cells of *Asparagus officinales* L. Proc Natl Acad Sci U S A 93:7623–7627

Matsubayashi Y, Takagi L, Sakagami Y (1997) Phytosulfokine-a, a sulfated pentapeptide, stimulates the proliferation of rice cells by means of specific high- and low-affinity binding sites. Proc Natl Acad Sci U S A 94:13357–13362

Matsubayashi Y, Takagi L, Omura N, Morita A, Sakagami Y (1999) The endogenous sulfated pentapeptide phytosulfokine-a stimulates tracheary element differentiation of isolated mesophyll cells of Zinnia. Plant Physiol 120:1043–1048

Matsubayashi Y, Ogawa M, Morita A, Sakagami Y (2002) An LRR receptor kinase involved in perception of a peptide plant hormone, phytosulfokine. Science 296:1470–1472

Matsubayashi Y, Ogawa M, Kihara H, Niwa M, Sakagami Y (2006) Disruption and overexpression of Arabidopsis phytosulokine receptor gene affects cellular longevity and potential for growth. Plant Physiol 142:45–53

Matsuzaki Y, Ogawa-Ohnishi M, Mori A, Matsubayashi Y (2010) Secreted peptide signals required for the maintenance of root stem cell niche in *Arabidopsis*. Science 329:1065–1067

McGurl B, Pearce G, Orozco-Cardenas M, Ryan CA (1992) Structure, expression, and antisense inhibition of the systemin precursor gene. Science 255:1570–1573

McGurl B, Orozco-Cardenas M, Pearce G, Ryan CA (1994) Overexpression of the prosystemin gene in transgenic tomato plants generates a systemic signal that constitutively induces proteinase inhibitor synthesis. Proc Natl Acad Sci U S A 91:9799–9802

Mishima M, Takayama S, Sasaki K, Jee J, Kojima C, Isogari A, Shirakawa M (2003) Structure of the male determinant factor for *Brassica* self-incompatibility. J Biol Chem 278:36389–36395

Miwa H, Kinoshita A, Fukuda H, Sawa S (2009) Plant meristems: CLAVATA3/ESR-related signaling in the shoot apical meristem and the root apical meristem. J Plant Res 122:31–39

Morse M, Pironcheva G, Gehring C (2004) AtPNP-A is a systemically mobile natriuretic peptide immunoanalogue with a role in *Arabidopsis thaliana* cell volume regulation. FEBS Lett 556:99–103

Motose H, Iwamoto K, Endo S, Demura T, Sakagami Y, Matsubayashi Y, Moore KI, Fukuda H (2009) Involvement of phytosulfokine in the attenuation of stress response during the transdifferentiation of Zinnia mesophyll cells into tracheary elements. Plant Physiol 150:437–447

Müller R, Bleckmann A, Simon R (2008) The receptor kinase CORYNE of Arabidopsis transmits the stem cell-limiting signal CLAVATA3 independently of CLAVATA1. Plant Cell 20:934–946

Narita NN, Moore S, Horiguchi G, Kubo M, Demura T, Fukuda H, Goodrich J, Tsukaya H (2004) Overexpression of a novel small peptide ROTUNIFOLIA4 decreases cel proliferation and alters leaf shape in *Arabidopsis thaliana*. Plant J 38:699–713

Nembaware V, Seoighe C, Sayed M, Gehring C (2004) A plant natriuretic peptide-like gene in the bacterial pathogen *Xanthomonas axonopodis* may induce hyper-hydration in the plant host: a hypothesis of molecular mimicry. BMC Evol Biol 4:10

Ni J, Clark SE (2006) Evidence for functional conservation, sufficiency, and proteolytic processing of the CLAVATA3 CLE domain. Plant Physiol 140:726–733

Oelkers K, Goffard N, Weiller GF, Gresshoff PM, Mathesius U, Frickey T (2008) Bioinformatic analysis of the CLE signaling peptide family. BMC Plant Biol 8:1. doi:10.1186/1471-2229-1188-1181

Ogawa M, Shinohara H, Sakagami Y, Matsubayashi Y (2008) *Arabidopsis* CLV3 peptide directly binds CLV1 ectodomain. Science 319:294

Ohyama K, Ogawa M, Matsubayashi Y (2008) Identification of a biologically active, small secreted peptide in Arabidopsis by in silico gene screening, followed by LC-MS-based structure analysis. Plant J 55:152–160

Okuda S, Tsutsui H, Shiina K, Sprunck S, Takeuchi H, Yui R, Kasahara RD, Hamamura Y, Mizukami A, Susaki D, Kawano N, Sakakibara T, Namiki S, Itoh K, Otsuka K, Matsuzaki M, Nozaki H, Kuroiwa T, Nakano A, Kanaoka MM, Dresselhaus T, Sasaki N, Higashiyama T (2009) Defensin-like polypeptide LUREs are pollen tube attractants secreted from synergid cells. Nature 458:357–361

Olsen AN, Skriver K (2003) Ligand mimicry? Plant-parasitic nematode polypeptide with similarity to CLAVATA3. Trends Plant Sci 8:55–57

Olsen AN, Mundy J, Skriver K (2002) Peptomics, identification of novel cationic Arabidopsis peptides with conserved sequence motifs. In Silico Biol 2:441–451

Opshal-Ferstad HG, Le Deunff E, Dumas C, Rogowsky PM (1997) *ZmESR*, a novel endosperm-specific gene expressed in a restricted region around the maize embryo. Plant J 12:235–246

Ozaki K, Leonard WJ (2002) Cytokine and cytokine receptor pleiotrophy and redundancy. J Biol Chem 277:29355–29358

Pearce G, Strydom D, Johnson S, Ryan CA (1991) A polypeptide from tomato leaves induces wound-inducible proteinase inhibitor proteins. Science 253:895–898

Pearce G, Moura DS, Stratmann J, Ryan CA (2001) RALF, a 5-KDa ubiquitous polypeptide in plants, arrests growth and development. Proc Natl Acad Sci U S A 98:12843–12847

Pearce G, Yamaguchi Y, Munske G, Ryan CA (2008) Structure-activity studies of AtPep1, a plant peptide signal involved in the innate immune response. Peptides 29:2083–2089

Peterson KM, Rychel AL, Torii KU (2010) Out of the mouths of plants: the molecular basis of the evolution and diversity of stomatal development. Plant Cell 22:296–306

Pharmawati M, Billington T, Gehring CA (1998) Stomatal guard cell responses to kinetin and natriuretic peptides are cGMP dependent. Cell Mol Life Sci 54:272–276

Qi Z, Verma R, Gehring C, Yamaguchi Y, Zhao Y, Ryan CA, Berkowitz GA (2010) Ca^{2+} signaling by plant *Arabidopsis thaliana* Pep peptides depends on AtPepR1, a receptor with guanylyl cyclase activity, and cGMP-activated Ca^{2+} channels. Proc Natl Acad Sci U S A 107:21193–21198

Rafudeen S, Gxaba G, Makgoke G, Bradley G, Pironcheva G, Raitt L, Irving H, Gehring C (2003) A role for plant natriuretic peptide immuno-analogues in NaCl- and drought-stress responses. Physiol Plant 119:554–562

Rojo E, Sharma VK, Kovaleva V, Raikhel NV, Fletcher JC (2002) CLV3 is localized to the extracellular space, where it activates the Arabidopsis CLAVATA stem cell signaling pathway. Plant Cell 14:969–977

Ruzvidzo O, Donaldson L, Valentine A, Gehring C (2011) The *Arabidopsis thaliana* natriuretic peptide AtPNP-A is a systemic regulator of leaf dark respiration and signals via the phloem. J Plant Physiol 168:1710–1714

Ryan CA, Pearce G (1998) Systemin: a polypeptide signal for plant defensive genes. Annu Rev Cell Dev Biol 14:1–17

Rychel AL, Peterson KM, Torii KU (2010) Plant twitter: ligands under 140 amino acids enforcing stomatal patterning. J Plant Res 123:275–280

Schilmiller AL, Howe GA (2005) Systemic signaling in the wound response. Curr Opin Plant Biol 8:369–377

Schopfer CR, Nasrallah ME, Nasrallah JB (1999) The male determinant of self-incompatibility in *Brassica*. Science 286:1697–1700

Silverstein KAT, Moskal WJ Jr, Wu HC, Underwood BA, Graham MA, Town CD, VandenBosch KA (2007) Small cysteine-rich peptides resembling antimicrobial peptides have been underpredicted in plants. Plant J 51:262–280

Srivastava R, Liu J-X, Howell SH (2008) Proteolytic processing of a precursor protein for a growth-promoting peptide by a subtilsin serine protease in Arabidopsis. Plant J 56:219–227

Srivastava R, Liu J-X, Guo H, Yin Y, Howell SH (2009) Regulation and processing of a plant peptide hormone, AtRALF23, in Aradidopsis. Plant J 59:930–939

Stenvik G-E, Butenko MA, Rae Urbanowicz B, Rose JKC, Aalen RB (2006) Overexpression of *INFLORESCENCE DEFICIENT IN ABSCISSION* activates cell separation in vestigial abscission zones in *Arabidopsis*. Plant Cell 18:1467–1476

Stenvik G-E, Tandstad NM, Guo Y, Shi C-L, Kristiansen W, Holmgren A, Clark SE, Aalen RB, Butenko MA (2008) The EPIP peptide of *INFLORESCENCE DEFICIENT IN ABSCISSION* is sufficient to induce abscission in *Arabidopsis* through the receptor-like kinases HASEA and HASEA-LIKE2. Plant Cell 20:1805–1817

Thoendel M, Horswill AR (2010) Biosynthesis of peptide signals in Gram-positive bacteria. Adv Appl Microbiol 71:91–112

Tör M, Lotze MT, Holton N (2009) Receptor-mediated signalling in plants: molecular patterns and programmes. J Exp Bot 60:3645–3654

Vanoosthuyse V, Miege C, Dumas C, Cock JM (2001) Two large *Arabidopsis thaliana* gene families are homologous to the *Brassica* gene superfamily that encodes pollen coat proteins and the male component of the self-incompatibility locus. Plant Mol Biol 16:17–34

Vesely DL, Giordano AT (1991) Atrial natriuretic peptide hormonal system in plants. Biochem Biophys Res Commun 179:695–700

Wang YH, Gehring C, Cahill DM, Irving HR (2007) Plant natriuretic peptide active site determination and effects on cGMP and cell volume regulation. Funct Plant Biol 34:645–653

Wang YH, Gehring C, Irving HR (2011) Plant natriuretic peptides are apoplastic and paracrine stress response molecule. Plant Cell Physiol 52:837–850

Wen J, Lease KA, Walker JC (2004) DVL, a novel class of small polypeptides: overexpression alters *Arabidopsis* development. Plant J 37:668–677

Wheeler JI, Irving HR (2010) Evolutionary advantages of secreted peptide signalling molecules. Funct Plant Biol 37:382–394

Whitford R, Fernandez A, De Groodt R, Ortega E, Hilson P (2008) Plant CLE peptides from two distinct functional classes synergistically induce division of vascular cells. Proc Natl Acad Sci U S A 105:18625–18630

Wu J, Kurten EL, Monshausen G, Hummel GM, Gilroy S, Baldwin IT (2007) NaRALF, a peptide signal essential for the regulation of root hair tip apoplastic pH in *Nicotiana attenuata*, is required for root hair development and plant growth in native soils. Plant J 52:877–890

Yamaguchi Y, Pearce G, Ryan CA (2006) The cell surface leucine-rich repeat receptor for AtPep1, an endogenous peptide elicitor in Arabidopsis, is functional in transgenic tobacco cells. Proc Natl Acad Sci U S A 103:10104–10109

Yamaoka K, Saharinen P, Pesu M, Holt VET III, Silvennoinen O, O'Shea JO (2004) The Janus kinases (Jaks). Genome Biol 5:253

Yang H, Matsubayashi Y, Nakamura K, Sakagami Y (2001) Diversity of Arabidopsis genes encoding precursors for phytosulfokine, a peptide growth factor. Plant Physiol 127:842–851

Yang S-L, Xie L-F, Mao H-Z, Puah CS, Yang W-C, Jiang L, Sundaresan V (2003) *TAPETUM DETERMINANT1* is required for cell specialization in the Arabidopsis anther. Plant Cell 15:2792–2804

Zhao X, de Palma J, Oane R, Gamuyao R, Luo M, Chaudhury A, Hervé P, Xue Q, Bennett J (2008) OsTDL1A binds to the LRR domain of rice receptor kinase MSP1, and is required to limit sporocyte numbers. Plant J 54:375–387

Peptides Regulating Apical Meristem Development

Marc Somssich and Rüdiger Simon

Abstract Small signaling peptides are involved in numerous aspects of plant development. A well-studied group of peptides is the CLE family with its most prominent member CLAVATA 3 (CLV3), a key regulator of apical meristem maintenance during growth and development. Members of the CLE peptide family share a high sequence similarity, but are capable of triggering independent pathways. Recent work has revealed complex networks of interacting receptors that recognize different peptides, as well as various posttranslational modifications of the peptides which could account for the signaling specificities within the family. With the CLV pathway well studied in *Arabidopsis*, the insight gained in this model plant can now be transferred to other plant species to control meristem development.

1 Peptides Regulating Apical Meristem Development

The organization, development, and maintenance of multicellular organisms depend on signaling networks that coordinate and integrate cells and tissue responses. Short- and long-range communication between different parts within an organism utilizes direct cell–cell communication, in part based on peptide signaling. In plants, secreted peptides signal via receptors that are mostly in the plasma membrane. The signaling peptides include molecules with diverse biological functions such as systemin, AtPep1, phytosulfokines, and CLAVATA3 (CLV3), a member of the large CLE peptide family. The focus of this chapter is the role of signaling peptides in mediating meristem cell communication and determination of cell fate.

M. Somssich (✉) • R. Simon
Institut für Entwicklungsgenetik, Heinrich-Heine-Universität Düsseldorf,
40225 Düsseldorf, Germany
e-mail: marc.somssich@uni-duesseldorf.de

H.R. Irving and C. Gehring (eds.), *Plant Signaling Peptides*,
Signaling and Communication in Plants 16, DOI 10.1007/978-3-642-27603-3_2,

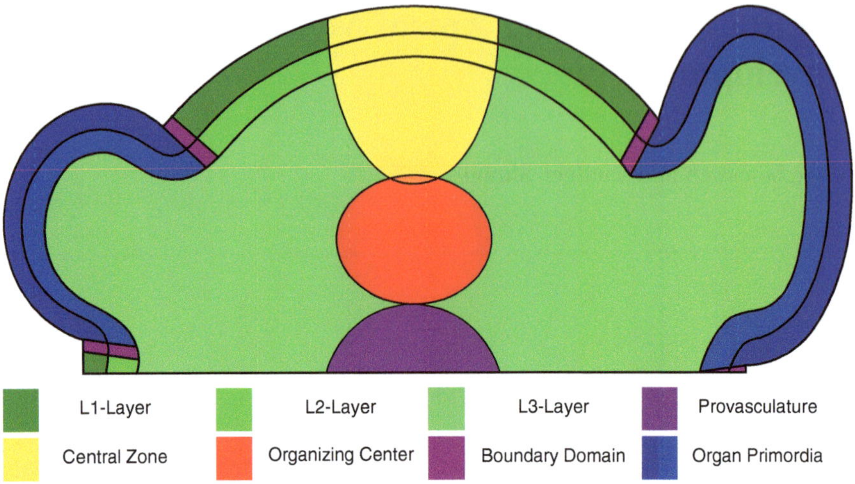

Fig. 1 Functional domains in the shoot apical meristem. Model of the *Arabidopsis* shoot apical meristem. The different subdomains are *color-coded*, based on different functions, gene expression profiles, and cell division rates

2 Structure of the Shoot Apical Meristem

The apical meristems are above-ground stem cell tissues. The main meristem, the shoot apical meristem (SAM), is located at the tip of the shoot and initiates the formation of all above-ground organs and consists of a dome-shaped group of cells that can be divided into specific subdomains based on their functions and gene expression profiles (see Fig. 1). The stem cell population is located at the center of the SAM apex, in the central zone (CZ). Stem cells are undifferentiated and divide slowly. Their daughter cells are displaced toward the peripheral zone (PZ) that encloses the central zone. In the peripheral zone, the cells divide more rapidly and give rise to daughter cells that can enter differentiation pathways or can become part of newly formed organs at the boundary of the meristem. Thus, the peripheral zone serves as a transit-amplifying zone.

The SAM also shows an organization into three distinct layers (L1–L3), which represent individual clones. The outermost layer (L1) contains the cells that will form the epidermis, while the cells of the L2 will form subepidermal tissue and the gametes. Cells in these two layers divide only anticlinally, that is, perpendicular to the meristem surface. All remaining cells form the L3 layer. Their cell divisions are not oriented, and they form the main body of the plant.

The stem cell containing central zone comprises cells of all three layers. Directly underneath these cells is a small population of cells with a low division rate that form the organizing center (OC), which in turn is necessary to initiate and maintain the stem cells of the central zone.

When a plant enters the reproductive stage of development, shoot meristems are transformed into inflorescence meristems (IM) that give rise to a number of lateral floral meristems (FM) that generate the floral organs. Controlled activity of meristems requires communication between stem cells and cells of neighboring tissues.

3 Genes Controlling Stem Cell Activity in the Apical Meristems

In *Arabidopsis thaliana*, the *fully fasciated* (*fuf*) mutants, *fasciata1* and *2* (*fas1/2*), as well as *clavata 1* (*clv1*) were among the first reported mutants that exhibited an enlarged meristem (Leyser and Furner 1992; Medford et al. 1992). Among them, *clv1* was genetically mapped as early as 1983. In 1989, *clv1* was described to have additional inner floral organs, resulting in a club-shaped pistil and silique (clava = club, *latin*) (Koornneef et al. 1983; Okada et al. 1989). In subsequent work, *CLV1* was established as a major regulator of SAM and floral meristem development. In addition to the generally enlarged meristems, caused by the accumulation of stem cells in the central zone, the mutants also exhibit more floral organs than the wild type (Clark et al. 1993). *CLV1* encodes a membrane-associated leucine-rich repeat (LRR) receptor-like kinase (RLK) with the LRR receptor domain on the extracellular and a kinase domain on the intracellular side of the membrane. It is first expressed in the shoot meristem during embryogenesis and later both in shoot and floral meristems. Within the meristem, it is expressed in L2 and L3 layer cells of the central zone, as well as cells of the organizing center (OC) (Clark et al. 1997).

In 1998, Kayes et al. described the *clv2* mutant, which resembles a weak *clv1* mutant with respect to the effects on meristem size and organ number. Additionally, *clv2* mutants develop abnormal organs, such as reduced anthers and elongated pedicels, indicating that CLV2 acts together with CLV1 in a pathway regulating the meristem, but independently with regard to organ development (Kayes and Clark 1998). *CLV2* encodes a membrane-associated LRR receptor-like protein (RLP), which lacks an intracellular kinase domain. While *CLV2* is prominently expressed in the shoot apex, RNA was detected in all tissues examined (Jeong et al. 1999).

CLV1 and CLV2 were expected to act together, possibly as a heterodimer, in a pathway regulating stem cell number and overall meristem size. In 2008, Müller et al. identified *CORYNE* (*CRN*), a membrane-associated protein kinase (PK) *crn* mutants are phenotypically weaker than *clv1*, but very similar to *clv2* mutants, also showing stem cell accumulation and organ defects. While CLV1 represents a full RLK, CLV2, which lacks the kinase domain, and CRN, which lacks a receptor domain, could complement each other. Because genetic data further revealed additive effects in *clv1/clv2* and *clv1/crn* double mutants, a new model was proposed with CLV1 functioning in parallel to a complex consisting of CLV2 and CRN. These two distinct pathways may, however, be interconnected through

Table 1 Genes involved in peptide-related meristem maintenance

Name	Annotation	Protein type	Expression in SAM	Proposed function
Peptides				
CLV3	*AT2G27250*	Peptide	Stem cells	Promotes stem cell differentiation
CLE42	*AT2G34925*	Peptide	Entire SAM	Could promote stem cell fate
Receptor/kinases				
CLV1	*AT1G75820*	LRR-RLK	CZ (L2,L3), OC	Promote stem cell differentiation by downregulating WUS. CLV1, and probably CLV2, can bind the CLV3 peptide
CLV2	*AT1G65380*	LRR-RLP	Entire SAM	
CRN	*AT5G13290*	Kinase	Entire SAM	
RPK2	*AT3G02130*	LRR-RLK	Periphery, OC	
BAM1	*AT5G65700*	LRR-RLK	Periphery	Can bind CLE peptides, partially suppress or enhance *clv* phenotypes
BAM2	*AT3G49670*	LRR-RLK	Periphery	
BAM3	*AT4G20270*	LRR-RLK	Periphery	
TDR/PXY	*AT5G61480*	LRR-RLK	Entire SAM	Could promote stem cell fate
Other factors				
WUS	*AT2G17950*	Homeodomain transcription factor	OC	Promotes stem cell fate
POL	*AT2G46920*	Protein phosphatase 2C	Entire SAM	Negative regulators of CLV1, and possibly CRN signaling
KAPP	*AT5G19280*	Protein phosphatase 2C	Entire SAM	

Genes, ATG numbers, predicted protein types, expression domains, and expected functions of genes related to peptide signaling in the *Arabidopsis* shoot apical meristem.

the formation of larger protein complexes that contain CLV1, CLV2, and CRN (Müller et al. 2008; Bleckmann et al. 2010; also see chapter "Peptides Regulating Root Growth" and "Peptides Regulating Plant Vascular Development"). The genes involved in *Arabidopsis* SAM regulation are listed in Table 1.

4 Feedback Regulation of Stem Cell Fate in Shoot Meristems

The *CLV3* gene was identified as a specific regulator of shoot and floral meristems in 1995. The *clv3* mutants, like *clv1* mutants, have strongly enlarged meristems with a broader stem cell domain and flowers with additional organs. Interestingly, *clv3* mutants are phenotypically stronger than *clv1* mutants, an observation that was later attributed to the second receptor complex, CLV2/CRN, still active in *clv1* mutants (Clark et al. 1995). In 1999, Fletcher et al. identified CLV3 as a small signaling peptide. *CLV3* is specifically expressed in the L1–L3 stem cell layers of the central zone, while *CLV1* is only expressed in deeper layers of the L3. This led to the conclusion that the CLV3 peptide could be secreted from the stem cells where it is produced and bind to the CLV1 or CLV2/CRN receptor complexes in L3 cells (Fletcher et al. 1999). Signaling through the kinase domains of CLV1 or CRN would then lead to the regulation of target genes, thereby restricting stem cell fate.

Since all of the above-mentioned mutants exhibit higher stem cell numbers, the respective proteins are thought to restrict stem cell fate in the wild type. A factor promoting stem cell fate was described in 1996 when it was shown that plants carrying mutations in the *WUSCHEL* (*WUS*) gene terminate growth prematurely and develop flowers that lack the inner organs due to an aberrant differentiation of stem cells (Laux et al. 1996). WUS is a homeodomain transcription factor that is expressed in the organizing center (OC), but confers stem cell fate to the cells above in the central zone (Mayer et al. 1998). A later study suggested that WUS does not directly influence the expression domain of *CLV3*, but regulates the CLV3 expression level, probably together with SHOOTMERISTEMLESS (STM), another homeodomain transcription factor (Brand et al. 2002).

Genetic interaction studies finally connected *CLV* with *WUS* function and allowed establishment of a model for stem cell maintenance in the SAM and floral meristem. Double mutants of *wus* with *clv1* or *clv3* do not maintain an active shoot meristem, thus resembling single *wus* mutants. The flowers of *wus clv1*, *wus clv2*, or *wus clv3* double mutants also resemble *wus* single mutants, indicating epistatic relationship of *wus* toward *clv1*, *clv2*, and *clv3*. *WUS* expression is drastically altered in the *clv* mutants; in all three mutants, the *WUS* expression domain is broader and extends into the lower layer of the central zone. This indicates that the CLV proteins normally function to restrict *WUS* expression from the stem cell domain. Accordingly, *WUS* overexpression causes meristem overgrowth, resulting in plants that resemble *clv* mutants.

Based on these results, it was suggested that the CLV3 peptide is secreted from stem cells and recognized by the CLV receptors in the lower organizing center (OC) cells to restrict *WUS* expression. Reduced WUS activity results in fewer stem cells being maintained, thereby lowering CLV3 signal intensity and consequentially strengthening *WUS* expression. Thus, a regulatory feedback loop is established that regulates stem cell number (Schoof et al. 2000). This model is based on the idea that CLV3 is being secreted, a reasonable assumption given the signal peptide in its N-terminal region (also see chapters "Peptides Regulating Plant Vascular Development," "Processing of Peptides," and "Methods to Identify New Partners of Plant Signaling Peptides"). The functionality of this signal was tested by Rojo et al. who showed that a full-length version of CLV3 is localized to the extracellular space, while a deletion construct, lacking the signal peptide, remains in the cytoplasm. Accordingly, only the full-length CLV3 version was able to rescue a *clv3* mutant, while the truncated version was not (Rojo et al. 2002). Once secreted, the CLV3 peptide is supposed to diffuse freely in the meristem and is bound by CLV1 and possibly other related RLKs (see below). One immediate consequence of CLV3 binding is a relocation of the CLV1 receptor to the vacuole and receptor degradation. Whether this relocation contributes to the signaling mechanism or serves to attenuate the signaling is currently not known (Nimchuk et al. 2011).

5 Peptides from the CLE Family Control Diverse Developmental Processes

The CLV3 protein, first described in 1999 by Fletcher et al. was predicted to consist of 96 amino acids and carry an N-terminal 18-amino-acid hydrophobic region that may act as a signal peptide, targeting CLV3 into the secretory pathway (Fletcher et al. 1999). The binding of CLV3 to the CLV1 extracellular domain was first shown in vitro, using membrane fractions of plant protoplasts overexpressing *CLV1* (lacking the kinase domain) and with isotope-labeled CLV3 peptide (Ogawa et al. 2008). In 2010, it was shown that this binding of CLV3 actually leads to phosphorylation of the CLV1 kinase (Betsuyaku et al. 2011). More than 400 RLKs are encoded in the *Arabidopsis* genome, and the majority of these represent orphan receptors that have not yet been paired with a ligand. First bioinformatic searches by Cock and McCormick in 2001 uncovered that the EMBRYO SURROUNDING REGION (ESR) proteins from *Zea mays* share a 14-amino-acid conserved region with CLV3. ESR proteins are also secreted peptides, and two *clv3* mutations described previously carry mutations inside the 14-amino-acid conserved region, pointing to a functional role of this domain (Fletcher et al. 1999). The conserved 14 amino acids were later termed the CLE domain (for *CLAVATA3/ESR-related)* or CLE-box that turned out to be conserved among family members in diverse plant species (Cock and McCormick 2001). Based on root growth assays, the CLE peptides were further classified into two classes, named A-type or B-type. For this, *Arabidopsis* seedlings were grown on media, containing one of 22 different *Arabidopsis* CLE peptides. Out of these 22, 18 arrested root growth, while 4, CLE41–44, did not. Based on this, CLE41–44 were classified B-type CLE peptides, and all the others as A-type, even though most subsequent studies focused only on a few peptides from each class (CLE3, CLE6, CLE19, and CLE40 representing the A-type and CLE41 representing the B-type). Concerning the SAM, treatment with A-type CLEs led to stem cell differentiation, while treatment with CLE41 did not, hinting at the possibility that A-type CLEs might regulate primary meristem formation (see chapter "Peptides Regulating Root Growth"), while B-type CLEs control other processes such as vascular differentiation (Whitford et al. 2008; see chapter "Peptides Regulating Plant Vascular Development"). In a follow-up study, overexpression of CLE1, CLE6, and CLE7 was shown to lead to stem cell differentiation in the SAM only, while CLE14 led to the termination of both SAM and root meristem (Meng and Feldman 2010).

CLE42 and tracheary element differentiation inhibitory factor (TDIF) (CLE41/44) have an almost identical amino acid sequence and are believed to be involved in the development of tracheary elements (Ito et al. 2006; see chapters "Peptides

Regulating Plant Vascular Development" and "Methods to Identify New Partners of Plant Signaling Peptides"). A recent study has uncovered that both peptides can promote the initiation of auxiliary meristems. Furthermore, CLE42 but not TDIF is expressed also in the SAM. The receptor for CLE42 is TDIF RECEPTOR (TDR), also known as *PHLOEM INTERCALATED WITH XYLEM* (PXY). The *tdr* mutants are insensitive to CLE42 treatment, and TDR too is expressed in the center of the SAM, overlapping with the CLE42 expression domain. This may indicate that TDR might act as receptor for CLE42 not only in the vascular system but also in the SAM. Furthermore, overexpression of CLE42 and TDIF led to increased and ectopic STM expression, a marker for stem cells. This observation implies that CLE42 signaling also contributes to the establishment and maintenance of shoot meristems (Yaginuma et al. 2011).

6 Posttranslational Processing of CLE Peptides

In 2006, matrix-assisted laser desorption/ionization time-of-flight mass spectrometry (MALDI-TOF-MS) from CLV3 overexpressing calli allowed the identification of a 12-amino-acid peptide as the active form of CLV3. This peptide was shown to be hydroxyprolinated on two of its three proline residues, revealing the mature CLV3 peptide sequence to be RTVP^{h}SGP^{h}DPLHH. The functionality of this peptide was tested by growing *Arabidopsis* seedlings on growth media containing 1 μM of the peptide and resulted in severely reduced root growth, reduced SAM size, and phyllotactic defects resembling those of *CLV3* overexpressing lines or *wus* mutants. Synthetic peptides consisting of the 14 amino acid CLE-box, or versions containing the 14-amino-acid conserved region plus up to 14 additional amino acids at the C-terminal end, did lead to the same result; however, an 11 amino acid version seemed inactive. Treatment with 12 amino acid versions of CLE40 and CLE19 led to similar results, showing that these peptides might function in a manner similar to that of CLV3. Proposed functions for the hydroxyprolination included peptide stabilization, trafficking, or storage (Kondo et al. 2006). Further analysis of the posttranslational modifications of CLV3 was performed in 2009 by Ohyama et al. who used nano-LC-MS/MS to purify CLE peptides from plant tissue and detected a 13-amino-acid form of CLV3. The authors confirmed the hydroxyl modification of at least one proline residue and detected a sugar chain linked to the proline at position 7. This sugar side chain was found to consist of three L-arabinose residues, making CLV3 a glycopeptide. Functional testing of these peptides and binding affinity to CLV1 showed that it was more active than the other versions of CLV3, in that treatment of *clv3* mutants with this modified glycopeptide led to a full rescue in concentrations in which the other forms could not fully restore the wild type. From this, it was concluded that this version of CLV3 is the candidate closest to the natural form. As glycosylation is known to affect protein conformation, binding specificity, and stability, it is conceivable that these modifications of CLV3 are essential for biological function and/or protect the mature peptide from active degradation by proteolysis (Ohyama et al. 2009; also see chapters "Peptides Regulating Plant Vascular Development" and "Processing of Peptides"). Meng et al. investigated how CLE peptides obtain their functional specificity and found evidence that regions outside the *CLE*-box are also important for CLE peptide activity (Meng et al. 2010).

7 Receptor Redundancies and Additional CLE-Dependent Pathways

BAM receptors were identified as additional receptor kinases contributing to stem cell specification. When searching for genes with sequence similarity to *CLV1*, DeYoung et al. identified the three *BARELY ANY MERISTEM* (*BAM1-3*) genes. Despite the high similarity between the *BAM* and *CLV1* genes, *bam* mutants show opposite effects to *clv1* mutants in that their meristems are reduced and terminate early and their flowers have less organs. Since CLV1 genes and BAM genes can at least partially replace each other if overexpressed, these differences in function must be caused by different expression patterns. Consistent with this, *BAM* genes are preferentially expressed in the flanks of the meristem, and only faintly in the *CLV1*-expressing domain (DeYoung et al. 2006). Genetic analysis revealed that the *bam1 bam2* mutations enhanced the *clv1* phenotype, but had no effect on the *clv2* phenotype and suppressed the *clv3* phenotype. In a first attempt to explain these effects, a previously described sequestration model was invoked. It was proposed that the BAM receptors protect the central zone from CLE peptides that are expressed in the lateral meristem or young organs by binding and sequestering them. In *bam* mutants, these peptides would be able to reach the center of the meristem, where they could activate *CLV1*, leading to a smaller meristem. These peptides would partially substitute for a lack of CLV3 in clv3 mutants, and this is consistent with a partial suppression of *clv3* phenotypes by *bam* mutants. The enhancement of the *clv1* phenotype by mutations in *bam1* and *bam2* was explained by a low expression of the BAM receptors in the center of the meristem where they would act redundantly with CLV1 (DeYoung and Clark 2008). This last point was later substantiated when interactions between the BAM and CLV1 receptors were shown in co-immunoprecipitation experiments. Furthermore, it was shown that BAM1 and BAM2 are capable of binding a variety of different CLE peptides using in vitro binding assays, including CLV3 (Guo et al. 2010). It should be noted that the *bam1 bam2 clv1* triple mutants revealed phenotypic traits like strong dwarfism and filamentous floral organs that were not observed in any of the single mutants. This supports the notion that *BAM* and *CLV1* might also act synergistically in other developmental pathways (DeYoung and Clark 2008).

RECEPTOR-LIKE PROTEIN KINASE 2/TOADSTOOL 2 (RPK2/TOAD2) was identified in a genetic screen for suppressors of CLE-induced shoot meristem growth arrest. The *rpk2* mutants resemble weak *clv* mutants and generate an enlarged shoot meristem due to stem cell proliferation and the formation of additional floral organs. Since the effects of *rpk2* are additive to *clv1* or *clv2*, RPK2 probably functions in an independent pathway in parallel to those controlled by CLV1 and CLV2/CRN. This idea is supported by the finding that RPK2 does not interact with CLV1 or CLV2/CRN in transient expression experiments in *Nicotiana benthamiana*, but only forms homomers (Kinoshita et al. 2010). However, co-immunoprecipitation experiments suggested weak interactions between CLV1 and RPK2 (Betsuyaku et al. 2011). Whether RPK2 also binds CLV3 has not yet been determined, but *rpk2* plants seem to be insensitive to the effects of exogenous CLV3 application, such as the reduction of SAM size and root meristem arrest (Kinoshita et al. 2010). Interestingly, RPK2 appears to be depleted from the central zone of the meristem, suggesting that similar to CLV1, the binding of CLV3 to RPK2 may trigger receptor degradation.

8 Downstream Signaling Components

The signaling components acting downstream of the CLV3 perceiving receptor kinases are mostly unknown. Genetic screens have so far identified two different types of phosphatases that are involved in signal attenuation or transduction. In 1997, the kinase-associated protein phosphatase (KAPP) was identified and reported to bind to and dephosphorylate the CLV1 kinase, raising the possibility of KAPP as a negative regulator of CLV signaling. This was supported by the observation that plants overexpressing *KAPP* resemble *clv1* mutants (Williams et al. 1997). However, KAPP is likely involved in a number of different signaling pathways and does not act only upon CLV1. In 2000, the *poltergeist* (*pol*) mutant was identified, and while *pol* single mutants do not have an obvious phenotype, they suppress even strong *clv1*, *clv2*, and *clv3* phenotypes, placing the POL gene function downstream of the CLVs (Yu et al. 2000). Additionally, *pol* enhances the *wus* phenotype. The *pol wus* double mutants show further reduced embryonic organ formation and an overall reduction of floral organ numbers, placing POL in a pathway parallel to the stem cell promoting function of WUS. Triple mutants of *pol wus* and *clv1* resemble *clv1 wus* double mutants, supporting the notion that POL might function in parallel to WUS, but requires CLV1. Interestingly, POL overexpression is sufficient to stimulate stem cell proliferation and maintenance. POL and the related gene POLTERGEIST-LIKE 1 (PLL1) encode membrane-localized protein phosphatases that were shown to be activated by phospholipids (Yu et al. 2003; Song and Clark 2005; Gagne and Clark 2010). The current model suggests that activation of the CLV receptors could antagonize the phospholipid-dependent stimulation of POL and/or PLL1.

The main target gene subject to regulation by the CLV pathway is WUS. This transcription factor acts non-cell-autonomously from the organizing center (OC) to promote stem cell fate in the central zone of the SAM. WUS acts in concert with the transcriptional corepressor TOPLESS (TPL) (Kieffer et al. 2006; Long et al. 2006). Direct target genes downregulated by WUS are the A-type *ARABIDOPSIS RESPONSE REGULATOR* (*ARR*) genes *ARR5*, *6*, *7*, and *15*, which allow signaling connection through the CLV pathway with cytokinin signaling, since these ARR genes are negative regulators of cytokinin signaling. Of these four related genes, at least ARR7 and 15 were shown to be involved in SAM regulation (Leibfried et al. 2005). Thus, WUS ultimately promotes cytokinin signaling in the meristem. The phytohormone auxin again promotes expression of at least *ARR7* and *15*, thereby antagonizing the functions of cytokinin and WUS (Zhao et al. 2010).

9 Evolutionary Conservation of Peptides and Signaling Pathways Directing Shoot Meristem Maintenance

9.1 CLV-Like Pathways in Dicotyledonous Plants

In 2002, the *Petunia hybrid WUS* homolog *TERMINATOR* (*TER*) was described. Just like *wus* mutants in *Arabidopsis*, *ter* mutants in *Petunia* do not maintain their stem cells. In wild type, *TER* is expressed in the central zone underneath the stem cells of the SAM. TER acts in parallel with a second transcription factor HAIRY MERISTEM (HAM) that belongs to the GRAS domain family. In *ham* mutants, the meristems cannot be maintained postembryonically and therefore terminate. Interestingly, *HAM* is only expressed in differentiated cells of organ primordia, sites of organ initiation, and the provascular tissue, but not in the meristem itself. Thus, HAM must act non-cell-autonomously. Since mutations in both genes led to meristem termination, the relationship between these transcription factors was analyzed. The *ham ter* double mutants exhibit a hybrid phenotype of traits that lead to the conclusion that they are likely to function in parallel to maintain stem cell activity in the meristem (Stuurman et al. 2002). To date, the identification of the highly similar *Arabidopsis WUS* homolog *PhWUS* (*TER*) is the only indication of a possible conservation of the CLAVATA pathway also in *Petunia*.

A *WUS* homolog for *Antirrhinum*, *ROSULATA* (*ROA*), was identified in 2006 and the *roa* mutants initiate meristems, but fail to maintain them. Like *WUS* in *Arabidopsis*, *ROA* is expressed below the stem cells (Kieffer et al. 2006). Next to the homeodomain, which is 83% identical between WUS and ROA and 100% identical between PhWUS (TER) and ROA, sequence analysis revealed three more conserved domains: an acidic domain (Mayer et al. 1998), the WUS box, and an ASLELXLN motif (Stuurman et al. 2002). Deletion of these three elements in the *roa-2* mutant resulted in a dominant-negative mutation. Expression of a similarly truncated version of WUS in *Arabidopsis* also caused dominant-negative effects, i.e., loss of meristem maintenance. Since the ASLELXLN motif is similar to known motifs of transcriptional repressors (e.g., the EAR motif), it was proposed that WUS and its homologs function as repressors that recruit transcriptional corepressors via their conserved C-terminal domain. If this repression is relieved by deleting the C-terminal part of WUS or ROA, stem cells are no longer prevented from entering differentiation, and the meristem terminates (Kieffer et al. 2006).

9.2 CLV-Like Pathways in Monocotyledonous Plants

An *in silico* screen for CLV3 homologs by Cock and McCormick (2001) brought up the *embryo-surrounding region* (*ESR*) genes from *Z. mays*. The sequence homology was confined to the 14 amino acid CLE domain and the functional signal peptide at the N terminus. Since ESR peptides are specifically expressed between the endosperm and embryo, it was suggested that they act as signaling molecules to coordinate development in zygotic and maternal tissue (Opsahl-Ferstad et al. 1997). In *Oryza sativa* (rice), Chu et al. (2006) described FLORAL ORGAN NUMBER 4 (FON4) as an ortholog of the *Arabidopsis* CLV3 peptide. The *fon4* mutants have enlarged shoot and floral meristems that produce more organs and, in the case of the floral meristem, do not terminate after the initiation of the carpel primordia. FON4, like CLV3, contains a CLE motif and a signal peptide for secretion. *FON4* is expressed only in the central apex of the meristem. The functional conservation of the CLV3 CLE motif was further tested by treating *Arabidopsis* and rice seedlings with CLV3 or FON4 peptide. CLV3 peptide induced the premature differentiation and growth arrest of both SAM and root meristem, while FON4 arrested SAM development, but had no effect in the root (Chu et al. 2006).

Homologs of CLV1 or CLV2 were found in *Z. mays* in 2001 and 2005, respectively. *FASCIATED EAR 2* (*FEA2*) was identified in a mutant screen, because *fea2* mutants exhibit larger inflorescences and floral meristems. *FEA2* encodes a LRR receptor-like protein with high sequence similarity to AtCLV2 that localizes to the plasma membrane (Taguchi-Shiobara et al. 2001). The THICK TASSEL DWARF1 (TD1) gene of maize was shown to restrict meristem size, and the *td1* mutants show enlarged inflorescence and floral meristems. *TD1* encodes a LRR RLK that is expressed more broadly in the plant compared to *CLV1* of *Arabidopsis*. Because *td1 fea2* double mutants show additive effects, it is possible that both CLV1 and CLV2 orthologs act in parallel pathways in meristem regulation, just like they do in *Arabidopsis* (Bommert et al. 2005). The CLV1 ortholog in *O. sativa* is *FLORAL ORGAN NUMBER 1* (*FON1*). While *CLV1* is expressed specifically in the central region of the meristem, *FON1* is expressed throughout the SAM and the floral meristem, which could indicate that regional specificity of FON1 function depends on a localized cofactor. So far, a CLV2 homolog has not been identified in rice (Suzaki et al. 2009).

WUS homologs were identified in *Z. mays* (*ZmWUS1* and *ZmWUS2)* and in *O. sativa* (*OsWUS*). In contrast to *Arabidopsis WUS*, the maize and rice *WUS* genes are expressed in the meristems in a pattern overlapping with that of the *CLV1* homolog *TD1*. *OsWUS* is expressed in leaf primordia and transiently in the SAM. *ZmWUS1* is expressed in the SAM, while *ZmWUS2* is expressed only in young leaves. Thus, *OsWUS* combines the expression patterns of both *ZmWUS* paralogs, which indicates that the two maize paralogs probably evolved from a single gene through duplication (Nardmann and Werr 2006).

10 Conclusion

The cells in the SAM are ultimately responsible for the generation of the majority of the plant body. Their growth and division rates and the identity and differentiation of daughter cells generated at the shoot meristem must be integrated and adapted to the requirements of the plant. Growth patterns will vary with changing temperatures, and nutrient or water availability, but also upon induction of flower formation. Thus, the meristem itself, but also the separate functional regions within a meristem, will need to perceive and respond to both extrinsic and intrinsic signals. Across the meristem, the responses of cell groups to these signals need to be integrated. Peptide signals have so far been shown to coordinate the development of the organizing center with the stem cell zone at the meristem tip. The responsible signaling pathways have been only partially elucidated, and we are just beginning to unravel the cross talk with hormonal signaling pathways. For example, CLV3 signaling from the stem cells can confine WUS expression, and thereby repress cytokinin signaling. While the generation of the CLV3 signal is strictly dependent on stem cells, cytokinin is both generated and perceived by many plant cells and feeds back upon CLV signaling by regulating the expression of the CLV receptors. We expect that other growth-regulating hormones, the prime candidate being auxin, will also modify the CLV communication pathway. How stem cell behavior in the shoot meristem is adjusted to plant growth requirements is not yet understood. Another layer of complexity in peptide signaling can be expected since a range of different peptides is generated in the shoot meristem. These peptides can be proteolytically processed and further modified by glycosylation, thus generating a large array of different signaling molecules. Furthermore, several hundred RLKs are expressed in meristems. Those that were already analyzed revealed the capacity for complex formation of hetero- or homomers, suggesting that even cells expressing only a limited number of receptors can form a dazzling number of different potential receptor complexes. We can expect that meristem cells employ a large number of peptide-controlled signaling pathways for communication between meristem domains. Redundant or parallel acting pathways are likely to complicate their genetic dissection, and we will need to improve our biochemical toolbox to unravel how specific peptides regulate diverse receptor complexes in meristem domains and how they control plant growth and differentiation.

References

Betsuyaku S, Takahashi F, Kinoshita A, Miwa H, Shinozaki K, Fukuda H, Sawa S (2011) Mitogen-activated protein kinase regulated by the CLAVATA receptors contributes to shoot apical meristem homeostasis. Plant Cell Physiol 52:14–29

Bleckmann A, Weidtkamp-Peters S, Seidel CA, Simon R (2010) Stem cell signaling in Arabidopsis requires CRN to localize CLV2 to the plasma membrane. Plant Physiol 152:166–176

Bommert P, Lunde C, Nardmann J, Vollbrecht E, Running M, Jackson D, Hake S, Werr W (2005) thick tassel dwarf1 encodes a putative maize ortholog of the Arabidopsis CLAVATA1 leucine-rich repeat receptor-like kinase. Development 132:1235–1245

Brand U, Grunewald M, Hobe M, Simon R (2002) Regulation of CLV3 expression by two homeobox genes in Arabidopsis. Plant Physiol 129:565–575

Chu H, Qian Q, Liang W, Yin C, Tan H, Yao X, Yuan Z, Yang J, Huang H, Luo D, Ma H, Zhang D (2006) The floral organ number4 gene encoding a putative ortholog of Arabidopsis CLAVATA3 regulates apical meristem size in rice. Plant Physiol 142:1039–1052

Clark SE, Running MP, Meyerowitz EM (1993) CLAVATA1, a regulator of meristem and flower development in Arabidopsis. Development 119:397–418

Clark SE, Running MP, Meyerowitz EM (1995) CLAVATA3 is a specific regulator of shoot and floral meristem development affecting the same processes as CLAVATA1. Development 121:2057–2067

Clark SE, Williams RW, Meyerowitz EM (1997) The CLAVATA1 gene encodes a putative receptor kinase that controls shoot and floral meristem size in Arabidopsis. Cell 89:575–585

Cock JM, McCormick S (2001) A large family of genes that share homology with CLAVATA3. Plant Physiol 126:939–942

DeYoung BJ, Clark SE (2008) BAM receptors regulate stem cell specification and organ development through complex interactions with CLAVATA signaling. Genetics 180:895–904

DeYoung BJ, Bickle KL, Schrage KJ, Muskett P, Patel K, Clark SE (2006) The CLAVATA1-related BAM1, BAM2 and BAM3 receptor kinase-like proteins are required for meristem function in Arabidopsis. Plant J 45:1–16

Fletcher JC, Brand U, Running MP, Simon R, Meyerowitz EM (1999) Signaling of cell fate decisions by CLAVATA3 in Arabidopsis shoot meristems. Science 283:1911–1914

Gagne JM, Clark SE (2010) The Arabidopsis stem cell factor POLTERGEIST is membrane localized and phospholipid stimulated. Plant Cell 22:729–743

Guo Y, Han L, Hymes M, Denver R, Clark SE (2010) CLAVATA2 forms a distinct CLE-binding receptor complex regulating Arabidopsis stem cell specification. Plant J 63:889–900

Ito Y, Nakanomyo I, Motose H, Iwamoto K, Sawa S, Dohmae N, Fukuda H (2006) Dodeca-CLE peptides as suppressors of plant stem cell differentiation. Science 313:842–845

Jeong S, Trotochaud AE, Clark SE (1999) The Arabidopsis CLAVATA2 gene encodes a receptor-like protein required for the stability of the CLAVATA1 receptor-like kinase. Plant Cell 11:1925–1934

Kayes JM, Clark SE (1998) CLAVATA2, a regulator of meristem and organ development in Arabidopsis. Development 125:3843–3851

Kieffer M, Stern Y, Cook H, Clerici E, Maulbetsch C, Laux T, Davies B (2006) Analysis of the transcription factor WUSCHEL and its functional homologue in Antirrhinum reveals a potential mechanism for their roles in meristem maintenance. Plant Cell 18:560–573

Kinoshita A, Betsuyaku S, Osakabe Y, Mizuno S, Nagawa S, Stahl Y, Simon R, Yamaguchi-Shinozaki K, Fukuda H, Sawa S (2010) RPK2 is an essential receptor-like kinase that transmits the CLV3 signal in Arabidopsis. Development 137:3911–3920

Kondo T, Sawa S, Kinoshita A, Mizuno S, Kakimoto T, Fukuda H, Sakagami Y (2006) A plant peptide encoded by CLV3 identified by in situ MALDI-TOF MS analysis. Science 313:845–848

Koornneef M, van Eden J, Hanhart CJ, Stam P, Braaksma FJ, Feenstra WJ (1983) Linkage map of *Arabidopsis thaliana*. J Hered 74:265–272

Laux T, Mayer KF, Berger J, Jurgens G (1996) The WUSCHEL gene is required for shoot and floral meristem integrity in Arabidopsis. Development 122:87–96

Leibfried A, To JP, Busch W, Stehling S, Kehle A, Demar M, Kieber JJ, Lohmann JU (2005) WUSCHEL controls meristem function by direct regulation of cytokinin-inducible response regulators. Nature 438:1172–1175

Leyser HMO, Furner IJ (1992) Characterisation of three shoot apical meristem mutants of *Arabidopsis thaliana*. Development 116:397–403

Long JA, Ohno C, Smith ZR, Meyerowitz EM (2006) TOPLESS regulates apical embryonic fate in Arabidopsis. Science 312:1520–1523

Mayer KF, Schoof H, Haecker A, Lenhard M, Jurgens G, Laux T (1998) Role of WUSCHEL in regulating stem cell fate in the Arabidopsis shoot meristem. Cell 95:805–815

Medford JI, Behringer FJ, Callos JD, Feldmann KA (1992) Normal and abnormal development in the Arabidopsis vegetative shoot apex. Plant Cell 4:631–643

Meng L, Feldman LJ (2010) CLE14/CLE20 peptides may interact with CLAVATA2/CORYNE receptor-like kinases to irreversibly inhibit cell division in the root meristem of Arabidopsis. Planta 232:1061–1074

Meng L, Ruth KC, Fletcher JC, Feldman L (2010) The roles of different CLE domains in Arabidopsis CLE polypeptide activity and functional specificity. Mol Plant 3:760–772

Müller R, Bleckmann A, Simon R (2008) The receptor kinase CORYNE of Arabidopsis transmits the stem cell-limiting signal CLAVATA3 independently of CLAVATA1. Plant Cell 20:934–946

Nardmann J, Werr W (2006) The shoot stem cell niche in angiosperms: expression patterns of WUS orthologues in rice and maize imply major modifications in the course of mono- and dicot evolution. Mol Biol Evol 23:2492–2504

Nimchuk ZL, Tarr PT, Ohno C, Qu X, Meyerowitz EM (2011) Plant stem cell signaling involves ligand-dependent trafficking of the CLAVATA1 receptor kinase. Curr Biol 21:345–352

Ogawa M, Shinohara H, Sakagami Y, Matsubayashi Y (2008) Arabidopsis CLV3 peptide directly binds CLV1 ectodomain. Science 319:294

Ohyama K, Shinohara H, Ogawa-Ohnishi M, Matsubayashi Y (2009) A glycopeptide regulating stem cell fate in *Arabidopsis thaliana*. Nat Chem Biol 5:578–580

Okada K, Komaki MK, Shimura Y (1989) Mutational analysis of pistil structure and development of *Arabidopsis thaliana*. Cell Differ Dev 28:27–37

Opsahl-Ferstad HG, Le Deunff E, Dumas C, Rogowsky PM (1997) ZmEsr, a novel endosperm-specific gene expressed in a restricted region around the maize embryo. Plant J 12:235–246

Rojo E, Sharma VK, Kovaleva V, Raikhel NV, Fletcher JC (2002) CLV3 is localized to the extracellular space, where it activates the Arabidopsis CLAVATA stem cell signaling pathway. Plant Cell 14:969–977

Schoof H, Lenhard M, Haecker A, Mayer KF, Jurgens G, Laux T (2000) The stem cell population of Arabidopsis shoot meristems in maintained by a regulatory loop between the CLAVATA and WUSCHEL genes. Cell 100:635–644

Song SK, Clark SE (2005) POL and related phosphatases are dosage-sensitive regulators of meristem and organ development in Arabidopsis. Dev Biol 285:272–284

Stuurman J, Jaggi F, Kuhlemeier C (2002) Shoot meristem maintenance is controlled by a GRAS-gene mediated signal from differentiating cells. Genes Dev 16:2213–2218

Suzaki T, Ohneda M, Toriba T, Yoshida A, Hirano HY (2009) FON2 SPARE1 redundantly regulates floral meristem maintenance with FLORAL ORGAN NUMBER2 in rice. PLoS Genet 5:e1000693

Taguchi-Shiobara F, Yuan Z, Hake S, Jackson D (2001) The fasciated ear2 gene encodes a leucine-rich repeat receptor-like protein that regulates shoot meristem proliferation in maize. Genes Dev 15:2755–2766

Whitford R, Fernandez A, De Groodt R, Ortega E, Hilson P (2008) Plant CLE peptides from two distinct functional classes synergistically induce division of vascular cells. Proc Natl Acad Sci USA 105:18625–18630

Williams RW, Wilson JM, Meyerowitz EM (1997) A possible role for kinase-associated protein phosphatase in the Arabidopsis CLAVATA1 signaling pathway. Proc Natl Acad Sci USA 94:10467–10472

Yaginuma H, Hirakawa Y, Kondo Y, Ohashi-Ito K, Fukuda H (2011) A novel function of TDIF-related peptides: promotion of axillary bud formation. Plant Cell Physiol 52:1354–1364

Yu LP, Simon EJ, Trotochaud AE, Clark SE (2000) POLTERGEIST functions to regulate meristem development downstream of the CLAVATA loci. Development 127:1661–1670

Yu LP, Miller AK, Clark SE (2003) POLTERGEIST encodes a protein phosphatase 2C that regulates CLAVATA pathways controlling stem cell identity at Arabidopsis shoot and flower meristems. Curr Biol 13:179–188

Zhao Z, Andersen SU, Ljung K, Dolezal K, Miotk A, Schultheiss SJ, Lohmann JU (2010) Hormonal control of the shoot stem-cell niche. Nature 465:1089–1092

Peptides Regulating Root Growth

Margret Sauter

Abstract The root apex drives root growth and development. It harbors an apical meristem with a central organizing center known as the quiescent center (QC). The QC cells divide rarely to produce a stem cell and a cell that retains QC cell identity. Stem cells divide more frequently to produce descendant cells which develop into defined cell types by lineage. Proximal and lateral stem cells produce vasculature, cortex, and epidermis. Distal stem cells produce the root cap. Cell–cell communication at the root apex is required to maintain proper cell identity and to promote cell type specification. Peptides are particularly suited for short-distance communication and have been identified in past years as key regulators of the root apical meristem. They are water-soluble, secreted molecules that are perceived by plasma membrane-localized receptors. And they are many. This chapter gives an overview of peptides with a known function in roots and summarizes key findings on how they contribute to the organization of the root structure and to root growth.

1 The Root Apical Meristem

In the root apical meristem (RAM) of advanced dicotylous plants, all cells are derived from specific tiers of stem cells by lineage, an organization known as the closed type of root meristem (Rost 2011). The stem cell region at the root apex is made up of functionally different zones: the quiescent center (QC), the proximal (toward the shoot), the lateral, and the distal (toward the root cap) stem cells (Stahl and Simon 2010; Tucker and Laux 2007). In *Arabidopsis thaliana*, the QC is made

M. Sauter (✉)
Plant Developmental Biology and Plant Physiology, University of Kiel, Am Botanischen Garten 5, 24118 Kiel, Germany
e-mail: msauter@bot.uni-kiel.de

H.R. Irving and C. Gehring (eds.), *Plant Signaling Peptides*,
Signaling and Communication in Plants 16, DOI 10.1007/978-3-642-27603-3_3,

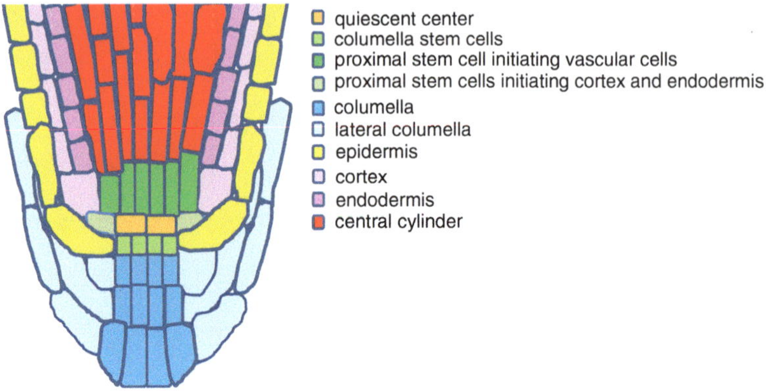

Fig. 1 The root apex of Arabidopsis. Schematic diagram of the stem cells and their descendant cells. The quiescent center (QC) cells act as the organizing center and are surrounded by the stem cells which are in immediate contact with QC cells. Columella cells originate from the distal stem cells. Lateral stem cells produce the lateral root cap, the epidermis, cortex, and endodermis. Distal stem cells produce the root cap

up of four cells which divide rarely and act as the organizing center. Stem cells, also termed initial cells, divide continuously to produce meristematic cells. The stem cells surround the QC and are thus in direct contact with the organizing center (Fig. 1). Stem cells divide asymmetrically to produce one cell that remains in contact with the QC and retains its stem cell identity and one meristematic cell that can continue to divide for a finite number of times before it differentiates. Proximal stem cells produce the vascular cylinder including the pericycle, while the endodermis, cortex, epidermis, and lateral root cap originate from lateral stem cells. The root cap of columella cells is derived from distal stem cells (Stahl et al. 2009). It protects the root tip. In addition, the starch granules, which are characteristic for columella cells, help to sense gravity.

The RAM hence encompasses two or even three stem cell regions: the proximal, the distal, and the lateral stem cells which originate from the same organizing center, the QC, but have vastly different tasks. The RAM is maintained and restricted by the interaction of signaling molecules that act over short distances. Among these, peptide signals have been identified as prominent regulators of meristem maintenance and root growth. Oftentimes, these signaling peptides are posttranslationally modified to enhance water solubility and stability or to provide binding specificity. All peptides identified so far are secreted into the apoplast where they are perceived by plasma membrane-localized leucine-rich repeat (LRR) receptors. This review will provide an overview of the peptides and peptide families which are known to participate in signaling of root growth and differentiation.

2 CLE Peptides Regulate Root Growth and Differentiation

2.1 CLE Peptides

The CLV3 (CLAVATA 3)/ENDOSPERM SURROUNDING REGION (ESR)-related (CLE) peptides are encoded as precursors by a gene family which consists of 32 members in *Arabidopsis*. All precursor proteins share a conserved region of 14 amino acids, the CLE motif, which is localized at or near the C terminus. The precursor proteins are further characterized by a signal peptide that targets the protein for secretion and a variable region between the signal peptide and the CLE motif. The best characterized member of the CLE family, CLV3, was shown to be a 13-amino-acid glycopeptide which is arabinosylated at the hydroxyproline residue Hyp7, a modification that is essential for its function (Ohyama et al. 2008). CLV3 controls the size of the stem cell niche in the shoot (Stahl and Simon 2010; also see chapter "Peptides Regulating Apical Meristem Development"). It is expressed in the central zone of the stem cell niche and limits stem cell proliferation. CLV3 peptide is perceived by the LRR kinase complex CLAVATA1 (CLV1)/CLAVATA2 (CLV2) and by the receptor-like kinases CLV2/CORYNE (CRN; also called SOL2). CLV2/CRN, but not CLV1/CLV2, also functions in root meristem development. *CLV1* knockout results in an enlarged shoot apical meristem (SAM). In the SAM, CLV3 acts by downregulating the expression of WUSCHEL (WUS), a homeobox transcription factor expressed in the organizing center of the SAM that maintains stem cells in an undifferentiated and proliferating state. *WUS* knockout results in terminal differentiation of stem cells and loss of the SAM.

CLE peptides also affect root growth. In *Arabidopsis*, some CLE peptides promote, whereas others inhibit, root growth. *CLE* genes have been classified according to their ability to affect root or SAMs as A-type *CLE* genes in contrast to B-type *CLE* genes which do not. A comprehensive promoter/GUS study showed that many *CLE* genes are active in roots with differential expression patterns. *CLE* genes are generally expressed throughout the plant and throughout development such that most likely almost any plant cell expresses one or more *CLE* genes at a time (Jun et al. 2010). Mostly, knockout mutants of single *CLE* genes are phenotypically normal, indicating some degree of redundancy in function. Overexpression of single *CLE* genes can result in root growth inhibition as shown for *Arabidopsis CLV3*, *CLV9*, *CLV10*, *CLV11*, and *CLV13* (Strabala et al. 2006). By contrast, overexpression of *CLE5*, *CLE6*, *CLE7*, *CLE18*, *CLE25*, *CLE26*, and, to a minor degree, *CLE2* and *CLE4* promoted root growth. Overexpression of *CLE* genes or application of synthetic CLE peptides has been found useful to decipher CLE signaling in roots. Even though the in planta function of specific CLE peptides is in most cases not understood, such studies provided insight into general regulatory abilities of CLE peptide signaling in roots.

CLV3 and *CLE19* are not expressed in roots or root meristems, respectively, but ectopic expression in roots reduces the root meristem size as does overexpression of *CLE40* or application of synthetic CLE19 or CLE40 peptide (Fiers et al. 2004; Wang and Fiers 2010). By contrast, the B-type CLE peptides CLE41 and CLE44 do not affect meristem size but promote proliferation of vascular cells and suppress differentiation of procambial cells (Ito et al. 2006). Cauliflower mosaic virus *35S (35S):CLE19* or *35S:CLE40* overexpression results in progressive root growth retardation (Hobe et al. 2003; Fiers et al. 2004). The RAM becomes progressively smaller because meristem cells differentiate prematurely until the RAM disappears completely 2 weeks after germination, resulting in root growth arrest. A similar consumption of the RAM was observed in plants overexpressing CLE14 or CLE20 (Meng et al. 2010). Transient application of CLE14 or CLE20 peptide to young seedlings demonstrated that the loss in meristematic activity and root growth was not restored after removal of peptides, indicating that loss of growth capacity induced by ectopic CLE peptide activity in the root is permanent (Meng and Feldman 2010). Premature differentiation of RAM cells is indicated by root hair formation close to the root tip.

While in *35S:CLE19* seedlings the RAM was consumed, the columella persisted and the roots showed a normal gravitropic response, indicating that cell differentiation from the proximal and distal stem cell niches is regulated through independent mechanisms. Native *CLE19* expression in roots is limited to pericycle cells facing the xylem poles. As neither knockout nor overexpression altered lateral root initiation, the function of the pericycle cell-specific expression of *CLE19* remains unexplored (Fiers et al. 2004). Expression of *CLE14* in epidermal cells, the root cap, and in root hairs may indicate a function in promoting differentiation of these cell types. Likewise, expression of *CLE20* in protoxylem cells may indicate a native function in vascular cell differentiation (Meng and Feldman 2010).

CLE peptides play a role in keeping a balance between stem cells and root cell differentiation. CLE peptides that inhibit root growth promote cell differentiation and inhibit stem cell identity. This likely occurs independently of auxin signaling. Analysis of auxin activity by *DR5:GUS* staining did not reveal altered auxin responsiveness in *CLE19* overexpressors except as a consequence of disappearing stem cells and QC cells.

2.2 CLE Peptide Perception

CLE peptides are perceived by LRR receptor-like kinases (LRR-RLKs) at the plasma membrane. The receptor pair CLV2/CRN has been implicated in CLE perception in roots. *CRN* is expressed in the QC and surrounding stem cells, and expression extends into the proximal root meristem (Müller et al. 2008). In loss-of-function mutants of *CLV2*, roots look phenotypically normal, but the *clv2* knockout mutant is no longer responsive to applied CLE19, CLE20, and CLE40 peptide (Birnboim et al. 2003; Fiers et al. 2005; Miwa et al. 2008). Similar to the *clv2* mutants, the *crn-1* mutant is per se aphenotypic, but it is insensitive to root growth inhibition by excess CLE40 peptide, indicating that CLE40 and possibly other CLE peptides are perceived by CRN (Müller et al. 2008). CLV2, which lacks a functional kinase domain, interacts with the receptor-like kinase CRN which has a very small extracellular domain, likely forming heterodimers or heterotetramers that are predicted to possess binding pockets for CLE14, CLE19, and CLE20 peptides (Bleckmann et al. 2010; Meng and Feldman 2010). As *crn-1* is insensitive to growth inhibition by CLE40, binding of this peptide likely also occurs through CRN/CLV2. The indigenous CLE-CRN/CLV2 pairs that signal root growth in the RAM have yet to be identified. Using fluorescently labeled proteins, Bleckmann et al. (2010) showed that CLV2 and CRN are mutually dependent on each other for transport from the endoplasmic reticulum to the plasma membrane. The observation that the *clv2* and *crn-1* loss-of-function mutants are aphenotypic indicates that redundant receptor(s) exist which can compensate for loss of CLE perception by the CRN/CLV2 complex. CRN/CLV2 receptors appear to be specific for higher plants, whereas CLV1, the receptor for CLV3 in the SAM, is also found in pteridophytes and in moss.

Arabidopsis CRINKLY4 (ACR4) is a membrane-localized receptor-like kinase which was shown to be required for the differentiation of columella cells. During lateral root development, *ACR4* is expressed as early as in the first pericycle cells formed after the first asymmetric cell divisions (De Smet et al. 2008). *acr4* produced 19% more lateral root meristems per centimeter than wild type, with disturbed spacing of adjacent lateral primordia, that is, close to each other and positioned at opposite poles. As many aberrant primordia do not emerge, lateral root density in *acr4* is lower than in wild type. In the RAM, *ACR4* is expressed in columella stem cells and in columella cells but not in the QC and not in the proximal stem cell region. In accord with this expression pattern, *acr4* mutants frequently displayed a distorted root cap, indicating that ACR4 is required for synchronized divisions of columella stem cells responsible for tiers of columella cells being formed (De Smet et al. 2008). ACR4 prevents extra asymmetric divisions, resulting in extra lateral root primordia or in unstructured root cap formation.

2.3 CLE Signaling

The WUS-related homeobox 5 transcription factor *WOX5* is expressed in the QC of the RAM. Loss of WOX5 results in premature differentiation of columella stem cells as indicated by starch accumulation, indicating that WOX5 acts in a non-cell-autonomous manner from the QC on columella stem cells to keep them in a meristematic state (Sarkar et al. 2007). The development of the proximal stem cell region appears to be normal in *wox5*. Ectopic *WOX5* (*35S:WOX*) expression blocks differentiation of columella stem cells into columella cells (Sarkar et al. 2007; Meng et al. 2010). The localized expression of *WOX5* in the QC is under control of *CLE40* which is expressed in differentiated root columella cells and hence acts in a non-cell-autonomous manner (Stahl et al. 2009). Loss of CLE40 leads to expanded *WOX5* expression and to an expanded columella stem cell region, indicating that CLE40 promotes premature columella cell differentiation via

WOX5 (Stahl et al. 2009). Treatment with CLE40 peptide shifts *WOX5* expression from the QC to cells proximal to the QC. The LRR-RLK ACR4 represses and limits *WOX5* expression in a similar manner as CLE40, indicating that CLE40 peptide acts through ACR4 to regulate *WOX5* expression (De Smet et al. 2008).

CLE40 promotes columella cell differentiation also in a *WOX5*-independent manner as *wox5* roots treated with CLE40 peptide displayed not only prematurely differentiated columella stem cells but in addition displayed differentiated columella cells at the site of the QC (Stahl et al. 2009). Taken together, CLE40 activity maintains the balance between columella stem cells and columella cell differentiation by controlling the expression of *WOX5* in the QC of the stem cell niche.

The SCR (SCARECROW), SHR (SHORT-ROOT), and PLT (PLETHORA) transcription factors regulate stem cell maintenance and QC establishment. High PLT levels promote QC fate, whereas low PLT levels induce mitotic activity of stem cells, and lowest PLT levels favor cell differentiation (Stahl and Simon 2010). Loss of WOX5 is phenotypically obvious only in the *shr*, *scr*, or *plt1/plt2* mutant backgrounds where it results in a smaller meristem (Sarkar et al. 2007), indicating that WOX5 contributes to the regulation of the proximal stem cell region in a redundant manner. Auxin regulates *PLT* expression but auxin does not restore mitotic activity or root growth inhibition by CLE peptide (Meng and Feldman 2010).

2.4 Nematodes Producing CLE Peptides

Interestingly, CLE peptide precursor genes were identified in plant-pathogenic cyst nematodes such as *Heterodera glycines* and *Globodera rostochiensis* which invade soybean and potato, respectively (Olsen and Skriver 2003; Lu et al. 2009; also see chapter "The Role of Plant Peptides in Symbiotic Interactions"). Functionality of the peptide encoded by the *H. glycines CLE* gene was shown by rescue of the *clv3* mutant phenotype in *Arabidopsis*. Furthermore, overexpression of CLE genes from *H. glycines* and *G. rostochiensis* in *Arabidopsis* caused consumption of the SAM and short roots identifying nematode *CLE* genes as A type (Wang et al. 2005; Lu et al. 2009). Four of the five *CLE* genes identified in *G. rostochiensis* possess multiple CLE motifs, possibly indicating functional diversification (Lu et al. 2009). It is also conceivable that multiple CLE peptide products that are produced from one transcript result in signal amplification. *Gr-CLE* transcripts were detected specifically in the dorsal gland cell of infective juvenile *G. rostochiensis* which penetrates a root cell with its stylet to eventually transform the plant tissue into a syncytial structure from which it feeds (Lu et al. 2009). Likewise, expression of the *H. glycines CLE* gene *Hg-SYV46* (also termed *HgCLE1*) was specific to the dorsal gland (Bakhetia et al. 2007). The nematode Gr-CLE1 precursor protein is processed to active peptide by host plant proteases. Hence, nematodes secrete CLE precursor proteins from their glands into the plant root cell from where they are secreted into the apoplast targeted by a signal that was shown to reside in the variable domain of the HgCLE2 propeptide sequence (Wang et al. 2010). CLE precursors are then processed to active CLE peptides in the apoplast by plant-specific protease and perceived by plant-intrinsic CLE receptors (Guo et al. 2011). In *Arabidopsis*, the CLV2/CRN receptors were shown to be required for nematode CLE signaling. *CLV2* and *CRN* are expressed in syncytia and, when mutated, result in reduced syncytium size (Replogle et al. 2011). Furthermore, *clv2* mutants lacking in addition either BARELY ANY MERISTEM 1 (BAM1) or BAM2 receptor showed strong resistance to root growth inhibition by Gr-CLE1 peptide, indicating that plant CLV/CRN/BAM1/BAM2 receptors can perceive nematode CLE peptides (Guo et al. 2011). BAM1 and BAM2 are close homologs of CLV1 and were previously shown to participate in CLV3 binding and in control of the SAM (DeYoung et al. 2006; Guo et al. 2010; also see chapter "Peptides Regulating Apical Meristem Development") and appear to also participate in root growth regulation.

Taken together, cyst nematodes successfully adopted plant CLE peptide signaling to manipulate plant cell fate to serve their own need (Mitchum et al. 2008). *H. glycines* selects a potentially meristematic pericycle cell for infection, while *G. rostochiensis* manipulates a cortical cell to develop into a syncytium (Lu et al. 2009). The multinucleate syncytia develop by extensive cell wall degradation and fusion of adjacent protoplasts. Elucidating the signaling pathways that are employed in these processes may provide insights into CLE functions not only in nematode–host interaction but may reveal novel CLE functions in planta. As *HgCLE1* and *HgCLE2* transcripts are high not only during syncytium formation but remain elevated during syncytium maintenance, CLEs may be required to promote and maintain differentiation of root cells into feeder cells rather than for maintaining meristematic activity. Such a function is reminiscent of CLE peptides that promote differentiation of root cap cells from columella stem cells and of premature differentiation of proximal stem cells. As the properties of differentiated syncytium are distinct from any single plant cell, additional factors likely exist that recruit plant cell pathways to define feeder cell identity in the syncytium. To which extent CLE peptide signaling contributes to cell-type specification and hence reprogramming of plant cells by parasitic nematodes remains to be elucidated. In any case, CLE peptides have proven useful tools to manipulate plant tissue with a high spatial resolution.

3 Rapid Alkalinization Factor

Rapid alkalinization factor (RALF) was first isolated from tobacco (*Nicotiana tabacum*) leaves. When added to a tobacco cell culture at nanomolar concentrations, it promoted rapid media alkalization and rapid activation of a mitogen-activated protein kinase (Pearce et al. 2001). From the partial N-terminal amino acid sequence, a cDNA was identified which encodes a 115-amino-acid precursor. RALFs are ubiquitous in plants, with *Arabidopsis* possessing 34 *RALF* genes (Olsen et al. 2002; also see chapters "Plant Peptide Signaling: An Evolutionary Adaptation" and "Peptides and the Regulation of Plant Homeostasis"). Chemically synthesized RALF from tomato required oxidation and subsequent disulfide bridge formation between Cys18 and Cys28 and between Cys41 and Cys47 for biological activity (Pearce et al. 2001). RALFs are synthesized from a preproprotein which contains an N-terminal signal sequence in accord with the localization of the mature peptide in the apoplast (Pearce et al. 2010). In *Nicotiana benthamiana*, a fusion protein library screen was performed to assign subcellular localization of proteins expressed after viral infection of leaves. In this screen, a RALF precursor protein fused to green fluorescent protein (GFP) was localized first to the endoplasmic reticulum and was subsequently detected in the cell wall, confirming that RALF synthesis and transport occurs via the secretory pathway (Escobar et al. 2003).

The conserved 49-amino-acid RALF peptide sequence is harbored at the C terminus of the preproprotein and is preceded by a conserved dibasic site which has been implicated in protein processing. An arginine69 to alanine (R69A) mutation at the second residue of the dibasic site resulted in the accumulation of a 10-kDa protein which corresponds to the size of the proprotein lacking the signal sequence and in loss of the processed 5 kDa RALF peptide (Matos et al. 2008). In *Arabidopsis*, *35S:AtRALF1* overexpression resulted in a semidwarf phenotype which was lost in the overexpression mutant *35S:AtRALF1(R69A)*. *35S:AtRALF1(R69A)* plants showed a wild-type phenotype. Arginine-arginine (RR) or lysine-arginine (KR) dibasic sites are generally recognized by a convertase or kexin-like subtilisin protease (Jiang and Rogers 1999). Mutation of both arginines of the predicted dibasic protease cleavage site to glycine in *Arabidopsis* AtRALF23 prevented release of a myc-tagged RALF peptide from the precursor protein. AtS1P was identified as the subtilase responsible for this proteolytic cleavage (Srivastava et al. 2009a, b). The *Ats1p* knockout mutant was incapable of processing the overexpressed proprotein, and the semidwarf phenotype of *35S:AtRALF23-myc*$_{ox}$ plants was lost in the *35S:AtRALF23-myc*$_{ox}$*/Atsip1* double mutant. AtS1P was predominately localized in the Golgi apparatus, suggesting proteolytic processing in this compartment. Other peptidases are presumably not involved as the predicted AtS1P cleavage site lies directly N-terminal of the RALF sequence (also see chapter "Processing of Peptides").

A 120-kDa and a 25-kDa membrane protein were shown to be cross-linked to ^{125}I-azido-*Le*RALF in tobacco suspension cells, indicating that they specifically bind RALF (Scheer et al. 2005). Binding was exclusively found in the membrane fraction. A 120-kDa RALF binding protein was also identified from tobacco and alfalfa suspension-cultured cells, indicating that RALF peptide signaling may be mediated via interaction with a plasma membrane-localized receptor. The molecular nature of RALF receptors and downstream signaling are still elusive.

While other secreted alkalization factors are involved in plant defense, RALFs were shown to affect plant growth and development. Inhibition of root growth in tomato and *Arabidopsis*, of root hair growth in *Arabidopsis*, regulation of nodulation in *Medicago truncatula* (see chapter "The Role of Plant Peptides in Symbiotic Interactions"), and cell elongation in microcalli of sugarcane suspension cells have been described. Addition of a chemically synthesized RALF from tomato led to inhibition of root growth in tomato and *Arabidopsis* seedlings (Pearce et al. 2001). Similarly, overexpression of AtRALF23 in *35S:AtRALF23-myc*$_{ox}$ *Arabidopsis* plants resulted in retarded root growth and in reduced hypocotyl elongation of seedlings (Srivastava et al. 2009a, b; see chapter "Processing of Peptides"). Conversely, gene silencing in *Nicotiana attenuata* in *irRALF* lines promoted root growth rates (Wu et al. 2007). *AtRALF23* expression was downregulated in response to treatment with brassinolide. Brassinosteroids promote growth, indicating that RALF23 counteracts brassinosteroid growth-promoting activity and is in turn subject to downregulation by brassinosteroids (Srivastava et al. 2009a, b). *AtRALF1* is not downregulated by brassinolide but also acts as a negative growth regulator (Matos et al. 2008).

RALFs are characterized by their ability to promote alkalization. Alkalization of *Solanum peruvianum* suspension cell media by RALF was dose dependent and effective between 0.25 and 250 nM, with a half maximal response obtained at approximately 1 nM (Pearce et al. 2001, 2010). *Arabidopsis* wild-type seedlings grown on unbuffered media showed acidification around their roots. *35S:AtRALF23-myc*$_{ox}$ seedlings had lost the ability to acidify their rhizosphere, indicating that AtRALF23 acts to promote alkalization in planta (Srivastava et al. 2009a, b). By contrast, silencing of *N. attenuata* RALF in an *irRALF* line resulted in an inability to acidify the media as measured by a pH-sensitive dye (Wu et al. 2007). Surface pH measurements according to an unpublished method revealed a lower frequency of short-term pH changes and peaks at higher pH. An elevated pH is unlike what is expected from loss of an alkalinization factor. Local acidification was shown to be required for root hair initiation and elongation (Bibikova et al. 1998) and indeed the *irRALF* lines had fewer and shorter root hairs (Wu et al. 2007). It remains unclear how these observations conform with the ability of other RALFs to increase extracellular pH. Changes in apoplastic pH have been associated with regulation of cell expansion. The mechanisms by which RALFs control cell wall pH are not described nor can it be excluded that other cellular mechanisms translate RALF activity into a growth response (also see chapter "Peptides and the Regulation of Plant Homeostasis").

4 POLARIS

POLARIS (*PLS*) was identified in a promoter trap screen in *Arabidopsis* as a gene expressed in the embryonic root and in root tips. Loss of *PLS* function results in shorter roots with radially expanded cells that are shorter and likely involves reduced cell division rates (Casson et al. 2002). Furthermore, *pls* seedlings have fewer lateral roots, altered vascular development, and loss of trichome formation in the Columbia background. The short *PLS* transcript encodes three peptides: the 36-amino-acid PLS peptide, a 9-amino-acid peptide, and, overlapping with the PLS sequence, a third open reading frame of 8 amino acids. Complementation with the 36-amino-acid PLS open reading frame partially restored the root growth phenotype, whereas a mutated start codon of the 36-amino-acid peptide transcript did not. PLS is thus an active peptide even though one or more additional factors appear to be required for the full response. *pls* seedlings were hypersensitive to growth inhibition by cytokinin, indicating that it negatively regulates cytokinin responsiveness or synthesis. Auxin enhanced expression of *PLS*, and *pls* seedlings were less sensitive to growth inhibition by auxin. Etiolated *pls* seedlings displayed a triple-response phenotype reminiscent of *ethylene overproducer 1* (*eto1*) and of the *constitutive ethylene response1* (*ctr1*) mutants. Ethylene synthesis is not altered and inhibition of ethylene signaling restored the wild-type triple-response phenotype (Chilley et al. 2006), indicating that PLS is a negative regulator of ethylene signaling. *PLS* expression is in turn suppressed by ethylene. Ethylene is known to inhibit auxin transport, thus pointing to a link between PLS, ethylene, and auxin activities. *pls* plants have a lower indole acetic acid (auxin, IAA) content than wild type and displayed strongly reduced transport of IAA, whereas *PLS* overexpressors have a higher free IAA content. Both lower IAA content and IAA transport in *pls* resulted from enhanced ethylene signaling. It was proposed that PLS expression is enhanced at the root tip by high levels of auxin to promote growth and is repressed by ethylene in the proximal root zones to allow for lateral root formation.

5 CEP1

The C-terminally encoded peptide (CEP) is encoded by a small gene family of five members in *Arabidopsis*. The CEP peptides are contained at the C terminus of precursor proteins of 82–126 amino acids in length. Processed CEP1 has 15 amino acids and contains two hydroxyproline residues (Ohyama et al. 2008). Synthetic CEP1 peptide at 10^{-6}–10^{-7} M concentration as well as constitutive overexpression of *CEP1* resulted in root growth inhibition in line with the finding that three of four *CEP* genes were expressed predominately in roots. The growth defect was explained in terms of a smaller meristem and shorter final cell lengths. More detailed studies are required to provide mechanistic insights into CEP function. Interestingly, histochemical P_{CEP1}:*GUS* analysis revealed localized expression at the SAM and at the base of emerging lateral roots, with no expression detected at the root apex. This expression pattern is not compatible with a proposed role in root meristem activity except in a non-cell-autonomous manner. A possible role of CEP1 in lateral root formation has yet to be shown.

6 The Sulfated Peptides PSY1, PSK, and RGF

6.1 PSY1

PSY1 (Plant peptide containing Sulfated tYrosine1) is an 18-amino-acid sulfopeptide with Pro16 and Pro17 being hydroxylated. Mass spectrometry analysis suggested that the peptide is not only sulfated but also glycosylated with three

L-arabinose moieties being attached to Hyp16 (Amano et al. 2007). *Arabidopsis* was reported to possess three PSY precursor genes. *PSY1* is expressed at higher levels in the shoot than in the root, with distinct expression in the SAM but not in the RAM. Nonetheless, constitutive overexpression of *PSY1* resulted in longer roots and seemingly in longer hypocotyls in light-grown seedlings. Both peptide modifications, that is, glycosylation and sulfation, were required for full root growth-promoting activity. The PSY1 receptor, here called PSYR, is an LRR receptor-like kinase closely related to the PSK receptors PSKR1 and PSKR2. *PSKR1* and *PSYR* have overlapping expression patterns. Complementation of the *pskr1/pskr2/psyr* triple mutant with P_{PSKR1}:PSYR restored the wild-type phenotype, indicating that the *PSKR1* promoter can drive proper *PSYR* expression. This may be owed to the fact that *PSKR1* appears to be ubiquitously expressed, albeit at low levels (Kutschmar et al. 2009; Stührwohldt et al. 2011). Despite overlapping expression patterns of *PSKR1*, *PSKR2*, and *PSYR*, the growth inhibitory effects of receptor knockout mutants appeared to be additive, indicating that the sulfated PSK and PSY peptides perform distinct roles in the control of root elongation.

6.2 *Phytosulfokine*

Phytosulfokine-α (PSK) was the first sulfated bioactive peptide discovered in plants (Matsubayashi and Sakagami 1996). PSK is a disulfated pentapeptide of the sequence [Y(SO_3H)IY(SO_3H)TQ] first isolated from the media of suspension cells of asparagus (Matsubayashi and Sakagami 1996), maize, and rice (Matsubayashi et al. 1997) where it was shown to promote division of cells grown at low density. The PSK precursor proteins of 80–120 amino acids are encoded by small gene families with five members in *Arabidopsis* and appear to be ubiquitous in higher plants. No precursor genes were identified in the genome of the moss *Physcomitrella patens*, but the tyrosylprotein sulfotransferase (TPST) responsible for peptide sulfation is present in moss, indicating that sulfated peptide signaling may have evolved early on in plant evolution (Zhou et al. 2010). PSK preproproteins possess a signal sequence for secretion and the conserved PSK pentapeptide backbone YIYTQ at the C terminus (Yang et al. 2001; Lorbiecke and Sauter 2002). The four to six amino acids at the very C terminus are mostly basic in nature, and the PSK pentapeptide is preceded by an aspartic acid residue conserved in all preproproteins, which is essential for PSK sulfation (Hanai et al. 2000). Two conserved arginine residues seven to eight amino acids N-terminal of the YIYTQ sequence are characteristic of protease binding sites. Striking conservation is found at the level of secondary structure where the post-signal sequence is predicted to form three α-helices with a turn between helices 1 and 2 (Lorbiecke and Sauter 2002). These features may as well be important for correct processing of the precursor. Synthesis of the disulfated pentapeptide is achieved by at least one protease and by TPST which also catalyzes sulfation of the PSY1 peptide (Komori et al. 2009; Zhou et al. 2010). TPST is localized to the *cis*-Golgi. Overexpression of TPST did not result in growth promotion (Zhou et al. 2010), indicating that sulfation is not a limiting step in active peptide synthesis. Sulfation of both tyrosine side chains of PSK is required for biological activity (Matsubayashi et al. 1996). The subtilase AtSBT1.1 was shown to specifically cleave the AtPSK4 precursor protein with lower or no activity toward other *Arabidopsis* precursor homologs (Srivastava et al. 2008; also see chapter "Processing of Peptides"). AtSBT1.1 was localized to the extracellular space, indicating that final processing of the active peptide occurs in the apoplast similar to nematode CLE peptide cleavage.

PSK is perceived by LRR-RLK receptors with two genes encoding PSK receptors in *Arabidopsis* (Matsubayashi et al. 2006). The extracellular 21 LRRs are separated by a 36-amino-acid island domain intercepted in LRR 18 which was shown to be the site of PSK binding (Shinohara et al. 2007). Root elongation growth of *Arabidopsis tpst-1* seedlings which is deficient in endogenous active PSK was promoted by as little as 0.3 nM PSK, indicative of high-affinity ligand binding (Stührwohldt et al. 2011). The cytoplasmic protein sequence of PSKR conforms to serine/threonine protein kinases and was shown to possess serine/threonine kinase activity in vitro (Kwezi et al. 2011). In addition, a guanylate cyclase catalytic center is present at the C terminus of PSKR1, and the cytoplasmic region of PSKR1 was shown to possess cyclase activity in protoplasts overexpressing PSKR1 (Kwezi et al. 2011). Cyclase activity was promoted by PSK but was still comparably low possibly due to the lack of additional cofactors. cGMP in turn had an inhibitory effect on PSKR1 kinase activity, suggesting an autoregulatory signaling loop. How these activities contribute to downstream PSK or to (auto-) regulatory pathways has yet to be established.

The sulfated peptides PSK, PSY1, and root growth factor (RGF) are all involved in the regulation of postembryonic root growth. Effector treatment as well as genetic evidence indicated that PSK signaling contributes to root elongation growth. Addition of PSK peptide or overexpression of the *PSK4* precursor gene resulted in longer roots in *Arabidopsis*, whereas PSK receptor knockout lines *pskr1* and *pskr2* had significantly shorter roots (Matsubayashi et al. 2006; Kutschmar et al. 2009). While the single receptor gene knockout lines were still responsive to PSK, *pskr1-3/pskr2-1* double knockout seedlings had completely lost their ability to respond to exogenously supplied PSK. Epidermal cell lengths of the *pskr1-3* mutant were shorter, indicating that regulation of root growth by PSK signaling was mainly the result of altered cell elongation (Kutschmar et al. 2009). Reduced epidermal cell length was also observed in *tpst-1* which lacks the ability for tyrosine sulfation of PSK, PSY1, and of other sulfated signal peptides (Komori et al. 2009). In *tpst-1* seedlings, reduced epidermal cell length was restored by exogenous application of PSK and was partially restored by PSY1 peptide (Matsuzaki et al. 2010). Zhou et al. (2010) reported that final lengths of cortex cells were not altered in *tpst-1* as compared to wild type, indicating that PSK and PSY1 specifically act on the epidermal cell layer.

The role of sulfated peptides in root growth is not confined to the regulation of epidermal cell size. *tpst-1* seedlings have much more severely stunted roots than seedlings devoid of PSK receptor, and they have a distorted root meristem (Fig. 2).

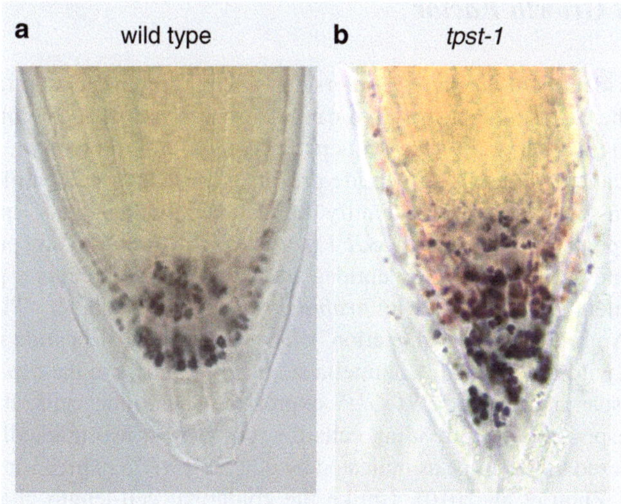

Fig. 2 Root tips of (**a**) wild-type and (**b**) *tpst-1* seedlings stained with Lugol solution to visualize starch granules of columella cells. *tpst-1* seedlings which are deficient in sulfated peptides have a distorted organization of the root apical meristem and elevated numbers of columella cell (Courtesy T. Nöthel)

Histochemical GUS analysis in *Arabidopsis* revealed that *TPST* is expressed as early as in the heart stage where it is localized to the hypophysis of embryos from which part of the root and the root cap are derived. Despite expression of *TPST* during embryo development, proper RAM formation during embryogenesis is not dependent on TPST. *TPST* expression persists to the postembryonic stage and is localized to the QC, surrounding RAM cells and in differentiated columella cells of root apices.

Deficiency in sulfated peptides in *tpst-1* caused disorganization of the RAM. The QC was expanded with extranumerary cells as also indicated by the QC marker gene *WOX5* which showed a similarly expanded expression pattern (Matsuzaki et al. 2010). The extra QC cells contained starch granules indicative of columella cell differentiation. QC cells of *tpst-1* roots were mitotically more active than their wild-type counterparts, with higher apparent activity also observed in the distal RAM of the columella. In accord with the higher mitotic activity, the columella stem cell layer was expanded. Thus, the columella stem cell region was larger, but columella stem cells at the same time were not able to maintain their undifferentiated state. Columella cells originating from the hyperactive columella stem cell region were distorted. This phenotype indicated that sulfated peptide(s) are required for the confinement of the QC and to maintain columella stem cell fate. Exogenous application of PSK or PSY1 did not restore the stem cell defects of the *tpst-1* mutant, indicating that additional active sulfated peptide(s) participate in the maintenance of QC and columella stem cell identity.

6.3 Root Growth Factor

In order to identify the sulfated factors that were responsible for the stem cell defects of the *tpst-1* mutant, putative candidates were searched bioinformatically based on prediction of protein sequences required for proteolytic processing. A putative candidate family of peptide precursor proteins was identified, and one precursor was overexpressed to identify the mature peptide by mass spectrometry. The mature peptide secreted from *RGF1* (*root growth factor1*)-overexpressing cells was 13 amino acids long. RGF peptides are encoded by a precursor gene family with nine members, *RGF1–9*, in *Arabidopsis* (also see chapter "Plant Peptide Signaling: An Evolutionary Adaptation"). The sulfated RGF1 peptide was capable of restoring wild-type QC and columella stem cell numbers in the *tpst-1* mutant.

Despite seemingly low *TPST:GUS* expression in mitotic cells, the proximal meristem responsible for providing cells for root growth was markedly shorter in *tpst-1* compared to wild type as indicated by *CYCB1;1:GUS* expression (Matsuzaki et al. 2010; Zhou et al. 2010). Unlike the epidermal cell-length phenotype, the short-meristem phenotype was not restored by combined PSK and PSY1 treatment. Treatment with RGF1 increased meristematic activity by about 70%, and a combination of RGF1, PSY1, and PSK fully restored root growth. PSY1 appeared to have the least root growth-promoting effect. *RGF1* gene transcription was exclusively detected in QC and columella stem cells, while RGF1 and cross-reactive peptides were localized by immunostaining to the whole meristematic region, indicating that RGFs likely are diffusible peptide signals. The *rgf1 rgf2 rgf3* triple knockout mutant displayed reduced meristematic activity which was complemented by overexpression of RGF1. RGF precursor genes were capable of restoring root growth of the *tpst-1* mutant to varying degrees, with the exception of *RGF8* which was ineffective (Matsuzaki et al. 2010). RGFs thus redundantly function in the maintenance of the postembryonic root stem cell niche.

While auxin does not regulate expression of *RGF1-4* (Matsuzaki et al. 2010), expression of the auxin reporter *DR5:GUS* in the root tip was expanded in *tpst-1* in parallel with higher free auxin levels likely as a result of altered synthesis and transport (Zhou et al. 2010). *tpst-1* seedlings showed reduced expression of the auxin efflux transporters *PIN4* (*PIN-FORMED4*) in the RAM and of *PIN3* and *PIN7* in QC and columella cells. In addition, the auxin synthesis genes *YUCCA2*, a flavin monooxygenase, *ASA1* (ANTHRANILATE SYNTHASE alpha1), and *ASB1* (ANTHRANILATE SYNTHASE beta1) were strongly downregulated in the RAM. *PIN1* expression and *PIN2* expression were not altered: *PIN2* is expressed in the epidermis and lateral root cap and is responsible for basipetal, that is, shootward, auxin transport and for gravitropic bending.

Taken together, sulfated peptide signaling, likely by RGFs, maintains synthesis and polar auxin transport in the RAM. Lack of RGF in *tpst-1* seedlings results in a short-meristem phenotype conceivably as a result of altered auxin activity. *PLETHORA* (*PLT*) genes show high expression in the root stem cell niche and a gradient of expression levels extending into the growth region of the root. *tpst-1* seedlings have strongly reduced *PLT1*

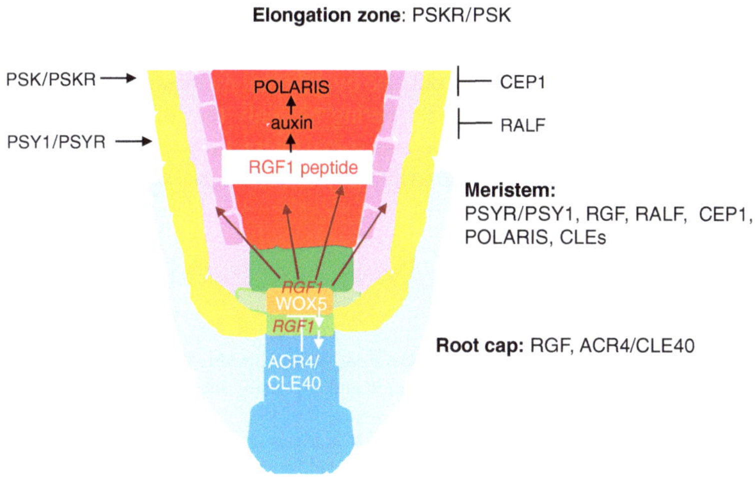

Fig. 3 Activities of peptide signals and their cognate receptors in the root apex

and *PLT2* expression which was restored by application of RGF1 peptide (Matsuzaki et al. 2010). The loss-of-function double mutant *plt1-4 plt2-2* which shows growth retardation was unresponsive to RGF1, indicating that PLT mediate RGF activity.

7 Conclusion

Plants have evolved signaling peptides which are oftentimes encoded by gene families indicative of varied and multiple functions. All signaling peptides that are active in roots are modified in one or more ways by hydroxylation of proline and subsequent glycosylation, sulfation of tyrosine residues, or disulfide bridge formation, modification which enhances water solubility, makes the peptides more stable, or provides binding specificity. All peptides identified so far are secreted into the apoplast where they are perceived by plasma membrane-LRR receptors. As such, they are excellent candidates for short-distance cell-to-cell signaling and have, at least in the case of plant-pathogenic nematodes, been successfully hijacked to manipulate their host to very locally differentiate a highly specified cell type, the feeder cell.

The endogenous functions of plant peptide signals appear to lie in maintaining the identity either of organizing center (QC) cells or of root stem cells as in the case of CLE and RGF control of root stem cell identities (Fig. 3). Or, peptides such as PSK and possibly RALFs contribute to root cell size regulation. Both mechanisms contribute to root growth control. Our understanding of how these different aspects of growth regulation in the root are achieved at the cellular or molecular level is still limited, and signaling networks, for instance, with hormonal pathways, have yet to be explored for most part. The RAM is complex in that it maintains two stem cell niches from one organizing center: the proximal stem cells which control root growth and the distal stem cells which are responsible for root cap formation. Study of peptide signaling revealed that these two stem cell niches are regulated differentially by different types of peptides such as CLEs and RGFs but also by different members of a peptide family such as the CLE peptide family. Deciphering not only the functions of individual peptides but the intricate network of peptide signal interactions and the links that likely exist to hormonal pathways is a formidable task ahead.

References

Amano Y, Tsubouchi H, Shinohara H, Ogawa M, Matsubayashi Y (2007) Tyrosine-sulfated glycopeptide involved in cellular proliferation and expansion in Arabidopsis. Proc Natl Acad Sci U S A 104:18333–18338

Bakhetia M, Urwin PE, Atkinson HJ (2007) qPCR analysis and RNAi define pharyngeal gland cell-expressed genes of *Heterodera glycines* required for initial interactions with the host. Mol Plant Microbe Interact 20:306–312

Bibikova TN, Jacob T, Dahse I, Gilroy S. (1998) Localized changes in apoplastic and cytoplasmic pH are associated with root hair development in Arabidopsis thaliana. Development 125:2925–2934

Birnboim K, Shasha DE, Wang JY, Jung JW, Lambert GM, Galbraith DW, Benfey PN (2003) A gene expression map of the Arabidopsis root. Science 302:1956–1960

Bleckmann A, Weidkamp-Peters SCAM, Simon R (2010) Stem cell signaling in Arabidopsis requires CRN to localize CLV2 to the plasma membrane. Plant Physiol 152:166–176

Casson S, Chilley PM, Topping JF, Evans M, Souter MA, Lindsey K (2002) The POLARIS gene of Arabidopsis encodes a predicted peptide for correct root growth and leaf vascular patterning. Plant Cell 14:1705–1721

Chilley PM, Casson SA, Tarkowski P, Hawkins N, Wang KL-C, Hussey PJ, Beale M, Ecker JR, Sandberg GK, Lindsey K (2006) The POLARIS peptide of *Arabidopsis* regulates auxin transport and root growth via effects on ethylene signaling. Plant Cell 18:3058–3072

De Smet I, Vassileva V, De Rybel B, Levesque MP, Grunewald W, Van Damme D, Van Noorden G, Naudts M, Van Isterdael G, De Clercq R, Wang JY, Meuli N, Vanneste S, Friml J, Hilson P, Jürgens G, Ingram GC, Inzé D, Benfey PN, Beeckman T (2008) Receptor-like kinase ACR4 restricts formative cell divisions in the *Arabidopsis* root. Science 322:594–597

DeYoung BJ, Bickle KL, Schrage KJ, Muskett P, Patel K, Clark SE (2006) The CLAVATA1-related BAM1, BAM2 and BAM3 receptor kinase-like proteins are required for meristem function in Arabidopsis. Plant J 45:1–16

Escobar NM, Haupt S, Thow G, Boevink P, Champman S, Oparka K (2003) High-throughput viral expression of cDNA-green fluorescent protein fusions reveals novel subcellular addresses and identifies unique proteins that interact with plasmodesmata. Plant Cell 15:1507–1523

Fiers M, Hause G, Boutilier K, Casamitjana-Martinez E, Weijers D, Offringa R, van der Geest L, van Lookeren CM, Liu C-M (2004) Mis-expression of the CLV3/ESR-like gene *CLE19* in *Arabidopsis* leads to a consumption of root meristem. Gene 327:37–49

Fiers M, Golemiec E, Xu J, van der Geest L, Heidstra R, Stiekema W, Liu C-M (2005) The 14-amino acid CLV3, CLE19, and CLE40 peptides trigger consumption of the root meristem in Arabidopsis through a *CLAVATA2*-dependent pathway. Plant Cell 17:2542–2553

Guo Y, Han L, Hymes M, Denver R, Clark SE (2010) Clavata2 forms a distinct CLE-binding receptor complex regulating Arabidopsis stem cell specification. Plant J 63:889–900

Guo Y, Ni J, Denver R, Wang X, Clark SE (2011) Mechanisms of molecular mimicry of plant CLE peptide ligands by the parasitic nematode *Globodera rostochiensis*. Plant Physiol 157:476–484

Hanai H, Nakayama D, Yang HP, Matsubayashi Y, Hirota Y, Sakagami Y (2000) Existence of a plant tyrosylprotein sulfotransferase: novel plant enzyme catalyzing tyrosine *O*-sulfation of preprophytosulfokine variants in vitro. FEBS Lett 470:97–101

Hobe M, Müller R, Grünewald M, Brand U, Simon R (2003) Loss of CLE40, a protein functionally equivalent to the stem cell restricting signal CLV3, enhances root waving in *Arabidopsis*. Dev Genes Evol 213:371–381
Ito Y, Nakanomyo I, Motose H, Iwamoto K, Sawa S, Dohmae N, Fukuda H (2006) Dodeca-CLE peptides as suppressors of plant stem cell differentiation. Science 313:842–845
Jiang LW, Rogers JC (1999) Functional analysis of a Golgi-localized Kex2p-like protease in tobacco suspension culture cells. Plant J 18:23–32
Jun J, Fiume E, Roeder AHK, Meng L, Sharma VK, Osmont KS, Baker C, Ha CM, Meyerowitz EM, Feldman LJ, Fletcher JC (2010) Comprehensive analysis of *CLE* polypeptide signaling gene expression and overexpression activity in *Arabidopsis*. Plant Physiol 154:1721–1736
Komori R, Amano Y, Ogawa-Ohnishi M, Matsubayashi Y (2009) Identification of tyrosyl sulfotransferase in *Arabidopsis*. Proc Natl Acad Sci U S A 106:15067–15072
Kutschmar A, Rzewuski G, Stührwohldt N, Beemster GTS, Inzé D, Sauter M (2009) PSK-α promotes root growth in Arabidopsis. New Phytol 181:820–831
Kwezi L, Ruzvidzo O, Wheeler JI, Govender K, Iacuone S, Thopson PE, Gehring C, Irving HR (2011) The phytosulfokine (PSK) receptor is capable of guanylate cyclase activity and enabling cyclic GMP-dependent signalling in plants. J Biol Chem 286:22580–22588
Lorbiecke R, Sauter M (2002) Comparative analysis of PSK peptide growth factor precursor homologs. Plant Sci 163:321–332
Lu S-W, Chen S, Wang J, Yu H, Chronis D, Mitchum MG, Wang X (2009) Structural and functional diversity of *CLAVATA3/ESR* (*CLE*)-like genes from the potato cyst nematode *Globodera rostochinensis*. Mol Plant Microbe Interact 22:1128–1142
Matos JL, Fiori CS, Silva-Filho MC, Moura DS (2008) A conserved dibasic site is essential for correct processing of the peptide hormone AtRALF1 in *Arabidopsis thaliana*. FEBS Lett 582:3343–3347
Matsubayashi Y , Hanai H, Hara O, Sakagami Y (1996) Active fragments and analogs of the plant growth factor, phytosulfokine: structure-activity relationships. Biochem Biophys Res Commun 225:209–214
Matsubayashi Y, Sakagami Y (1996) Phytosulfokine, sulfated peptides that induce the proliferation of single mesophyll cells of *Asparagus officinalis*. Proc Natl Acad Sci U S A 93:7623–7627
Matsubayashi Y, Takagi L, Sakagami Y (1997) Phytosulfokine-alpha, a sulfated pentapeptide, stimulates the proliferation of rice cells by means of specific high- and low-affinity binding sites. Proc Natl Acad Sci U S A 94:13357–13362
Matsubayashi Y, Ogawa M, Kihara H, Niwa M, Sakagami Y (2006) Disruption and overexpression of Arabidopsis phytosulfokine receptor gene affects cellular longevity and potential for growth. Plant Physiol 142:45–53
Matsuzaki Y, Ogawa-Ohnishi M, Mori A, Matsubayashi Y (2010) Secreted peptide signals required for maintenance of root stem cell niche in Arabidopsis. Science 329:1065–1067
Meng L, Feldman LJ (2010) CLE14/CLE20 peptides may interact with CLAVATA2/CORYNE receptor-like kinases to irreversibly inhibit cell division in the root meristem of Arabidopsis. Planta 232:1061–1074
Meng L, Ruth KC, Fletcher JC, Feldman LJ (2010) The roles of different CLE domains in *Arabidopsis* CLE polyepeptide activity and functional specificity. Mol Plant 3:760–772
Mitchum MG, Wang X, Davis EL (2008) Diverse and conserved roles of CLE peptides. Curr Opin Plant Biol 11:75–81
Miwa H, Betsuyaku S, Iwamoto K, Kinoshita A, Fukuda H, Sawa S (2008) The receptor-like kinase SOL2 mediates CLE signaling in Arabidopsis. Plant Cell Physiol 49:1752–1757
Müller R, Bleckmann A, Simon R (2008) The receptor kinase CORYNE of *Arabidopsis* transmits the stem cell-limiting signal CLAVATA3 independently of CLAVATA1. Plant Cell 20:934–946
Ohyama K, Ogawa M, Matsubayashi Y (2008) Identification of a biologically active, small, secreted peptide in Arabidopsis by in silico gene screening, followed by LC-MS-based structure analysis. Plant J 55:152–160
Olsen AN, Skriver K (2003) Ligand mimicry? Plant-parasitic nematode polypeptide with similarity to CLAVATA3. Trends Plant Sci 8:55–57

Olsen AN, Mundy J, Skriver K (2002) Peptomics, identification of novel cationic *Arabidopsis* peptides with conserved sequence motifs. In Silico Biol 2:441–451
Pearce G, Moura DS, Stratmann J, Ryan CA (2001) RALF, a 5-kDa ubiquitous polypeptide in plants, arrests root growth and development. Proc Natl Acad Sci USA 98:12843–12847
Pearce G, Yamaguchi Y, Munske G, Ryan CA (2010) Structure-activity studies of RALF, Rapid alkalization factor, reveal an essential – YISY – motif. Peptides 31:1973–1977
Replogle A, Wang J, Bleckmann A, Hussey RS, Baum TJ, Sawa S, Davies EL, Wang X, Simon R, Mitchum MG (2011) Nematode CLE signaling in Arabidopsis requires CLAVATA2 and CORYNE. Plant J 65:430–440
Rost TL (2011) The organization of roots of dicotyledonous plants and the positions of control points. Ann Bot 107:1213–1222
Sarkar AK, Luijten M, Miyamashi S, Lenhard M, Hashimoto T, Nakajima K, Scheres B, Heidstra R, Laux T (2007) Conserved factors regulate signalling in *Arabidopsis thaliana* shoot and root stem cell organizers. Nature 446:811–814
Scheer JM, Pearce G, Ryan CA (2005) *Le*RALF, a plant peptide that regulates root growth and development, specifically binds to 25 and 120 kDa cell surface membrane proteins of *Lycopersicon peruvianum*. Planta 221:667–674
Shinohara H, Ogawa M, Sakagami Y, Matsubayashi Y (2007) Identification of ligand binding site of phytosulfokine receptor by on-column photoaffinity labeling. J Biol Chem 282:124–131
Srivastava R, Liu JX, Howell SH (2008) Proteolytic processing of a precursor protein for a growth-promoting peptide by a subtilisin serine protease in Arabidopsis. Plant J 56:219–227
Srivastava R, Liu J-X, Howell S (2009a) Proteolytic processing of a precursor protein for a growth-promoting peptide by a subtilisin serine protease in Arabidopsis. Plant J 56:219–227
Srivastava R, Liu J-X, Guo H, Yin Y, Howell S (2009b) Regulation and processing of a plant peptide hormone, AtRALF23, in Arabidopsis. Plant J 59:930–939
Stahl Y, Simon R (2010) Plant primary meristems: shared functions and regulatory mechanisms. Curr Opin Plant Biol 13:53–58
Stahl Y, Wink RH, Ingram GC, Simon R (2009) A signalling module controlling the stem cell niche in *Arabidopsis* root meristems. Curr Biol 19:909–914
Strabala TJ, O'Donnell PJ, Smit A-M, Ampomah-Dwamena C, Martin EJ, Netzler N, Nieuwenhuizen NJ, Quinn BD, Foote HCC, Hudson KR (2006) Gain-of-function phenotypes of many *CLAVATA3/ESR* genes, including four new members, correlate with tandem variations in the conserved *CLAVATA3/ESR* domain. Plant Physiol 140:1331–1344
Stührwohldt N, Dahlke R, Steffens B, Johnson A, Sauter M (2011) Phytosulfokine-α controls hypocotyl length and cell expansion in *Arabidopsis thaliana* through phytosulfokine receptor 1. PLoS One 6:e21054
Tucker MR, Laux T (2007) Connecting the paths in plant stem cell regulation. Trends Cell Biol 17:403–410
Wang G, Fiers M (2010) CLE peptide signalling during plant development. Protoplasma 240:33–43
Wang X, Mitchum MG, Gao B, Li C, Diab H, Baum TJ, Hussey RS, Davies EL (2005) A parasitism gene from a plant-parasitic nematode with function similar to CLAVATA3/ESR (CLE) of *Arabidopsis thaliana*. Mol Plant Pathol 6:187–191
Wang J, Lee C, Replogle A, Joshi S, Korkin D, Hussey R, Baum TJ, Davis EL, Wang X, Mitchum MG (2010) Dual roles for the variable domain in protein trafficking and host-specific recognition of *Heterodera gylcines* CLE effector proteins. New Phytol 187:1003–1017
Wu J, Kurten EL, Monshausen G, Hummel GM, Gilroy S, Baldwin IT (2007) NaRALF, a peptide signal essential for the regulation of root hair tip apoplastic pH in *Nicotiana attenuata*, is required for root hair development and plant growth in native soils. Plant J 52:877–890
Yang H, Matsubayashi Y, Nakamura K, Sakagami Y (2001) Diversity of *Arabidopsis* genes encoding precursors for phytosulfokine, a peptide growth factor. Plant Physiol 127:842–851
Zhou W, Wei L, Xu J, Zhai Q, Jiang H, Chen R, Chen Q, Sun J, Chu J, Zhu L, Liu C-M, Li C (2010) *Arabidopsis* tyrosylprotein sulfotransferase acts in the auxin/PLETHORA pathway in regulating postembryonic maintenance of the root stem cell niche. Plant Cell 22:3692–3709

Peptides Regulating Plant Vascular Development

Hiroo Fukuda

Abstract Cell-to-cell communication is a fundamental mechanism for maintaining homeostasis of the plant meristems. Cellular proliferation and differentiation in the vascular meristem are regulated by hormonal signals and secreted peptides. Recent studies have highlighted the importance of an intercellular signaling pathway consisting of a CLE peptide TDIF and its receptor PXY/TDR (PHLOEM INTERCALATED WITH XYLEM/TDIF receptor) in determining the fates of procambial cells (vascular stem cells), namely self-renewal or commitment to differentiation. The comparison of the TDIF–TDR signaling pathways with the CLAVATA pathways revealed common and distinct features of homeostasis between these two meristems. Some other CLE peptides are recently found to function in xylem development via modification of plant hormone signaling. Here, emerging concepts of how intercellular peptide signals regulate vascular cell communities in global geometry are summarized.

1 Introduction

Distinct types of vascular cells are produced from the vascular meristem continuously derived from the apical and basal meristems in plants. Plant vascular tissues are composed of xylem, phloem, and their intervening procambium/cambium. Procambial/cambial cells in the vascular meristem retain meristematic activity even outside of the apical meristems and continue to provide cells to the two conductive vascular tissues, xylem and phloem. In this sense, procambial and cambial cells are considered as vascular stem cells (Hirakawa et al. 2011). Individual species of vascular plants form distinct radial patterns of vascular tissues in each

H. Fukuda (✉)
Department of Biological Sciences, Graduate School of Science, The University of Tokyo,
7-3-1 Hongo, Bunkyo-ku, Tokyo 113-0033, Japan
e-mail: fukuda@biol.s.u-tokyo.ac.jp

H.R. Irving and C. Gehring (eds.), *Plant Signaling Peptides*,
Signaling and Communication in Plants 16, DOI 10.1007/978-3-642-27603-3_4,

organ. These genetically defined radial patterns are governed by cell-to-cell communication between distinct vascular cells (Fukuda et al. 2007; Lehesranta et al. 2010; Hirakawa et al. 2010b, 2011). On the other hand, the vascular system forms a complex continuous network throughout the plant body, which is essential for long-distance transport of water, nutrients, and signaling molecules and for mechanical support. This continuous vascular formation is also fundamentally based on cell-to-cell communication, thus achieving a functional network of the vascular system. It has long been believed that mobile phytohormones such as auxins, cytokinins, and brassinosteroids mediate cell–cell communication in vascular development (Fukuda 2004). Recent work has revealed that novel signaling peptides play essential roles in regulating tissue patterning during vascular development. Therefore, this chapter reviews recent developments in studies of intercellular signaling peptides with a special focus on their cross talk with plant hormone signaling pathways.

2 CLE Peptides

2.1 *Tracheary Element Differentiation Inhibitory Factor*

2.1.1 Characteristics of TDIF

In vitro xylogenesis of *Zinnia elegans* L., in which isolated mesophyll cells transdifferentiate into xylem cells including tracheary elements, offers a useful experimental system for the analyses of intercellular signal molecules regulating xylem development (Fukuda and Komamine 1980; Fukuda 2004). Using this system as a bioassay, an extracellular factor suppressing tracheary element differentiation was isolated from cultured medium of *Zinnia* cells. This factor was identified as a dodeca-amino acid peptide of HEVHypSGHypNPISN with hydroxylation on two proline residues and was designated tracheary element differentiation inhibitory factor (TDIF) (Ito et al. 2006). Although some CLE peptides such as CLAVATA 3 (CLV3) and CLE2 are further modified with arabinosylation on a hydroxylated proline residue(s) (Ohyama et al. 2009), isolated TDIF contains no glycosylated residues. The full-length cDNA corresponding to TDIF encodes a protein of 132 amino acids, but only 12 from H_{120} to N_{131} match to the TDIF sequence (Fig. 1a). This result implies that TDIF is produced through the removal of both residues M_1–A_{119} at the N terminus and R_{132} at the C terminus. A homology search revealed that the C-terminal 12-amino-acid sequences of the *Arabidopsis* CLE41 and CLE44 were identical to the TDIF sequence and those of CLE42 and CLE46 are highly homologous to the TDIF sequence. Indeed, *cle41-1* mutation promoted xylem vessel differentiation (Hirakawa et al. 2010a). Therefore, *CLE41* must act as a gene responsible for TDIF activity in situ. Experiments using chemically synthesized 26 CLE dodeca-peptides with two hydroxyproline residues

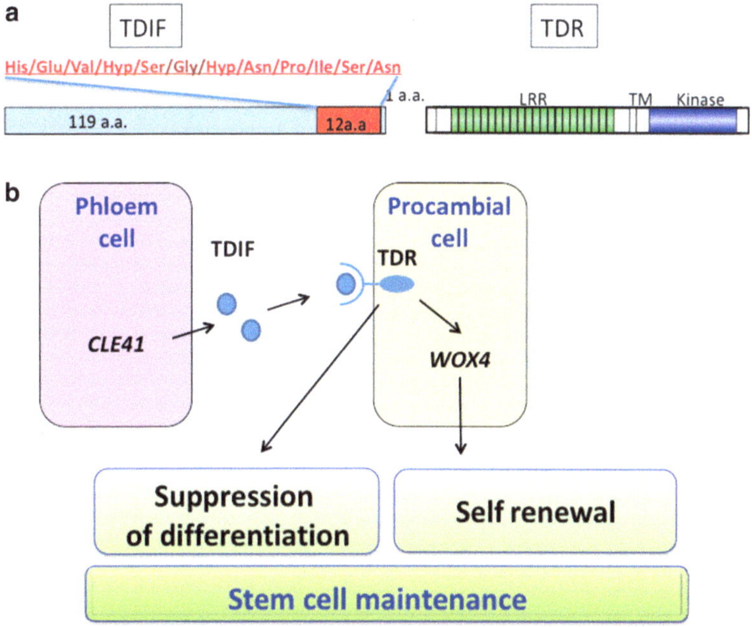

Fig. 1 Structures and signaling pathway of TDIF–TDR (**a**) Structures of TDIF and TDR. TDIF is produced as a dodeca-amino acid peptide by processing and hydroxylation. TDR is a receptor for TDIF and contains leucine-rich repeat (LRR), transmembrane (TM), and kinase domains. (**b**) Regulation of stem cell maintenance in procambial cells by the TDIF–TDR signaling pathway. TDIF is produced from *CLE41* in phloem cells. TDIF, which is secreted from phloem cells, is perceived by TDR in procambial cells. TDR signaling is diverged into two pathways, each of which promotes self-renewal or inhibits xylem differentiation from procambial cells. The former involves *WOX4* and the latter involves an uncharacterized factor(s). Both mechanisms function in stem cell maintenance in procambial cells

indicated that the dodeca-amino acid peptides from CLE42 as well as CLE41/CLE44 had strong TDIF activity, but the dodeca-amino acid peptides from the remaining CLE proteins did not exhibit TDIF activity (Table 1). Therefore, CLE42 and CLE41/CLE44 peptides function specifically in suppressing xylem vessel differentiation.

The synthetic TDIF peptide at as low as 30 pM inhibits tracheary element differentiation by 50% in *Zinnia* xylogenic culture (Ito et al. 2006), suggesting that an extremely low amount of TDIF is capable of functioning in vascular development in situ. Alanine-scanning mutagenesis of TDIF revealed that the G_6 and P_9, which are conserved in almost all *Arabidopsis* CLE peptides, and the slightly less conserved residues V_3, N_8, and N_{12} are indispensable for TDIF activity, and probably also for general activity of CLE peptides (Ito et al. 2006). In contrast, H_1 may confer a TDIF-specific biological activity because it is unique to the CLE peptides with TDIF activity. The 13-amino-acid peptide with an additional arginine at the C terminus reduced activity by 1/7 compared with the original 12-amino-acid

Table 1 Inhibition of xylem differentiation and root growth by the synthetic dodeca-amino acid CLE peptides with two hydroxyproline residues

	Inhibition		
	TE differentiation in *Zinnia*culture	Protoxylem formation in *Arabidopsis*	Root growth
No peptide	–	–	–
CLE1/3/4	–	–	–
CLE2	–	–	–
CLE5/6	–	–	–
CLE7	–	–	–
CLE41/44	+	–	–
CLE42	+	–	–
CLE46	–	–	–
CLV3	–	+	+
CLE8	–	+	+
CLE9/10	–	+	+
CLE11	–	+	+
CLE12	–	+	+
CLE13	–	+	+
CLE14	–	+	+
CLE16	–	+	+
CLE17	–	+	+
CLE18	–	+	+
CLE19	–	+	+
CLE20	–	+	+
CLE21	–	+	+
CLE22	–	+	+
CLE27	–	+	+
CLE40	–	+	+
CLE45	–	+	+
CLE25	–	–	+
CLE26	–	–	+

–, Nonfunction; +, inhibition.

TDIF peptide, indicating that the accurate cleavage at the C-termini is required for producing highly active TDIF peptide.

2.1.2 TDR, a TDIF Receptor

Loss-of-function mutations of three *CLAVATA* loci (*CLV1*, *CLV2*, and *CLV3*) confer a similar phenotype including fascinated stems, increased floral organ and whorl numbers, and disrupted phyllotaxis, which suggest that the three genes function in the same pathway (Haecker and Laux 2001). *CLV3* encodes a 13-amino-acid CLE peptide, which is modified by three arabinose moieties (Fletcher et al. 1999; Brand et al. 2000; Kondo et al. 2006; Ohyama et al. 2009). The *CLV1* gene encodes a member of leucine-rich repeat (LRR) receptor-like kinase (RLK) subfamily XI (Clark et al. 1997; Shiu and Bleecker 2001). *CLV2* encodes a receptor-like protein (LRR-RLP) that has extracellular LRR domains but lacks the kinase domain (Jeong et al. 1999). The CLV2 protein associates with CORYNE (CRN)/SUPPRESSOR OF LLP1 (SOL) 2, a membrane-associated protein kinase with a short extracellular domain, to establish a functional receptor complex that mediates transmembrane signal transduction (Müller et al. 2008; Zhu et al. 2010; Bleckmann et al. 2010). Recently, a novel LRR-RLK, RECEPTOR-LIKE PROTEIN KINASE 2 (RPK2)/TOADSTOOL 2 (TOAD2), was identified, and CLV1, CLV2-CRN/SOL2, and RPK2/TOAD2 are thought to be three major receptor complexes that transmit the CLV3 signal independently (Kinoshita et al. 2010; also see chapters "Peptides Regulating Apical Meristem Development," "Peptides Regulating Root Growth," and "The Role of Plant Peptides in Symbiotic Interactions").

A receptor for TDIF, designated TDIF receptor (TDR), was discovered by a screening of a collection of LRR-RLK T-DNA insertion lines for TDIF insensitivity (Hirakawa et al. 2008). TDIF binds in vitro specifically to the extracellular domain of TDR protein produced in tobacco cultured cells. TDR belongs to the subclass XI of LRR-RLK (Fig. 1a). Because the CLV3–CLV1 system is also comprised of a set of a CLE peptide and a member of the LRR-RLK XI subfamily, different sets of a CLE peptide and an LRR-RLK XI member may function in various developmental processes. Fisher and Turner (2007) reported that *pxy-1* (*phloem intercalated with xylem-1*), which may act as an antimorphic allele, induces smaller procambium and phloem adjacent to the xylem in flower stalks. This mutation occurred in the *TDR* locus.

A phylogenetic analysis of 26 members of the LRR-RLK XI subfamily based on their LRR domains, which CLE ligands would bind to, revealed a clade comprising *TDR/PXY*, *CLV1*, *BAM* (*barely any meristem*)*1*, *BAM2*, *BAM3*, *PXL* (*PXY-like*)*1*, and *PXL2* among 26 members of the subclass XI (Hirakawa et al. 2008). Mutation in *BAM* and *PXL* genes causes abnormality of the shoot apical meristem (SAM) or vascular tissues or both (Fisher and Turner 2007; DeYoung et al. 2006). Therefore, five genes other than *TDR* and *CLV1* in the same clade might also encode specific receptors for the respective CLE peptides, which play roles in vascular development as well as the SAM.

2.1.3 A Target Transcription Factor of the TDIF–TDR Signaling

WUSCHEL (WUS), a homeodomain transcription factor essential for stem cell maintenance in the SAM, is well established as a key target of the CLV signaling pathway (see chapter "Peptides Regulating Apical Meristem Development"). The CLV3 signal represses the expression domain of *WUS*, thereby restricting the stem cell population (Mayer et al. 1998; Brand et al. 2000; Schoof et al. 2000). The *Arabidopsis* genome contains 15 *WUSCHEL-related HOMEOBOX (WOX)* family members, which are classified into three subgroups (Haecker et al. 2004). Gene expression and genetic analyses showed *WOX4* to be a transcriptional target of the

TDR-dependent TDIF signaling pathway (Hirakawa et al. 2010a). In the SAM, activation of *CLV3* expression results in the reduction of the *WUS* expression within 3 h (Müller et al. 2006). *WOX4* expression is induced 1 h after TDIF application. This similarity in the timing of induction suggests that similar signaling components regulate *WOX* gene expression after the perception of the CLE ligand in both TDIF–TDR and CLV3–CLV1 signaling pathways, although their effects on the target gene expression level are opposite.

2.1.4 Function of the TDIF–TDR Signaling

TDR/PXY is expressed preferentially in procambium and cambium cells (Fisher and Turner 2007; Hirakawa et al. 2008). In contrast, *CLE41* and *CLE44* are expressed specifically in phloem and more widely in its neighbors, respectively (Hirakawa et al. 2008). Immunohistochemistry using an anti-TDIF antibody reveals that TDIF is secreted to the apoplast surrounding phloem cells (Hirakawa et al. 2008). These results, together with the fact that TDR works in procambial/cambial cells, indicate that TDIF produced in phloem and its neighbor regulates procambial cell fates in a non-cell-autonomous fashion (Fig. 1b). Indeed, overproduction of CLE41 in xylem cells results in a disordered vascular pattern (Etchells and Turner 2010).

The *tdr* mutant exhibits discontinuous formation of vessel strands without affecting phloem strand formation in *Arabidopsis* leaves (Hirakawa et al. 2008). This phenotype is consistent with the result that TDIF suppresses tracheary element differentiation in vitro (Ito et al. 2006). Exogenously supplied TDIF also enhances procambial/cambial cell proliferation in *Arabidopsis* hypocotyls and *Zinnia* cultured cells. In situ, the defects in *TDR* cause the exhaustion of procambial/cambial cells between phloem and xylem, resulting in formation of xylem vessels adjacent to phloem cells in *Arabidopsis* hypocotyls (Hirakawa et al. 2008, 2010a). This phenotype is also observed in *cre41*. Thus, the TDIF–TDR signaling pathway suppresses vessel differentiation and promotes self-renewal in procambial/cambial cells. Then, how are these two processes regulated in a procambial/cambial cell? In *wox4*, procambial cell proliferation is reduced. However, *wox4* has at least one cell layer of procambium between xylem and phloem cells, although in the *tdr* mutant, procambial cells have vanished, resulting in the formation of vessel cells adjacent to the phloem. Therefore, *WOX4* must be in charge of only procambial cell proliferation, one of the two events that the TDIF–TDR signaling pathway regulates, namely, proliferation and vessel differentiation of procambial cells (Fig. 1b). The transcriptional target governing the other event remains elusive.

2.1.5 Regulation of Stem Cell Fates by the CLE Peptide–LRR-RLK–WOX Signaling Pathway

Stem cells have two properties: self-renewal and commitment to differentiation. These properties are maintained in the specific microenvironment called stem cell

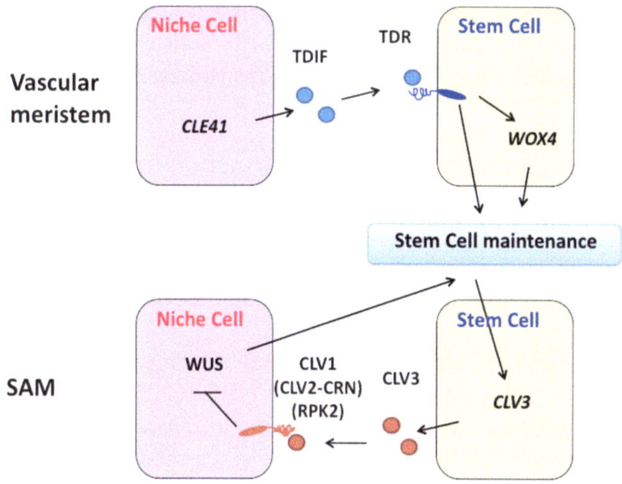

Fig. 2 Mode of action of the TDIF and the CLV3 signaling in stem cell maintenance. Components of the TDIF signaling and CLV3 signaling pathways are similar, and also both CLE peptides act as an intercellular signal between niche cells and stem cells. However, they differ in the mode of action in stem cell maintenance

niche, where stem cells perceive signals from the surrounding niche cells and continue to produce daughter cells that are to differentiate (Fig. 1b) (Stahl and Simon 2005; Scheres 2007; Haecker and Laux 2001). In *tdr* mutants, consistently, the loss of intervening procambial cells between phloem and xylem results in the cessation of secondary growth. In *wox4* mutants, however, procambial cell proliferation is reduced but still continues. The growth termination in cambium of adult *tdr* hypocotyls causes outgrowth of cells located outside of vascular cells to form cavities on the outline of xylem tissue. Thus, TDIF derived from phloem acts as a signal for stem cell maintenance in the vascular meristem by affecting both xylem fate commitment and proliferation. Interestingly, *wox4* mutation increases the size of the cavity in a *tdr* mutant background (Hirakawa et al. 2010a). This result suggests that WOX4 is also regulated by a signal other than TDIF.

The TDIF–TDR–WOX4 pathway is quite similar to the CLV3–CLV1–WUS pathway in the SAM (see chapter "Peptides Regulating Apical Meristem Development"), because both pathways are composed of a CLE peptide, an LRR-RLK XI, and a WOX transcriptional regulator and function in stem cell maintenance in meristems (Fig. 2). This fact means that the maintenance of vascular meristem is regulated by machinery similar to that of the SAM. However, there exist some critical differences (Fig. 2): CLV3 suppresses the expression of *WUS*, which in turn restricts stem cell population (Brand et al. 2000; Schoof et al. 2000), whereas TDIF promotes the expression of *WOX4* to enhance stem cell proliferation. In addition, the effect of *WUS* is non-cell-autonomous (Mayer et al. 1998), whereas that of *WOX4* is cell/tissue autonomous (Hirakawa et al. 2008). *WUS* affects *CLV3* to make a negative feedback loop (Brand et al. 2000; Schoof et al. 2000), whereas

WOX4 does not influence *CLE41* and *CLE44* (Hirakawa et al. 2008). To understand a cell communication network conserved in both meristems, further sophisticated analyses, in particular, on function of genes downstream of WUS and WOX4, are necessary.

2.2 *Other CLE Peptides*

2.2.1 Other CLE Peptides Related to Vascular Development

The *Arabidopsis* genome contains 32 *CLE* genes encoding at least 26 different CLE peptides (Ito et al. 2006; Jun et al. 2008). TDIF (CLE41/CLE44) and CLE42 peptide inhibit tracheary element differentiation in *Zinnia* culture cells, and function in vascular meristem cell maintenance by promoting its proliferation and suppressing its differentiation in *Arabidopsis* hypocotyls and leaves. However, these peptides do not affect significantly vascular development in roots (Kondo et al. 2011). Xylem in roots contains two different vessels: metaxylem and protoxylem. In contrast to TDIF and CLE42, treatment with some other 12-amino-acid CLE peptides at 1 μM inhibited formation of protoxylem vessels but not of metaxylem vessels, in both the primary root and lateral roots (Table 1). Previous studies have revealed that 19 out of the *Arabidopsis* 26 CLE dodeca-peptides inhibit root growth (Ito et al. 2006; Kinoshita et al. 2007). Seventeen out of these 19 CLE peptides inhibited protoxylem vessel formation without affecting metaxylem vessel formation. Interestingly, the other two peptides, CLE25 and CLE26, were peptides that most severely inhibited root growth (Kinoshita et al. 2007). This result suggests that protoxylem vessel inhibition by CLE peptides does not result indirectly from root growth inhibition. Although several synthesized CLE peptides did not affect any developmental aspect, some of them may not be active forms, because CLE2 becomes functional only after modification with three arabinose moieties in situ (Ohyama et al. 2009).

Whitford et al. (2008) reported that overexpression of some *CLE* genes (*CLE6*, *CLV3*, and *CLE19*) cooperates with TDIF to enhance procambium proliferation, although they could not affect vascular development alone. These *CLE* genes are known to reduce the size of the SAM and root apical meristem (RAM), when ectopically expressed in *Arabidopsis* plants. In *clv1* and *clv2* mutants, these CLE peptides fail to induce a reduction in meristem size, but they can enhance TDIF-induced procambium proliferation. These results imply that CLE6, CLV3, and CLE19 peptides cooperate with TDIF to enhance procambial cell proliferation through receptors different from CLV1 and CLV2. Therefore, one CLE ligand may bind to a range of receptors. Furthermore, the competition analysis using the mature CLV3 and CLE2 peptides indicated that a particular LRR-RLK may well recognize some different CLE ligands with different affinities (Ohyama et al. 2009). Because in cambium proliferation, WOX4 functions not only downstream of the TDIF–TDR signaling pathway but also downstream of an unknown signaling pathway(s) in cambium proliferation (Hirakawa et al. 2010a), it is plausible that CLE6, CLV3, and CLE19 peptides might act at procambial/cambial proliferation through WOX4 in roots.

An earlier report indicated that, in *Arabidopsis*, transcripts for some of the *CLE* genes (*CLE10*, *CLE26*, *CLE27*) increase during in vitro xylogenesis (Kubo et al. 2005). Kondo et al. (2011) reported that *pCLE9::GUS* and *pCLE10::GUS* are preferentially expressed in vascular tissues in roots. *Brassica napus CLE19* gene is expressed in a few pericycle cells facing the protoxylem poles as well as in organ primordia (Fiers et al. 2004). These results also support that various CLE peptides may play roles in regulating vascular development (also see chapters "Peptides Regulating Root Growth" and "The Role of Plant Peptides in Symbiotic Interactions").

In a *Zinnia* cell culture, a CLV3 peptide promotes tracheary element differentiation, although CLV3 cannot overcome the TDIF-suppressed tracheary element differentiation (Ito et al. 2006). Therefore, there may be the two counteracting pathways in the CLE signaling, one promotive and the other inhibitory in cell differentiation. The fact that BAM LRR-RLKs, which are the most homologous to CLV1 in *Arabidopsis*, function oppositely to CLV1 (DeYoung et al. 2006) may also suggest two counteracting pathways of the CLE signaling.

2.2.2 CLE Peptides That Inhibit Protoxylem Vessel Formation Through Activating Cytokinin Signaling

In the vascular system, the CLE peptide-enhanced procambial proliferation is positively regulated by auxin and negatively regulated by 1-naphthylphthalamic acid (NPA), which is an inhibitor of auxin transport (Whitford et al. 2008). This result suggests that auxin is necessary for enhancement of procambium proliferation by CLE peptides. Additionally, cellular proliferation in the procambium/cambium is also controlled by cytokinin and its signaling mediators in *Arabidopsis* and *Poplar* (Mähönen et al. 2006; Matsumoto-Kitano et al. 2008; Nieminen et al. 2008; Hejátko et al. 2009). These findings suggest that, in general, CLE peptides may regulate the fate of stem cells through cross talk with signaling pathways mediated by other plant hormones, such as auxin and cytokinin (also see chapters "Peptides Regulating Root Growth" and "The Role of Plant Peptides in Symbiotic Interactions").

CLE9 and *CLE10*, which encode the same CLE peptide, are preferentially expressed in vascular cells of roots (Kondo et al. 2011). The CLE9/CLE10 peptide inhibits protoxylem vessel formation without affecting metaxylem vessel formation in *Arabidopsis* roots. Microarray analysis revealed that the CLE9/CLE10 peptide specifically reduces expression of type A *ARABIDOPSIS RESPONSE REGULATORs* (*ARRS*), which are known as negative regulators of cytokinin signaling (Kiba et al. 2003; To et al. 2004, 2007), in particular, *ARR5* and *ARR6* (Kondo et al. 2011). Indeed, the *arr5 arr6* double mutants often exhibited incomplete protoxylem vessels in roots (Kondo et al. 2011). Consistently, a double mutant

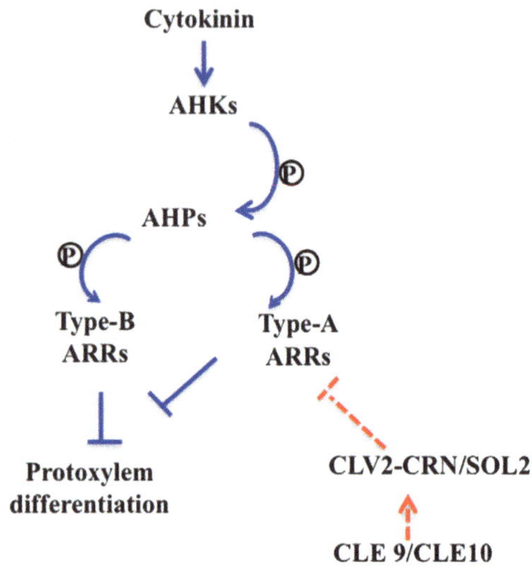

Fig. 3 A model illustrating the cross talk between cytokinin and CLE9/CLE10 peptide signaling pathways to regulate protoxylem vessel formation. CLE peptides positively regulate cytokinin signaling via suppressing type A ARRs (*ARABIDOPSIS RESPONSE REGULATORs*), resulting in the inhibition of protoxylem vessel formation

for two type B ARRs, *ARR10* and *ARR12*, which function positively in cytokinin signaling, is resistant to CLE9/CLE10 peptide in terms of protoxylem vessel formation. Interestingly, other combinations of type B ARR mutants such as *arr1arr12* and *arr1arr10* showed much weaker resistance against TDIF than *arr10arr12*. This result implies that *ARR10* and *ARR12* act as major type B ARRs at the cross talk with the TDIF signaling, and as expected, *arr10arr12* displayed ectopic protoxylem vessel formation. Thus, CLE9/CLE10 peptide activates cytokinin signaling through the repression of *ARR5* and *ARR6*, resulting in inhibition of protoxylem vessel formation (Fig. 3). Genetic analysis suggests that the CLV2 membrane receptor and its partner CRN/SOL2 kinase (Müller et al. 2008; Miwa et al. 2008) may act at protoxylem vessel formation downstream of the CLE9/CLE10 peptide signaling (Kondo et al. 2011). Further identification of downstream components of the CLE9/CLE10 peptide signaling will provide a direct molecular link between the CLE9/CLE10 and cytokinin signaling.

Similar cross talk between CLE signaling and type A ARRs has been reported in the SAM maintenance, in which the transcription factor WUS acts downstream of CLV3 signaling (Schoof et al. 2000). WUS directly represses the expression of several type A *ARR*s (*ARR5*, *ARR6*, *ARR7*, and *ARR15*) in the SAM (Leibfried et al. 2005). This suggests that CLE peptides may commonly regulate cytokinin signaling through type A ARRs during plant development, although signaling pathways of many other CLE peptides remain to be elucidated. WUS also upregulates *CLV3* expression in the SAM maintenance through negative feedback (Brand et al. 2002). In crown root development, OsWOX11, which is a rice homologue of *Arabidopsis* WOX11, directly represses *RR2*, a type A response regulator gene (Zhao et al. 2009). However, our GeneChip analysis revealed that no *CLE* or *WOX* gene was induced by CLE9/CLE10 peptide application. Therefore, although some type A ARRs may play a role as downstream factors of CLE signaling pathways, the mechanisms by which type A ARRs are induced may vary.

3 Other Peptides Involved in Signaling Between Vascular Cells

3.1 Phytosulfokine (PSK)

Matsubayashi and Sakagami (1996) isolated a disulfated pentapeptide named phytosulfokine [PSK; Y(SO_3H)IY(SO_3H)TQ] as a potent mitogenic factor from the conditioned medium of cultures of asparagus mesophyll cell culture. PSK was also discovered as a substance that stimulates cell proliferation or somatic embryogenesis from suspension cultures of rice, maize, and carrot (Matsubayashi and Sakagami 2006). Genes encoding ~80-aa precursors of PSK have also been isolated in various plant species (Matsubayashi and Sakagami 2006). Each preproprotein of PSK has a secretory signal sequence at the N-terminal and a PSK sequence near the C-terminal flanked by dibasic amino acid residues, implying proteolytic processing as that of animal peptide hormones. Tyrosine *O*-sulfation of PSK precursors, which is catalyzed by tyrosylprotein sulfotransferase in the Golgi apparatus (Komori et al. 2009), is essential for the biological activities of PSK (Matsubayashi and Sakagami 1996). After the processing and secretion, PSK binds to its LRR-RLK on the plasma membrane (Matsubayashi et al. 2002; also see chapters "Peptides Regulating Root Growth," "Peptides and the Regulation of Plant Homeostasis," and "Methods to Identify New Partners of Plant Signaling Peptides").

PSK was also isolated from cultured medium of a xylogenic *Zinnia* cells (Matsubayashi et al. 1999). Tracheary element differentiation was suppressed when *Zinnia* cells were cultured at low cell densities less than 1.0×10^4 cells mL^{-1} (Fukuda and Komamine 1980; Matsubayashi et al. 1999; Motose et al. 2001a). This suppression was restored by the supply of PSK. A *Zinnia* PSK gene, *ZePSK1*, is expressed with two peaks between 12 and 24 h and between 60 and 72 h during the xylogenic culture (Motose et al. 2009). The first and second peaks appear to be induced by wound stimuli and brassinosteroids, respectively. Wound-induced PSK downregulates defense response in *Zinnia* cells through transcriptional regulation of a particular gene set and heals wounding of them, which make them competent for tracheary element differentiation in vitro (Motose et al. 2009). However, the role of brassinosteroid-induced PSK remains to be elucidated.

3.2 POLARIS

The *POLARIS (PLS)* open reading frame is located within a short (500 nucleotides) auxin-inducible transcript and encodes a predicted polypeptide of 36 amino acid residues (Casson et al. 2002). The *pls* mutants exhibit a semidominant phenotype characterized by short-root phenotype, reduced leaf vascularization, and altered responses to exogenous auxin and cytokinin. The *pls* roots are hyperresponsive to exogenous cytokinins and show elevated expression of the cytokinin-inducible gene *ARR5/IBC6* compared with the wild type. In *pls* mutants, the short-root phenotype is partially rescued by a low concentrations of auxin. Further detailed analysis revealed that PLS negatively regulates ethylene responses to modulate cell division and expansion via downstream effects on microtubule dynamics and auxin signaling (Chilley et al. 2006). It is still unclear if this peptide functions in leaf vascular patterning directly, or indirectly through phytohoromes-dependent regulation of cell division and expansion.

3.3 ENOD40

ENOD40s are early plant symbiotic genes (nodulins) linked to the dedifferentiation and division of root cortical cells before nodule primordium formation (Batut et al. 2011; also see chapter "The Role of Plant Peptides in Symbiotic Interactions"). The ~0.7-kb *enod40* gene does not display any long open reading frame but has coding capacity for two short peptides, a strongly conserved 12–13-amino-acid-long peptide A and a 4–30-amino-acid-long peptide B. ENOD40 peptides are synthesized de novo but not resulting from proteolytic cleavage of larger precursors. Peptides A and B do not display signal sequences either and are thus predicted to act within the cell. However, due to their size, they may travel from cell to cell. Knockdown of *ENOD40*s leads to significant suppression of nodule formation in *Lotus japonica* (Kumagai et al. 2006). In addition to nodules, this gene is expressed preferentially in early pericyles during normal developmental of plants, even nonlegumes (Vleghels et al. 2003). In rice, transcription of *OsENOD40* is confined to parenchyma cells surrounding the protoxylem during the early stages of development of lateral vascular bundles that conjoin an emerging leaf, thus suggesting that ENOD40 peptides might function as a signal affecting xylem development (Kouchi et al. 1999).

3.4 Xylogen

Xylogen is not a small peptide but a relatively large protein, which functions as an intercellular signal molecule that promotes tracheary element differentiation in vascular tissues (Motose et al. 2004). Motose et al. (2001a, b) first demonstrated the involvement of promotive local intercellular communication during tracheary element differentiation in a xylogenic *Zinnia* cell culture. Novel bioassay systems with *Zinnia* cell culture resulted in the isolation of the mediator designated "xylogen," which is a hybrid-type molecule with both properties of arabinogalactan proteins and nonspecific lipid transfer proteins (Motose et al. 2004). Xylogen predominantly accumulated in the meristem, procambium, and xylem. In the xylem, xylogen was polarly localized in the cell walls of differentiating tracheary elements. The *Arabidopsis* genome contains *AtXYP1* and *AtXYP2*, two highly sequence-related homologues of the *Zinnia* xylogene gene. Proteins produced from *AtXYP1* and *AtXYP2* in tobacco cultured cells had xylogen activity. AtXYP2 is expressed preferentially in vascular cells in *Arabidopsis* plants (Kobayashi et al. 2011). Double knockouts of *AtXYP1* and *AtXYP2* exhibited defects in xylem continuity (Motose et al. 2004). These findings have suggested that polarly secreted xylogen draws neighboring cells into the pathway of xylem differentiation to achieve continuous xylem development.

3.5 Receptors for Putative Peptide Ligands Responsible for Vascular Development

There are more than 600 RLK genes in the *Arabidopsis* genome (Shiu et al. 2001). They may be candidates of receptors for peptide ligands. However, only a limited number of ligand–receptor combinations have been demonstrated. Comprehensive gene expression analyses revealed that a number of uncharacterized RLK genes are expressed preferentially during vascular development (Kubo et al. 2005; Ohashi-Ito et al. 2010; Agusti et al. 2011). Agusti et al. (2011) identified two LRR-RLKs, MORE LATERAL GROWTH 1 (MOL1), and REDUCED IN LATERAL GROWTH 1 (RUL1), both of which are expressed in the cambium. The *mol1* and *rul1* mutants displayed an enhanced and reduced cambium activity, respectively, indicating that they function as opposing regulators of cambium activity. Recently, a gene for an LRR-RLK, COE1 (COMMISSURAL VEIN EXCESSIVE1), was identified from rice as a regulator for commissural vein patterning (Sakaguchi et al. 2010). Interestingly, *COE1* functions in regulation of vein intervals downstream of both auxin and brassinosteroid signaling pathways. There are still a huge number of uncharacterized vascular tissue-related RLKs. Further analysis of signaling pathways regulated by these receptors must provide fruitful insight into a complex signaling network shaping the well-organized vascular tissue.

4 Conclusions and Perspectives

Recent findings mentioned above are summarized below:

1. Several CLE peptides play major roles in vascular pattern formation in non-cell-autonomous manner through their specific LRR-RLKs.
2. The TDIF–TDR–WOX4 signaling pathway regulates vascular stem cell maintenance. A robust similarity in the molecular components in SAM and vascular meristem suggests that the CLE peptides–LRR-RLKs–WOXs are a common machinery for regulation of stem cell maintenance in meristems.

3. CLE9/10 peptide represses protoxylem vessel formation via activation of cytokinin signaling through the transcriptional suppression of type A ARRs. Together with the fact that WUS regulates the expression of type A ARR, cross talk among plant hormones and peptide ligands may be involved in the regulation of vascular development.
4. Our knowledge about peptides other than CLE peptides in relation to vascular development is still very limited. However, preferential expression of many RLKs in distinct vascular cells suggests the involvement of novel peptides in vascular development.

These findings open a new window in the field of local cell–cell interaction governing vascular development. Now we are about to grasp a sophisticated network of cell–cell signaling pathways involving phytohormones and peptides, which governs the initiation and maintenance of vascular stem cells and their commitment to differentiation into xylem and phloem cells. Further intensive analyses are necessary for displaying the whole image of the signaling network. In particular, it should be essential to understand cross talks among signaling pathways of phytohormones and peptides. We also need to clarify peptide maturation process, non-cell-autonomous manner of function, and intracellular signaling pathways at molecular levels for each peptide signal.

Acknowledgments I thank S. Betsuyaku for critical reading of this manuscript. This work was supported in part by Grants-in-Aid from the Ministry of Education, Science, Sports and Culture of Japan (19060 009), from the Japan Society for the Promotion of Science (23227001), and from Bio-oriented Technology Research Advancement Institution (BRAIN).

References

Agusti J, Lichtenberger R, Schwarz M, Nehlin L, Greb T (2011) Characterization of transcriptome remodeling during cambium formation identifies MOL1 and RUL1 as opposing regulators of secondary growth. PLoS Genet 7:e1001312

Batut J, Mergaert P, Masson-Boivin C (2011) Peptide signalling in the rhizobium-legume symbiosis. Curr Opin Microbiol 14:181–187

Bleckmann A, Weidtkamp-Peters S, Seidel CA, Simon R (2010) Stem cell signaling in Arabidopsis requires CRN to localize CLV2 to the plasma membrane. Plant Physiol 152:166–176

Brand U, Fletcher JC, Hobe M, Meyerowitz EM, Simon R (2000) Dependence of stem cell fate in Arabidopsis on a feedback loop regulated by CLV3 activity. Science 289:617–619

Brand U, Grünewald M, Hobe M, Simon R (2002) Regulation of CLV3 expression by two homeobox genes in Arabidopsis. Plant Physiol 129:565–575

Casson SA, Chilley PM, Topping JF, Evans IM, Souter MA, Lindsey K (2002) The POLARIS gene of Arabidopsis encodes a predicted peptide required for correct root growth and leaf vascular patterning. Plant Cell 14:1705–1721

Chilley PM, Casson SA, Tarkowski P, Hawkins N, Wang KLC, Hussey PJ, Beale M, Ecker JR, Sandberg GK, Lindsey K (2006) The POLARIS peptide of Arabidopsis regulates auxin transport and root growth via effects on ethylene signaling. Plant Cell 18:3058–3072

Clark SE, Williams RW, Meyerowitz EM (1997) The CLAVATA1 gene encodes a putative receptor kinase that controls shoot and floral meristem size in Arabidopsis. Cell 89:575–585
DeYoung BJ, Bickle KL, Schrage KJ, Muskett P, Patel K, Clark SE (2006) The CLAVATA1-related BAM1, BAM2 and BAM3 receptor kinase-like proteins are required for meristem function in Arabidopsis. Plant J 45:1–16
Etchells JP, Turner SR (2010) The PXY-CLE41 receptor ligand pair defines a multifunctional pathway that controls the rate and orientation of vascular cell division. Development 137:767–774
Fiers M, Hause G, Boutilier K, Martinez EC, Wejers D, Offringa R, van der Geest L, van Lookren CM, Liu CM (2004) Mis-expression of the CLV3/ESR-like gene CLE19 in Arabidopsis leads to a consumption of root meristem. Gene 327:37–49
Fisher K, Turner S (2007) PXY, a receptor-like kinase essential for maintaining polarity during plant vascular-tissue development. Curr Biol 17:1061–1066
Fletcher JC, Brand U, Running MP, Simon R, Meyerowitz EM (1999) Signaling of cell fate decisions by CLAVATA3 in Arabidopsis shoot meristems. Science 283:1911–1914
Fukuda H (2004) Signals that control plant vascular cell differentiation. Nat Rev Mol Cell Biol 5:379–391
Fukuda H, Komamine A (1980) Establishment of an experimental system for the study of tracheary element differentiation from single cells isolated from the mesophyll of *Zinnia elegans*. Plant Physiol 65:57–60
Fukuda H, Hirakawa Y, Sawa S (2007) Peptide signaling in vascular development. Curr Opin Plant Biol 10:477–482
Haecker A, Laux T (2001) Cell-cell signaling in the shoot meristem. Curr Opin Plant Biol 4:441–446
Haecker A, Gross-Hardt R, Geiges B, Sarkar A, Breuninger H, Herrmann M, Laux T (2004) Expression dynamics of WOX genes mark cell fate decisions during early embryonic patterning in Arabidopsis thaliana. Development 131:657–668
Hejátko J, Ryu H, Kim GT, Dobesová R, Choi S, Choi SM, Choi SM, Soucek P, Horák J, Pekárová B, Palme K, Brzobohaty B, Hwang I (2009) The histidine kinases CYTOKININ-INDEPENDENT1 and ARABIDOPSIS HISTIDINE KINASE2 and 3 regulate vascular tissue development in Arabidopsis shoots. Plant Cell 21:2008–2021
Hirakawa Y, Shinohara H, Kondo Y, Inoue A, Nakanomyo I, Ogawa M, Sawa S, Ohashi-Ito K, Matsubayashi Y, Fukuda H (2008) Non-cell-autonomous control of vascular stem cell fate by a CLE peptide/receptor system. Proc Natl Acad Sci U S A 105:15208–15213
Hirakawa Y, Kondo Y, Fukuda H (2010a) TDIF peptide signaling regulates vascular stem cell proliferation via the WOX4 homeobox gene in Arabidopsis. Plant Cell 22:2618–2629
Hirakawa Y, Kondo Y, Fukuda H (2010b) Regulation of vascular development by CLE peptide-receptor systems. J Integr Plant Biol 52:8–16
Hirakawa Y, Kondo Y, Fukuda H (2011) Establishment and maintenance of vascular cell communities through local signaling. Curr Opin Plant Biol 14:17–23
Ito Y, Nakanomyo I, Motose H, Iwamoto K, Sawa S, Dohmae N, Fukuda H (2006) Dodeca-CLE peptides as suppressors of plant stem cell differentiation. Science 313:842–845
Jeong S, Trotochaud AE, Clark SE (1999) The Arabidopsis CLAVATA2 gene encodes a receptor-like protein required for the stability of the CLAVATA1 receptor-like kinase. Plant Cell 11:1925–1934
Jun JH, Fiume E, Fletcher JC (2008) The CLE family of plant polypeptide signaling molecules. Cell Mol Life Sci 65:743–755
Kiba T, Yamada H, Sato S, Kato T, Tabata S, Yamashino T, Mizuno T (2003) The type-A response regulator, ARR15, acts as a negative regulator in the cytokinin-mediated signal transduction in Arabidopsis thaliana. Plant Cell Physiol 44:868–874
Kinoshita A, Nakamura Y, Sasaki E, Kyozuka J, Fukuda H, Sawa S (2007) Gain-of-function phenotypes of chemically synthetic CLAVATA3/ESR-related (CLE) peptides in Arabidopsis thaliana and Oryza sativa. Plant Cell Physiol 48:1821–1825

Kinoshita A, Betsuyaku S, Osakabe Y, Mizuno S, Nagawa S, Stahl Y, Simon R, Yamaguchi-Shinozaki K, Fukuda H, Sawa S (2010) RPK2/TOAD2 is an essential receptor-like kinase transmitting the CLV3 signal in Arabidopsis. Development 137:3911–3920

Kobayashi Y, Motose H, Iwamoto K, Fukuda H (2011) Expression and genome-wide analysis of xylogen-type gene family. Plant Cell Physiol 52:1095–1106

Komori R, Amano Y, Ogawa-Ohnishi M, Matsubayashi Y (2009) Identification of tyrosylprotein sulfotransferase in Arabidopsis. Proc Natl Acad Sci U S A 106:15067–15072

Kondo T, Sawa S, Kinoshita A, Mizuno S, Kakimoto T, Fukuda H, Sakagami Y (2006) A plant peptide encoded by CLV3 identified by in situ MALDI TOF-MS analysis. Science 313:845–848

Kondo Y, Hirakawa Y, Fukuda H (2011) CLE peptides can negatively regulate protoxylem vessel formation via cytokinin signaling. Plant Cell Physiol 52:37–48

Kouchi H, Takane K, So RB, Ladha JK, Reddy PM (1999) Rice ENOD40: isolation and expression analysis in rice and transgenic soybean root nodules. Plant J 18:121–129

Kubo M, Udagawa M, Nishikubo N, Horiguchi G, Yamaguchi M, Ito J, Mimura T, Fukuda H, Demura T (2005) Transcription switches for protoxylem and metaxylem vessel formation. Genes Dev 19:1855–1860

Kumagai H, Kinoshita E, Ridge RW, Kouchi H (2006) RNAi knock-down of ENOD40s leads to significant suppression of nodule formation in Lotus japonicus. Plant Cell Physiol 47:1102–1111

Lehesranta SJ, Lichtenberger R, Helariutta Y (2010) Cell-to-cell communication in vascular morphogenesis. Curr Opin Plant Biol 13:59–65

Leibfried A, To JP, Busch W, Stehling S, Kehle A, Demar M et al (2005) WUSCHEL controls meristem function by direct regulation of cytokinin-inducible response regulators. Nature 438:1172–1175

Mähönen AP, Bishopp A, Higuchi M, Nieminen KM, Kinoshita K, Tormakangas K, Ikeda Y, Oka A, Kakimoto T, Helariutta Y (2006) Cytokinin signaling and its inhibitor AHP6 regulate cell fate during vascular development. Science 311:94–98

Matsubayashi Y, Sakagami Y (1996) Phytosulfokine, sulfated peptides that induce the proliferation of single mesophyll cells of Asparagus officinalis L. Proc Natl Acad Sci U S A 93:7623–7627

Matsubayashi Y, Sakagami Y (2006) Peptide hormones in plants. Annu Rev Plant Biol 57:649–674

Matsubayashi Y, Takagi L, Omura N, Morita A, Sakagami Y (1999) The endogenous sulfated pentapeptide phytosulfokine-alpha stimulates tracheary element differentiation of isolated mesophyll cells of zinnia. Plant Physiol 120:1043–1048

Matsubayashi Y, Ogawa M, Morita A, Sakagami Y (2002) An LRR receptor kinase involved in perception of a peptide plant hormone, phytosulfokine. Science 296:1470–1472

Matsumoto-Kitano M, Kusumoto T, Tarkowski P, Kinoshita-Tsujimura K, Václavíková K, Miyawaki K, Kakimoto T (2008) Cytokinins are central regulators of cambial activity. Proc Natl Acad Sci U S A 105:20027–20031

Mayer KF, Schoof H, Haecker A, Lenhard M, Jürgens G, Laux T (1998) Role of WUSCHEL in regulating stem cell fate in the Arabidopsis shoot meristem. Cell 95:805–815

Miwa H, Betsuyaku S, Iwamoto K, Kinoshita A, Fukuda H, Sawa S (2008) The receptor-like kinase SOL2 mediates CLE signaling in Arabidopsis. Plant Cell Physiol 49:1752–1757

Motose H, Fukuda H, Sugiyama M (2001a) Involvement of local intercellular communication in the differentiation of zinnia mesophyll cells into tracheary elements. Planta 213:121–131

Motose H, Sugiyama M, Fukuda H (2001b) An arabinogalactan protein(s) is a key component of a fraction that mediates local intercellular communication involved in tracheary element differentiation of zinnia mesophyll cells. Plant Cell Physiol 42:129–137

Motose H, Sugiyama M, Fukuda H (2004) A proteoglycan mediates inductive interaction during plant vascular development. Nature 429:873–878

Motose H, Iwamoto K, Endo S, Demura T, Sakagami Y, Matsubayashi Y, Moore KL, Fukuda H (2009) Involvement of phytosulfokine in the attenuation of stress response during the transdifferentiation of *Zinnia* mesophyll cells into tracheary elements. Plant Physiol 150:437–447

Müller R, Borghi L, Kwiatkowska D, Laufs P, Simon R (2006) Dynamic and compensatory responses of Arabidopsis shoot and floral meristems to CLV3 signaling. Plant Cell 18:1188–1198

Müller R, Bleckmann A, Simon R (2008) The receptor kinase CORYNE of Arabidopsis transmits the stem cell-limiting signal CLAVATA3 independently of CLAVATA1. Plant Cell 20:934–946

Nieminen K, Immanen J, Laxell M, Kauppinen L, Tarkowski P, Dolezal K, Tähtiharju S, Elo A, Decourteix M, Ljung K, Bhalerao R, Keinonen K, Albert VA, Helariutta Y (2008) Cytokinin signaling regulates cambial development in poplar. Proc Natl Acad Sci U S A 105:20032–20037

Ohashi-Ito K, Oda Y, Fukuda H (2010) *Arabidopsis* VASCULAR-RELATED NAC-DOMAIN6 directly regulates genes that govern programmed cell death and secondary wall formation in a coordinated way during xylem differentiation. Plant Cell 22:3461–3473

Ohyama K, Shinohara H, Ogawa-Ohnishi M, Matsubayashi Y (2009) A glycopeptide regulating stem cell fate in Arabidopsis thaliana. Nat Chem Biol 5:578–580

Sakaguchi J, Itoh JI, Ito Y, Nakamura A, Fukuda H, Sawa S (2010) COE1, an LRR-RLK responsible for commissural vein pattern formation in rice. Plant J 63:405–416

Scheres B (2007) Stem-cell niches: nursery rhymes across kingdoms. Nat Rev Mol Cell Biol 8:345–354

Schoof H, Lenhard M, Haecker A, Mayer KF, Jürgens G, Laux T (2000) The stem cell population of Arabidopsis shoot meristems in maintained by a regulatory loop between the CLAVATA and WUSCHEL genes. Cell 100:635–644

Shiu SH, Bleecker AB (2001) Receptor-like kinases from Arabidopsis form a monophyletic gene family related to animal receptor kinases. Proc Natl Acad Sci U S A 98:10763–10768

Stahl Y, Simon R (2005) Plant stem cell niches. Int J Dev Biol 49:479–489

To JP, Haberer G, Ferreira FJ, Deruère J, Mason MG, Schaller GE, Alonso JM, Ecker JR, Kieber JJ (2004) Type-A Arabidopsis response regulators are partially redundant negative regulators of cytokinin signaling. Plant Cell 16:658–671

To JP, Deruère J, Maxwell BB, Morris VF, Hutchison CE, Ferreira FJ, Schaller GE, Kieber JJ (2007) Cytokinin regulates type-A Arabidopsis Response Regulator activity and protein stability via two-component phosphorelay. Plant Cell 19:3901–3914

Vleghels I, Hontelez J, Ribeiro A, Fransz P, Bisseling T, Franssen H (2003) Expression of ENOD40 during tomato plant development. Planta 218:42–49

Whitford R, Fernandez A, De Groodt R, Ortega E, Hilson P (2008) Plant CLE peptides from two distinct functional classes synergistically induce division of vascular cells. Proc Natl Acad Sci U S A 105:18625–18630

Zhao Y, Hu Y, Dai M, Huang L, Zhou DX (2009) The WUSCHEL-related homeobox gene WOX11 is required to activate shoot-borne crown root development in rice. Plant Cell 21:736–748

Zhu Y, Wang Y, Li R, Song X, Wang Q, Huang S, Jin JB, Liu C-M, Lin J (2010) Analysis of interactions among the CLAVATA3 receptors reveals a direct interaction between CLAVATA2 and CORYNE in Arabidopsis. Plant J 61:223–233

The S-LOCUS CYSTEINE-RICH PROTEIN (SCR): A Small Peptide with A High Impact on the Evolution of Flowering Plants

Isabelle Fobis-Loisy, Rumen Ivanov, and Thierry Gaude

Abstract Self-incompatibility (SI) is an archetypal cell-to-cell communication system in which self-pollen is rejected to prevent inbreeding. In crucifers (or Brassicaceae family), the pollen SI determinant is a small peptide, the S-LOCUS CYSTEINE-RICH PROTEIN (SCR, also known as SP11). During self-pollination, SCR binds to the extracellular domain of its cognate stigmatic receptor, the S-LOCUS RECEPTOR KINASE (SRK). This initiates a signaling cascade leading to self-pollen inhibition. The genes encoding both SI determinants are tightly linked in a multiallelic genomic region defined as the *S*-locus. Among *S* alleles, *SCR* shows extreme variability, and few residues in the protein have been reported to be critical for its specificity. In a heterozygous situation, *SCR* alleles display complex dominance relationships based on the silencing of certain *SCR* genes. This chapter provides a summary of the role of SCR in SI, the relationships between *SCR* alleles, and the role SCR loss played in the model plant *Arabidopsis thaliana*.

1 Introduction

One characteristic of flowering plants is the very high frequency of hermaphroditic species in which male and female reproductive organs are found in close vicinity. This trait, which facilitates self-fertilization and hence inbreeding, could have been deleterious for the expansion of flowering plants. However, angiosperms have

---

IFL and RI contributed equally to this work

I. Fobis-Loisy • T. Gaude (✉)
Centre National de la Recherche Scientifique, Institut National de la Recherche Agronomique, Université Claude Bernard Lyon I, Ecole Normale Supérieure de Lyon, Lyon, France
e-mail: thierry.gaude@ens-lyon.fr

R. Ivanov
Department of Plant Biology, Saarland University, Campus A 2.4, 66123 Saarbruecken, Germany

H.R. Irving and C. Gehring (eds.), *Plant Signaling Peptides*,
Signaling and Communication in Plants 16, DOI 10.1007/978-3-642-27603-3_5,

conquered a large variety of environments and are the most successful groups of terrestrial flora. This success is generally thought to derive from the acquisition by the angiosperms of mechanisms that limit or even prevent self-fertilization within hermaphroditic flowers. For instance, difference between the maturity period of reproductive organs (i.e., protogyny and protandry) is an efficient means to avoid self-pollination. However, the most sophisticated mechanism to control self-fertilization is self-incompatibility (SI), which allows the pistil to recognize and reject self-pollen grains. Self-incompatible plants are not able to produce seeds when their flowers receive pollen from their own flowers or from flowers of genetically related individuals. Approximately half of the angiosperm species have acquired SI systems. In the *Brassicaceae*, self-pollen recognition is genetically controlled by a single multiallelic locus, the *S*-locus (de Nettancourt 2001; Gaude et al. 2006). Pollen rejection occurs when the same *S*-allele specificity is expressed by both the pollen and the pistil. Extensive analyses of SI systems clearly indicate that loci other than the *S*-locus are also required for a complete SI response (Gaude et al. 2006; Ivanov et al. 2010). In most SI species, the *S*-allele specificity carried by the pollen is determined by its own haploid genome. These species have a gametophytic SI system. By contrast, in the *Brassicaceae*, pollen behavior depends on the diploid (sporophytic) genome of the pollen parent plant. In that case, these species possess a sporophytic SI system. Because SI response requires specific male and female determinants, at least two distinct linked genes are expected to be located at the *S*-locus. This led to the use of the term "*S*-haplotypes" to designate variants of the *S*-locus (Boyes and Nasrallah 1993). The number of *S*-haplotypes is particularly large in *Brassica* species, with over 90 in *Brassica oleracea* (Ruffio-Chable and Gaude 2001).

In the early 1980s, with the emergence of molecular genetics and novel protein analysis methods, several groups focused their attention on the identification of the male and female determinants. Because self-pollen rejection occurs very early following pollination in *Brassicaceae* species, SI determinants were thought to be localized in the pollen wall and on the stigmatic surface, respectively. Indeed, cytological studies revealed that self-pollen fails to hydrate properly or, if it germinates, pollen tube growth is arrested at the surface of stigmatic papillar cells. In this chapter, after a brief description of how determinants of the SI response were discovered in the *Brassicaceae*, we present our current knowledge of the signaling pathway that leads to self-pollen recognition and rejection. In particular, we focus on the male determinant SCR, which functions as a ligand triggering the specific activation of the plasma membrane receptor kinase SRK.

2 Discovery of the Female and Male Determinants of the Self-Incompatibility Response in *Brassicaceae*

Two genes were first described at the *S*-locus to encode putative female determinants as their expression patterns coincided with the acquisition of SI by the maturing stigma. These were the *S-LOCUS GLYCOPROTEIN* (*SLG*) and

S-LOCUS RECEPTOR KINASE (*SRK*), respectively (Nasrallah et al. 1985; Stein et al. 1991). The precise role of SLG in SI still remains questionable as some self-incompatible *Brassica* plants do not express SLG and as the self-incompatible *Arabidopsis lyrata* species has no *SLG* gene (Okazaki et al. 1999; Nishio and Kusaba 2000; Kusaba et al. 2001). The *SRK* gene encodes a single-pass transmembrane serine/threonine kinase, which appears as the key player in the recognition of self-pollen. Indeed, *Brassica* plants mutated in the *SRK* gene and that do not express *SRK* transcripts or only truncated SRK transcripts were found to be self-compatible (Goring et al. 1993; Nasrallah et al. 1994). In addition, Takasaki et al. (2000) generated transgenic *Brassica* plants that exhibit a novel *S*-allele specificity by introducing an *SRK* allele different from the endogenous *SRK* alleles. These data argue in favor of SRK being the female determinant of SI.

In contrast to the female factor, the male *S*-determinant remained elusive for a long time before its identification in 1999 (Schopfer et al. 1999; Suzuki et al. 1999). The first evidence that molecules carried by the pollen might be related to SI response came from the work of Stephenson et al. (1997). These authors isolated proteins coating the cavities of the pollen wall and showed that, when mixed with pollen of a different *S*-haplotype, these could induce pollen rejection by the stigma in an *S*-haplotype-specific manner. Biochemical analysis of these pollen coat proteins (PCPs) combined with pollination bioassay revealed that the molecules responsible for this activity are small basic proteins of molecular mass under 10 kDa. Two years later, the group of Professor Hinata, in Japan, isolated the *S-locus protein 11* (*SP11*) gene in *Brassica rapa* (Suzuki et al. 1999). This gene, located in the *SLG-SRK* region of the S_9-haplotype, encodes a small cysteine-rich protein resembling the PCP described earlier by Stephenson et al. (1997). In a parallel study based on the systematic sequencing of the *S*-locus region of different *S*-haplotypes in *B. oleracea*, the group of June and Mike Nasrallah in the USA described the *S-LOCUS CYSTEINE-RICH PROTEIN (SCR)* gene, which corresponds to *SP11* (Schopfer et al. 1999). Analysis of the *Arabidopsis* genome showed that SP11/SCR proteins belong to a novel class of highly polymorphic peptides within a family of proteins named the PCP family (Vanoosthuyse et al. 2001). In the following sections, we will use the term SCR to designate SP11/SCR gene products. SCR sequence contains a signal peptide which is cleaved before secretion of the mature peptide by the tapetal cells of the anther. The mature SCR peptides are small (5.7–8.0 kDa), hydrophilic, basic (isoelectric points over 8.0) secreted peptides, which contain generally eight conserved cysteine residues and a conserved glycine (Schopfer et al. 1999; Takayama et al. 2000, 2001). Immunolocalization studies revealed SCRs to be expressed in the tapetum at early developmental stages of the flower buds, and then to migrate to the pollen wall surface at later stages, accumulating in the pollen coat of mature pollen grains (Shiba et al. 2001; Iwano et al. 2003). The role of SCR as the male determinant was clearly established by loss-of-function and gain-of-function experiments, which revealed that the *SCR* gene is necessary and sufficient for determining SI specificity in pollen of *Brassica* species (Schopfer et al. 1999; Shiba et al. 2001). More recently, the unequivocal implication of both SRK and SCR as essential factors for self-pollen recognition was demonstrated by the reintroduction of SI in the naturally self-compatible *Arabidopsis thaliana* species through transformation with a functional *SRK–SCR* gene pair from its close relative self-incompatible *A. lyrata* (Nasrallah et al. 2002, 2004).

3 Interaction Between SCR and SRK

3.1 Structural Prerequisites for the Interaction Between SCR and SRK

In order to explain self-pollen recognition on a more mechanistic level, it was predicted that it follows the typical receptor–ligand recognition model, where SCR serves as a ligand for SRK. This was supported by the expression pattern of SCR and its localization in the pollen coat (Shiba et al. 2001). Two independent studies published in 2001 confirmed this hypothesis. Kachroo et al. (2001) used recombinant SCR_6 expressed in tobacco (*Nicotiana benthamiana*) leaves to demonstrate that it can specifically interact with both recombinant extracellular domain of SRK_6 ($eSRK_6$) and native SRK_6 from stigma microsomal extracts. Takayama et al. (2001) used similar approach to demonstrate specific interaction of SCR_8 to recombinant $eSRK_8$ and native SRK_8. In addition, it was shown that the S_8 stigma microsomal extracts possess two binding activities for SCR_8, with one of these corresponding to the full-length receptor. A later study from the same group showed that the second one represented a form of SRK_8 that comprised the extracellular, transmembrane, and part of the juxtamembrane domain of the receptor (tSRK) (Shimosato et al. 2007). Interestingly, this form was documented also in the S_3-haplotype a decade earlier (Giranton et al. 1995).

The interaction between the receptor and its ligand occurs at the plasma membrane of the stigma papilla cell (Ivanov and Gaude 2009b) and therefore the pollen coat-located SCR needs to first traverse the papilla cell wall. Purified SCR proteins are not capable of eliciting self-incompatibility response on their own. A microscopical study showed that purified SCR_8 protein was unable to penetrate the cell wall of S_8 papillae and instead required the presence of a pollen grain (Iwano et al. 2003). In other studies, extracted pollen coat proteins are applied to facilitate this artificial self-incompatibility response (Kachroo et al. 2001). The identity of these facilitators remains unknown at present.

Mass spectrometric analysis of SCR_8, which was affinity-purified from pollen coats, suggested that SCR exists as a monomer with fully oxidized cysteine residues forming four intramolecular disulfide bonds (Takayama et al. 2001) (Fig. 1). This was later verified by investigating the tertiary structure of SCR_8 (Mishima et al. 2003). In addition, structure-based sequence alignments identified a loop which includes a hypervariable region (Fig. 1). It was proposed that this loop is responsible for the allele-specific interaction with SRK. As in SCRs from certain haplotypes

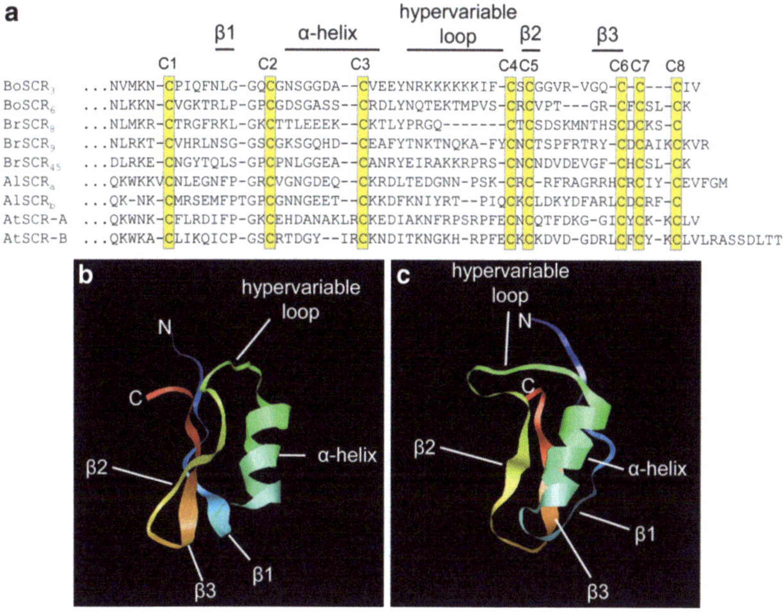

Fig. 1 Structure of SCR (**a**) Alignment of SCR sequences from *Brassica oleracea* (Bo), *Brassica rapa* (Br), *Arabidopsis lyrata* (Al), and *Arabidopsis thaliana* (At) excluding the known or predicted signal peptides. The eight cysteine residues (C1–C8) are *highlighted*. Regions corresponding to the alpha-helix (α-helix), the three beta sheets (β1–β3), and the hypervariable loop are indicated above the alignment. The GenBank accession numbers for the used sequences are as follows: *BoSCR*$_3$ (AJ278643), *BoSCR*$_6$ (AF195625), *BrSCR*$_8$ (AB035504), *BrSCR*$_9$ (AB022078), *BrSCR*$_{45}$ (AB039763), *AlSCR*$_a$ (AB052753), and *AlSCR*$_b$ (AB052754). Sequences for *AtSCR-B* and the restored *AtSCR-A* were taken from Tsuchimatsu et al. (2010). (**b**, **c**) Predicted three-dimensional structure of BoSCR$_3$. Amino- and carboxy-termini (N and C, respectively), alpha-helix (α-helix), the three beta sheets (β1–β3), and the hypervariable loop are indicated. The image in (**c**) represents the same structure as in (**b**) but after a 90° clockwise rotation about the vertical axis. The molecule was modeled using the EasyModeller software (Sali and Blundell 1993; Kuntal et al. 2010) on the basis of its homology to SCR$_8$ (Mishima et al. 2003)

this loop is relatively small, it was proposed that an additional non-allele-specific interaction surface exists (Mishima et al. 2003). In support of the latter, several subsequent studies have shown that interaction between SCR and SRK does not necessarily result in self-incompatibility reaction (Chookajorn et al. 2004; Kemp and Doughty 2007; Naithani et al. 2007).

In an attempt to identify the regions of SCR responsible for interaction with SRK, Chookajorn et al. (2004) swapped specific homologous domains between SCR from the S_6- and S_{13}-haplotype. On the basis of the SCR$_6$ backbone, the authors were able to convert its haplotype specificity to S_{13} using either domain swapping or only a few amino acid changes. In a similar approach, it was shown that swapping of the domains flanking the hypervariable region could change the haplotype specificity of very homologous SCR molecules (Sato et al. 2004).

Swapping of the hypervariable region did not result in a specificity change in any of the two studies, and together with the fact that SCR molecule is able to tolerate a high amount of mutations without having its properties as a ligand compromised, it could be concluded that the allele specificity of the interaction is dependent on the structural features of SCR rather than a specific amino acid sequence.

At the plasma membrane of the papilla cell, the full-length receptor can exist as a monomer and to a lesser extent as a dimer (Giranton et al. 2000; Naithani et al. 2007; Shimosato et al. 2007). Using artificially dimerized forms of eSRK, Shimosato et al. (2007) could demonstrate that binding to SCR can occur only when SRK is a dimer. There are several other proteins, such as eSRK (Ivanov and Gaude 2009a), SLG (Cabrillac et al. 1999), and perhaps tSRK, which might be part of the SRK complex at the plasma membrane; however, their role in ligand recognition and self-incompatibility is either not analyzed or highly debated (Nishio and Kusaba 2000).

In an in vitro approach, Kemp and Doughty (2007) used a peptide array to map the receptor sequence responsible for SCR binding and proposed this to be shared between two hypervariable regions within the extracellular part of SRK. Later, using an approach based on domain swapping and mutagenesis, Boggs et al. (2009a) identified the same regions as responsible for the in vivo ligand-specific activation of self-incompatibility. It was shown that similarly to the case with SCR, only a few amino acid residues are essential to the specific receptor activation. It was proposed that the domains form a pocket structure that can specifically accommodate the cognate ligand (Boggs et al. 2009a). Therefore, the successful interaction and SRK activation depends on the three-dimensional structure of both the ligand and receptor. The great tolerance of both SCR and extracellular SRK domain to mutations has been used as an argument for explaining the coevolution of new *S* alleles (Chookajorn et al. 2004). It has been proposed that despite the accumulation of mutations within SRK and SCR, they still preserve their allelic specificity. However, as a result, certain groups of SRK–SCR pairs may develop a stronger specificity to each other than to the corresponding proteins of their haplogroup. In case this results in a strong self-incompatibility response, this subgroup may be evolutionary favored and evolve independently of the original haplogroup, finally losing specificity to it. Thus, a new haplogroup can arise without compromising the self-incompatibility at any point and avoiding self-compatible intermediates (Chookajorn et al. 2004).

3.2 Postrecognition Events in the Stigma Papilla Cell

As SCR is delivered to a certain position on the papilla cell by the pollen, it can target only the SRK molecules that underlie the attachment site. This and also the limited distribution of SRK into patches, or "SI domains," on the plasma membrane ensure that receptor activation and self-incompatibility response is contained within a limited space (Fig. 2) (Ivanov and Gaude 2009b). As a result, if two or more

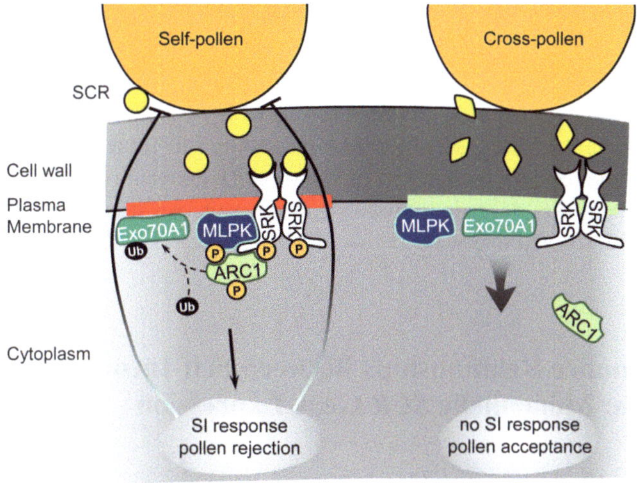

Fig. 2 Localized self-incompatibility response in *Brassicaceae*. The papilla cell responses to pollination are localized to the zone of contact. In an extreme case, a cell can respond adequately to simultaneous self- and cross-pollination. One of the factors that underlie this phenomenon is the unequal distribution of SRK in domains along the plasma membrane (indicated as *dark* and *light bars* along the plasma membrane). In case of self-pollination, activation will be limited to the contact domain (*dark bar*), leaving the neighboring domains (*light bar*) ready for an independent recognition process. Upon pollination, SCR is delivered to the papilla cell by the pollen grain. SCR traverses the cell wall and can in principle come in contact with the receptor kinase SRK at the papilla cell plasma membrane. In case of self-pollination (*left part of the image*), SCR (*yellow circles*) interacts with SRK. This results in several phosphorylation events and the formation of an active receptor complex that includes SRK, the M-LOCUS PROTEIN KINASE (MLPK), and the E3 ubiquitin ligase ARC1. Activated ARC1 ubiquitinates numerous targets within the cell, including the plasma membrane-localized member of the exocyst complex Exo70A1. Exo70A1 is then probably targeted for proteasome-dependent degradation. As a result of this cascade, the hydration and germination of the pollen grain are suppressed. If cross-pollination occurs (*right side of the image*), the noncognate SCR (*yellow rhomboids*) cannot activate the SRK and no SRK phosphorylation can be observed. Removal of Exo70A1 from the plasma membrane is required for pollen acceptance; however, the reason for this remains to be uncovered

pollen grains attach simultaneously to the same papilla cell, the decision whether they will be accepted or rejected will be independent in each case (Sarker et al. 1988). Recognition of the cognate ligand results in the phosphorylation of the kinase domain of SRK (Cabrillac et al. 2001). The activated complex includes also the M-LOCUS PROTEIN KINASE (MLPK) (Kakita et al. 2007) and the E3 ubiquitin ligase ARM REPEAT CONTAINING 1 (ARC1), which binds only the phosphorylated kinase domain of SRK (Gu et al. 1998). In vitro phosphorylation experiments indicate that ARC1 may be a substrate for both kinases (Samuel et al. 2008). Activation of ARC1 results in the ubiquitination of multiple targets in the cell (Stone et al. 2003). While for the most part their identity remains unknown, one of the ARC1 targets may be the component of the exocyst complex Exo70A1. Studies on *Brassica* and *A. thaliana* have shown that Exo70A1 is required for self-incompatibility response but also for pollen acceptance during compatible pollination (Samuel et al. 2009). Once the complex is activated, it dissociates and the different components may be degraded. ARC1 and MLPK are probably sent to the proteasome (Samuel et al. 2008). The fate of SCR is not known. A study using an antibody against SRK_3, which could functionally substitute for the native ligand, showed that the receptor complex is endocytosed (Ivanov and Gaude 2009b). As the activated receptor is degraded, probably in the vacuole, it can be speculated that SCR is also destroyed this way.

4 Dominance Relationships Between Self-Incompatibility Alleles: Monoallelic *SCR* Gene Expression

4.1 Classification of S*-Haplotypes*

As *Brassicaceae* self-incompatibility is sporophytic, the phenotypes of both the stigma and the pollen are determined by two, or more, copies of the *S*-locus. Extensive genetic analyses of *Brassica* plants, heterozygous for the *S*-locus, have demonstrated that there are dominant and recessive alleles. Cases of codominance can be also observed. These relationships between *S* alleles have been observed in both stigma and pollen.

Allelic interactions between two *S*-haplotypes can differ in stigma and pollen. In stigma, codominance is common whereas dominance/recessiveness is frequent in pollen (Thompson and Taylor 1966; Ockendon 1975). *Brassica S*-haplotypes have been separated into two classes, class I and class II, clearly separable based on sequence divergence in the S domain of the *SLG*/*SRK* genes (Nasrallah et al. 1991). The amino acid sequence similarity of SLG/SRK S domain is about 65% between classes and 80–90% within classes. A phylogenic tree of class I/class II SCR amino acid sequences shows that the two groups also form distinct clusters (Shiba et al. 2002). Class I are known as high-activity alleles because they exhibit a strong incompatibility phenotype in which no, or only few, pollen tubes could bypass the self-incompatibility barrier after self-pollination. Oppositely, class II alleles, also known as low-activity alleles, have a weak incompatibility phenotype with an average of 10–30 pollen tubes developing per self-pollinated stigma. In pollen, the class I *S*-haplotypes are known to be generally dominant over the class II *S*-haplotypes (Nasrallah et al. 1991).

The pollen-recessive *S* alleles remain "hidden" during the pollen–stigma interaction and can therefore overcome the self-incompatibility barrier. Consequently, these alleles occur at a higher frequency in the population and persist longer than the dominant alleles (Uyenoyama 2000). *A. lyrata* is a close relative to the highly self-fertile model *A. thaliana*, but it has functional self-incompatibility. While in cultivated *Brassica*, only two dominance types have been identified, studies in *A. lyrata* show more dominance classes and a high degree of complexity in dominance relationships. Synthesis of results from cross-pollinations supports a hypothesis of division of S-haplotypes into four dominance groups (A2 > A3 > B > A1) with codominance within classes in pistils and a strictly linear hierarchy in pollen (Schierup et al. 2001; Kusaba et al. 2002; Mable et al. 2003; Bechsgaard et al. 2004; Prigoda et al. 2005).

4.2 Recessive Allele Expression Depends on Their Allelic Partner

Brassica class II *SCR* genes exhibit an exclusively sporophytic expression in the anther tapetum and little or no gametophytic expression in the pollen, which is in contrast to the sporophytic/gametophytic expression pattern of class I *SCR*. Furthermore, class II *SCR* transcripts were detected at early stages of anther development, and the expression levels decrease during the later stages when tapetal degradation occurs. This is in contrast with class I *SCR* where the strength of expression is maintained throughout the anther development (Schopfer and Nasrallah 2000; Takayama et al. 2000; Shiba et al. 2002). Such a difference in spatial and temporal distribution of *SCR* transcripts has also been reported in *A. lyrata* (Kusaba et al. 2002). The expression of SCR_b (SCR_{20} in Prigoda et al. 2005), which belongs to the most dominant A2 class, was detected both in microspores and in tapetal cells throughout the anther development. On the other hand, transcripts of the more recessive SCR_a allele (SCR_{13} in Prigoda et al. 2005) were not detected in microspores and only at early stages of anther development.

The expression pattern of a dominant *SCR* allele remains unaffected in a heterozygous plant whatever the allelic partner. However, sporophytic expression of recessive *SCRs* is drastically reduced in combination with a dominant allele. This has been reported in *Brassica* species (Shiba et al. 2002; Kakizaki et al. 2003) as well as in *A. lyrata* (Kusaba et al. 2002). In combination with a more dominant allele, a reduction of at least 30-fold of *A. lyrata* SCR_a transcripts was deduced from quantitative reverse transcriptase-PCR experiments (Kusaba et al. 2002). The authors generated a population of S_aS_b and S_aS_a plants by forcing selfing of a heterozygote S_aS_b $(S_b>S_a)$. SCR_a expression level was always reduced in combination with S_b allele in the parent plant as well as in the progeny. Interestingly, when the S_a allele faced another S_a haplotype in the progeny, the expression level of SCR_a is restored to a normal level, despite being silenced in the heterozygote parent plant. The same set of experiments performed in *Brassica* showed in addition that recessive *SCR* expression in the progeny was reduced regardless of whether the S allele was of maternal or parental origin (Shiba et al. 2006). *SCR* repression represents a typical example of monoallelic gene silencing but different from the well-documented genomic imprinting as silencing is reversible and not influenced by parental origin of the S allele.

To gain insight into the mechanism that controls recessive *SCR* silencing, Fujimoto et al. (2006) analyzed the expression level of a *Brassica* class II *SCR* in combination with a nonfunctional Class I allele. *B. rapa* cultivar *yellow sarson* is self-compatible and carries the *S-f2* class I haplotype. The *S-f2* locus presents an insertion of a retrotransposon-like sequence in the first intron of *SRK* and 89-bp deletion in the promoter region of *SCR*. No transcripts of *SRK* and *SCR* were detected in *S-f2* homozygotes, suggesting that both mutated genes are not functional. Moreover, promoter analysis using transgenic plants indicated that the *SCR* promoter of *S-f2* has no activity. Interestingly, in heterozygote, *S-f2* is still able to suppress the expression of a recessive *SCR*. Thus, the presence of transcripts from the dominant allele is not necessary for the silencing of the recessive allele. Importantly, within the same class, an allele can be either dominant or recessive depending on the other *S*-haplotype present in the heterozygote (Kakizaki et al. 2003). This clearly demonstrates that the region responsible for the SCR silencing is located within the *S*-locus outside of the *SCR* gene and thus segregates with the *S*-haplotype.

4.3 Dominance Relationships Depend on DNA Methylation Triggered by a Trans-Acting Small RNA Found Within the S-Locus

Dominance among *SCR* alleles involves transcriptional silencing, and one major way to affect gene expression in eukaryotic cells is DNA methylation. Shiba et al. (2006) identified a 56-nucleotide region within the promoter of *Brassica SCR* genes which is differentially methylated depending on the allelic combination. Bisulfite sequencing analysis showed a significant increase in methylation level only in the recessive *SCR* promoter, but not in the dominant alleles and only in heterozygotes. In addition, *SCR* methylation occurs very early during anther development, before the initiation of *SCR* transcription, and is restricted to tapetal cells (Shiba et al. 2006). Thus, in a heterozygote, the expression of the recessive *SCR* allele is silenced as a result of de novo methylation of its promoter region induced, in *trans*, by the presence of a dominant *S* allele. The authors further speculated that this regulation might be mediated by small RNA (sRNA) and therefore searched for regions of identity between the dominant *S*-locus and the methylated region. A short sequence highly similar to the methylated target region in recessive *SCR* promoter was found within four *Brassica* class I *S*-loci (Tarutani et al. 2010). Moreover, a corresponding 24-nucleotide sRNA (designated *Smi* for *SCR methylation inducer*) which shares 18 bases with the recessive *SCR* promoter was found to accumulate in anthers at early stages of development. Experiments in transgenic plants expressing a dominant *SMI* region (*SMIdom*) in *S*-locus homozygotes demonstrated that the *Smidom* transgene is able to silence all the recessive *SCR* tested. As a result, plants with homozygous recessive *SCR* alleles became fully

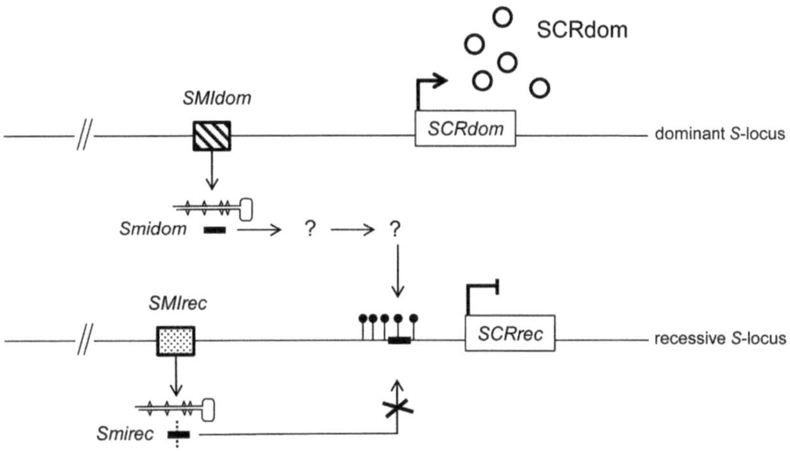

Fig. 3 Molecular basis for dominance relationships between *S* alleles in anthers. In a heterozygous situation, expression of each copy of the *SCR* gene in the anther depends on the genetic environment of the other *S*-locus. Dominant *S*-loci contain a *SMI* region (*SCR/SP11 Methylation Inducing, hatched box*) which encodes a noncoding *SMI* transcript predicted to form an imperfect stem loop encompassing the 24-mer (*Smidom thick black line*). Sequence of the *SMIdom* is highly similar (18 out of 24 nucleotides) to a short region within the promoter (*thick black line* in the recessive *S*-locus) of the recessive *SCR* gene (*SCRrec*). The *Smidom* is responsible directly, as suggested by Tarutani et al. (2010), or indirectly, as proposed by Finnegan et al. (2011), for the methylation of this homologous site as well as the surrounding region (symbol). Methylation blocks *SCRrec* transcription, which results in its silencing in the presence of *S*-locus with dominant *SCR* allele (*SCRdom*). The methylated target site is not found within the 5′ upstream region of the dominant SCR, which is normally expressed and translated into protein (SCRdom, *circles*). An *SMI* homologue region is also found within the recessive *S*-locus (*SMIrec, dotted box*), and *Smirec* small RNA can be detected in anther tissue. However, *Smirec* differs in sequence from *Smidom* by one base (*dotted line*). This mismatch is sufficient to prevent the methylation of the recessive *SCR*, and consequently, *Smirec* does not interfere with *SCR* expression

self-compatible. In addition, bisulfite sequencing revealed that the *SCR* promoter region methylated in a dominant/recessive *S*-heterozygote is also the target for methylation in transgenic plants (Tarutani et al. 2010). These results suggest that the *Smidom* directs promoter methylation leading to silencing of recessive *SCR* (Fig. 3). Surprisingly, a *SMI* region is also found in class II *S*-locus and *Smirec* (*Smi recessive*) to accumulate, albeit at a lower level, in anthers of a recessive *S* homozygote. The 24-nucleotide *Smirec* contains one base substitution in the middle of the region that matches with the recessive *SCR* promoter. Interestingly, when the same mismatch is introduced in *SMIdom*, the mutated sequence is unable to repress the recessive *SCR* in transgenic plants (Tarutani et al. 2010). Thus, one base pair mismatch is sufficient to abolish the process of targeted DNA methylation.

To date, the mechanistic link between accumulation of the *Smi* and the promoter methylation remains unsolved. Likewise, how one mismatch in *Smirec* can block this methylation is still enigmatic. Tarutani et al. (2010) propose a direct effect of the *Smi* on DNA methylation. Alternatively, Finnegan et al. (2011) suggest that the *Smi* acts as a miRNA to first cleave the noncoding transcript from the antisense strand of the SCR promoter. This cleaved transcript is then converted to double-stranded RNA that is further processed in small RNA, triggering direct promoter methylation. Additional experiments are needed to further support both models. Interestingly, Shiba et al. (2006) reported that methylation patterns of SCR promoters in heterozygous plants for two *Brassica* class II *S* alleles are similar to those observed in class I/class II heterozygotes. In fact, increase in methylation rate is only detected in the promoter region of the most recessive *SCR*. Thus, methylation seems to be a general process to repress expression of recessive *SCRs*. In the near future, further investigation will probably provide additional insight into the putative role of *Smi* in mediating dominance relationships within the same class of dominance especially in wild species such as *A. lyrata* for which *S* alleles show deeper divergences and more dominance classes than those in cultivated *Brassica*.

5 Stability of the *SCR* Gene and the Evolution of Self-Fertilization in *A. thaliana*

The switch from outcrossing to selfing is one of the major evolutionary transitions in flowering plants and is thought to have occurred independently in numerous lineages. Although selfing leads to inbreeding depression, it can be regarded as an advantage in certain conditions when reproductive partners or pollinators become rare or when new territories should be colonized (Busch and Schoen 2008). The model plant *A. thaliana* is highly self-fertile, and outcrossing has never been reported among the hundreds of its accessions collected from various geographical locations. Indeed, all accessions analyzed to date contain several independent mutations and rearrangements that disrupt the self-incompatibility system (Kusaba et al. 2001; Nasrallah et al. 2002; Liu et al. 2007; Sherman-Broyles et al. 2007; Tang et al. 2007; Boggs et al. 2009b; Tsuchimatsu et al. 2010). Transformation of *A. thaliana* with a pair of functional *SRK–SCR*, from the outcrossing relative *A. lyrata*, demonstrated that in some *A. thaliana* populations, direct inactivation of the *S*-locus is the primary cause of self-fertility, whereas in other accessions, presence of mutations, not only at the *S*-locus but also at "modifier" loci, is responsible for the breakdown of self-incompatibility (Nasrallah et al. 2004; Liu et al. 2007; Boggs et al. 2009b). Interestingly, four accessions, among them Wei-1, harbor functional alleles of *SRK* and other genes required for the female self-incompatibility reaction. However, a 213-bp inversion within the second exon of *SCR* results in a nonfunctional truncated SCR protein. Transgenic Wei-1 plants that express a restored *SCR* with a full-length open reading frame turn out to be self-incompatible (Tsuchimatsu et al. 2010). This strongly suggests that in Wei-1 and probably in the three other accessions, loss of self-incompatibility is linked to *SCR* disruption and that *SCR* inversion is the cause for the switch to selfing.

This observation supports the earlier hypothesis that, when disrupting self-incompatibility, mutations targeting the male component (*SCR*) will have advantage over those targeting the female component (*SRK*). This may be explained by the fact that *SCR* mutations will be spread by both pollen and seeds, while mutations in *SRK* will spread only through seeds (Busch and Schoen 2008).

6 Conclusion

Since the first report describing the isolation of the putative male determinants of SI (Stephenson et al. 1997) and the discovery of the SCR peptides (Schopfer et al. 1999; Suzuki et al. 1999), tremendous progress has been made in the comprehension of the SCR structure, function in SI, and regulation of their expression. Yet, some mysteries wait to be unraveled. Thus, it is puzzling that application of purified SCRs on the surface of papilla cells expressing the cognate SRK cannot activate the SI signaling pathway. How do SCR peptides travel across the thick cell wall of papilla cells, and what are the components required for a successful SRK–SCR interaction? Does this process involve posttranslational modifications of SCR? Once SCR associates with SRK, what is the dynamics of internalization of the receptor–ligand complex? Do SRK–SCR complexes derived from recessive *S*-haplotypes behave as dominant ones in terms of protein association strength and cytoplasmic targets? How can we explain at the molecular level the fact that recessive combinations lead to weak SI response, authorizing a few self-pollen tubes to grow into the stigmatic tissue?

For the near future, we may predict that the more general use of transgenic SI *A. thaliana* lines will greatly facilitate the study of SI and improve our knowledge of this female–male recognition mechanism that has played such a crucial role in the evolutionary success of flowering plants.

References

Bechsgaard J, Bataillon T, Schierup MH (2004) Uneven segregation of sporophytic self-incompatibility alleles in *Arabidopsis lyrata*. J Evol Biol 17:554–561

Boggs NA, Dwyer KG, Nasrallah ME, Nasrallah JB (2009a) In vivo detection of residues required for ligand-selective activation of the S-locus receptor in *Arabidopsis*. Curr Biol 19:786–791

Boggs NA, Nasrallah JB, Nasrallah ME (2009b) Independent *S*-locus mutations caused self-fertility in *Arabidopsis thaliana*. PLoS Genet 5:e1000426

Boyes DC, Nasrallah JB (1993) Physical linkage of the *SLG* and *SRK* genes at the self-incompatibility locus of *Brassica oleracea*. Mol Gen Genet 236:369–373

Busch JW, Schoen DJ (2008) The evolution of self-incompatibility when mates are limiting. Trends Plant Sci 13:128–136

Cabrillac D, Delorme V, Garin J, Ruffio-Chable V, Giranton JL, Dumas C, Gaude T, Cock JM (1999) The S_{15} self-incompatibility haplotype in *Brassica oleracea* includes three *S* gene family members expressed in stigmas. Plant Cell 11:971–986

Cabrillac D, Cock JM, Dumas C, Gaude T (2001) The S-locus receptor kinase is inhibited by thioredoxins and activated by pollen coat proteins. Nature 410:220–223

Chookajorn T, Kachroo A, Ripoll DR, Clark AG, Nasrallah JB (2004) Specificity determinants and diversification of the *Brassica* self-incompatibility pollen ligand. Proc Natl Acad Sci USA 101:911–917

de Nettancourt D (2001) Incompatibility and incongruity in wild and cultivated species. Springer-Verlag, Berlin

Finnegan EJ, Liang D, Wang MB (2011) Self-incompatibility: Smi silences through a novel sRNA pathway. Trends Plant Sci 16:238–241

Fujimoto R, Sugimura T, Fukai E, Nishio T (2006) Suppression of gene expression of a recessive *SP11/SCR* allele by an untranscribed *SP11/SCR* allele in *Brassica* self-incompatibility. Plant Mol Biol 61:577–587

Gaude T, Fobis-Loisy I, Miege C (2006) Control of fertilization by self-incompatibility mechanisms. In: Jordan B (ed) The molecular biology and biotechnology of flowering. CAB International, Wallingford, CT, pp 269–297

Giranton JL, Ariza MJ, Dumas C, Cock JM, Gaude T (1995) The *S locus receptor kinase* gene encodes a soluble glycoprotein corresponding to the SKR extracellular domain in *Brassica oleracea*. Plant J 8:827–834

Giranton JL, Dumas C, Cock JM, Gaude T (2000) The integral membrane S-locus receptor kinase of *Brassica* has serine/threonine kinase activity in a membranous environment and spontaneously forms oligomers in planta. Proc Natl Acad Sci USA 97:3759–3764

Goring DR, Glavin TL, Schafer U, Rothstein SJ (1993) An *S receptor kinase* gene in self-compatible *Brassica napus* has a 1-bp deletion. Plant Cell 5:531–539

Gu T, Mazzurco M, Sulaman W, Matias DD, Goring DR (1998) Binding of an arm repeat protein to the kinase domain of the S-locus receptor kinase. Proc Natl Acad Sci USA 95:382–387

Ivanov R, Gaude T (2009a) *Brassica* self-incompatibility: a glimpse below the surface. Plant Signal Behav 4:996–998

Ivanov R, Gaude T (2009b) Endocytosis and endosomal regulation of the S-receptor kinase during the self-incompatibility response in *Brassica oleracea*. Plant Cell 21:2107–2117

Ivanov R, Fobis-Loisy I, Gaude T (2010) When no means no: guide to *Brassicaceae* self-incompatibility. Trends Plant Sci 15:387–394

Iwano M, Shiba H, Funato M, Shimosato H, Takayama S, Isogai A (2003) Immunohistochemical studies on translocation of pollen *S*-haplotype determinant in self-incompatibility of *Brassica rapa*. Plant Cell Physiol 44:428–436

Kachroo A, Schopfer CR, Nasrallah ME, Nasrallah JB (2001) Allele-specific receptor-ligand interactions in *Brassica* self-incompatibility. Science 293:1824–1826

Kakita M, Murase K, Iwano M, Matsumoto T, Watanabe M, Shiba H, Isogai A, Takayama S (2007) Two distinct forms of M-locus protein kinase localize to the plasma membrane and interact directly with S-locus receptor kinase to transduce self-incompatibility signaling in *Brassica rapa*. Plant Cell 19:3961–3973

Kakizaki T, Takada Y, Ito A, Suzuki G, Shiba H, Takayama S, Isogai A, Watanabe M (2003) Linear dominance relationship among four class-II *S* haplotypes in pollen is determined by the expression of *SP11* in *Brassica* self-incompatibility. Plant Cell Physiol 44:70–75

Kemp BP, Doughty J (2007) S cysteine-rich (SCR) binding domain analysis of the *Brassica* self-incompatibility S-locus receptor kinase. New Phytol 175:619–629

Kuntal BK, Aparoy P, Reddanna P (2010) EasyModeller: a graphical interface to MODELLER. BMC Res Notes 3:226

Kusaba M, Dwyer K, Hendershot J, Vrebalov J, Nasrallah JB, Nasrallah ME (2001) Self-incompatibility in the genus *Arabidopsis*: characterization of the *S* locus in the outcrossing *A. lyrata* and its autogamous relative *A. thaliana*. Plant Cell 13:627–643

Kusaba M, Tung CW, Nasrallah ME, Nasrallah JB (2002) Monoallelic expression and dominance interactions in anthers of self-incompatible *Arabidopsis lyrata*. Plant Physiol 128:17–20

Liu P, Sherman-Broyles S, Nasrallah ME, Nasrallah JB (2007) A cryptic modifier causing transient self-incompatibility in *Arabidopsis thaliana*. Curr Biol 17:734–740

Mable BK, Schierup MH, Charlesworth D (2003) Estimating the number, frequency, and dominance of *S*-alleles in a natural population of *Arabidopsis lyrata* (*Brassicaceae*) with sporophytic control of self-incompatibility. Heredity 90:422–431

Mishima M, Takayama S, Sasaki K, Jee JG, Kojima C, Isogai A, Shirakawa M (2003) Structure of the male determinant factor for *Brassica* self-incompatibility. J Biol Chem 278:36389–36395

Naithani S, Chookajorn T, Ripoll DR, Nasrallah JB (2007) Structural modules for receptor dimerization in the S-locus receptor kinase extracellular domain. Proc Natl Acad Sci USA 104:12211–12216

Nasrallah JB, Kao TH, Goldberg ML, Nasrallah ME (1985) A cDNA clone encoding an S-locus-specific glycoprotein from *Brassica oleracea*. Nature 318:263–267

Nasrallah JB, Nishio T, Nasrallah ME (1991) The self-incompatibility genes of *Brassica*: expression and use in genetic ablation of floral tissues. Annu Rev Plant Physiol Plant Mol Biol 42:393–422

Nasrallah JB, Rundle SJ, Nasrallah ME (1994) Genetic evidence for the requirement of the *Brassica S-locus receptor kinase* gene in the self-incompatibility response. Plant J 5:373–384

Nasrallah ME, Liu P, Nasrallah JB (2002) Generation of self-incompatible *Arabidopsis thaliana* by transfer of two *S* locus genes from *A. lyrata*. Science 297:247–249

Nasrallah ME, Liu P, Sherman-Broyles S, Boggs NA, Nasrallah JB (2004) Natural variation in expression of self-incompatibility in *Arabidopsis thaliana*: implications for the evolution of selfing. Proc Natl Acad Sci USA 101:16070–16074

Nishio T, Kusaba M (2000) Sequence Diversity of *SLG* and *SRK* in *Brassica oleracea* L. Ann Bot 85:141–146

Ockendon DJ (1975) Dominance relationships between *S*-alleles in the stigmas of Brussels sprouts (*Brassica oleracea* var. *gemmifera*). Euphytica 24:165–172

Okazaki K, Kusaba M, Ockendon DJ, Nishio T (1999) Characterization of *S* tester lines in *Brassica oleracea*: polymorphism of restriction fragment length of *SLG* homologues and isoelectric points of S-locus glycoproteins. Theor Appl Genet 98:1329–1334

Prigoda NL, Nassuth A, Mable BK (2005) Phenotypic and genotypic expression of self-incompatibility haplotypes in *Arabidopsis lyrata* suggests unique origin of alleles in different dominance classes. Mol Biol Evol 22:1609–1620

Ruffio-Chable V, Gaude T (2001) *S*-haplotype polymorphism in *Brassica oleracea*. Acta Hortic 546:257–261

Sali A, Blundell TL (1993) Comparative protein modelling by satisfaction of spatial restraints. J Mol Biol 234:779–815

Samuel MA, Mudgil Y, Salt JN, Delmas F, Ramachandran S, Chilelli A, Goring DR (2008) Interactions between the S-domain receptor kinases and AtPUB-ARM E3 ubiquitin ligases suggest a conserved signaling pathway in *Arabidopsis*. Plant Physiol 147:2084–2095

Samuel MA, Chong YT, Haasen KE, Aldea-Brydges MG, Stone SL, Goring DR (2009) Cellular pathways regulating responses to compatible and self-incompatible pollen in *Brassica* and *Arabidopsis* stigmas intersect at Exo70A1, a putative component of the exocyst complex. Plant Cell 21:2655–2671

Sarker RH, Elleman CJ, Dickinson HG (1988) Control of pollen hydration in *Brassica* requires continued protein synthesis, and glycosylation is necessary for intraspecific incompatibility. Proc Natl Acad Sci USA 85:4340–4344

Sato Y, Okamoto S, Nishio T (2004) Diversification and alteration of recognition specificity of the pollen ligand SP11/SCR in self-incompatibility of *Brassica* and *Raphanus*. Plant Cell 16:3230–3241

Schierup MH, Mable BK, Awadalla P, Charlesworth D (2001) Identification and characterization of a polymorphic receptor kinase gene linked to the self-incompatibility locus of *Arabidopsis lyrata*. Genetics 158:387–399

Schopfer CR, Nasrallah JB (2000) Self-incompatibility. Prospects for a novel putative peptide-signaling molecule. Plant Physiol 124:935–940

Schopfer CR, Nasrallah ME, Nasrallah JB (1999) The male determinant of self-incompatibility in *Brassica*. Science 286:1697–1700

Sherman-Broyles S, Boggs N, Farkas A, Liu P, Vrebalov J, Nasrallah ME, Nasrallah JB (2007) *S* locus genes and the evolution of self-fertility in *Arabidopsis thaliana*. Plant Cell 19:94–106

Shiba H, Takayama S, Iwano M, Shimosato H, Funato M, Nakagawa T, Che FS, Suzuki G, Watanabe M, Hinata K, Isogai A (2001) A pollen coat protein, SP11/SCR, determines the pollen S-specificity in the self-incompatibility of *Brassica* species. Plant Physiol 125:2095–2103

Shiba H, Iwano M, Entani T, Ishimoto K, Shimosato H, Che FS, Satta Y, Ito A, Takada Y, Watanabe M, Isogai A, Takayama S (2002) The dominance of alleles controlling self-incompatibility in *Brassica* pollen is regulated at the RNA level. Plant Cell 14:491–504

Shiba H, Kakizaki T, Iwano M, Tarutani Y, Watanabe M, Isogai A, Takayama S (2006) Dominance relationships between self-incompatibility alleles controlled by DNA methylation. Nat Genet 38:297–299

Shimosato H, Yokota N, Shiba H, Iwano M, Entani T, Che FS, Watanabe M, Isogai A, Takayama S (2007) Characterization of the SP11/SCR high-affinity binding site involved in self/non self recognition in *Brassica* self-incompatibility. Plant Cell 19:107–117

Stein JC, Howlett B, Boyes DC, Nasrallah ME, Nasrallah JB (1991) Molecular cloning of a putative receptor protein kinase gene encoded at the self-incompatibility locus of *Brassica oleracea*. Proc Natl Acad Sci USA 88:8816–8820

Stephenson AG, Doughty J, Dixon S, Elleman C, Hiscock S, Dickinson HG (1997) The male determinant of self-incompatibility in *Brassica oleracea* is located in the pollen coating. Plant J 12:1351–1359

Stone SL, Anderson EM, Mullen RT, Goring DR (2003) ARC1 is an E3 ubiquitin ligase and promotes the ubiquitination of proteins during the rejection of self-incompatible *Brassica* pollen. Plant Cell 15:885–898

Suzuki G, Kai N, Hirose T, Fukui K, Nishio T, Takayama S, Isogai A, Watanabe M, Hinata K (1999) Genomic organization of the *S* locus: identification and characterization of genes in *SLG/SRK* region of *S(9)* haplotype of *Brassica campestris* (syn. *rapa*). Genetics 153:391–400

Takasaki T, Hatakeyama K, Suzuki G, Watanabe M, Isogai A, Hinata K (2000) The S receptor kinase determines self-incompatibility in *Brassica* stigma. Nature 403:913–916

Takayama S, Shiba H, Iwano M, Shimosato H, Che FS, Kai N, Watanabe M, Suzuki G, Hinata K, Isogai A (2000) The pollen determinant of self-incompatibility in *Brassica campestris*. Proc Natl Acad Sci USA 97:1920–1925

Takayama S, Shimosato H, Shiba H, Funato M, Che FS, Watanabe M, Iwano M, Isogai A (2001) Direct ligand-receptor complex interaction controls *Brassica* self-incompatibility. Nature 413:534–538

Tang C, Toomajian C, Sherman-Broyles S, Plagnol V, Guo YL, Hu TT, Clark RM, Nasrallah JB, Weigel D, Nordborg M (2007) The evolution of selfing in *Arabidopsis thaliana*. Science 317:1070–1072

Tarutani Y, Shiba H, Iwano M, Kakizaki T, Suzuki G, Watanabe M, Isogai A, Takayama S (2010) *Trans*-acting small RNA determines dominance relationships in *Brassica* self-incompatibility. Nature 466:983–986

Thompson KF, Taylor JP (1966) Non-linear dominance relationships between *S* alleles. Heredity 21:345–362

Tsuchimatsu T, Suwabe K, Shimizu-Inatsugi R, Isokawa S, Pavlidis P, Stadler T, Suzuki G, Takayama S, Watanabe M, Shimizu KK (2010) Evolution of self-compatibility in *Arabidopsis* by a mutation in the male specificity gene. Nature 464:1342–1346

Uyenoyama MK (2000) Evolutionary dynamics of self-incompatibility alleles in *Brassica*. Genetics 156:351–359

Vanoosthuyse V, Miege C, Dumas C, Cock JM (2001) Two large *Arabidopsis thaliana* gene families are homologous to the *Brassica* gene superfamily that encodes pollen coat proteins and the male component of the self-incompatibility response. Plant Mol Biol 46:17–34

Peptides Modulating Development of Specialized Cells

Lee Hunt and Julie E. Gray

Abstract Pairs of specialized guard cells on the leaf surface of higher plants form stomatal pores that control water release and carbon dioxide entry. The patterning of stomata is precisely controlled during leaf development to ensure the correct cellular spacing within the epidermis. The EPIDERMAL PATTERNING FACTOR (EPF) family of small secreted cysteine-rich peptides is important in this cell-to-cell signaling process as the EPF-like peptides can act as either positive or negative regulators of stomatal development. The properties of the EPF-like peptides and how they may interact with putative receptor components, such as TOO MANY MOUTHS (TMM) and ERECTA family proteins, and compete to activate or inactivate the downstream kinase cascade are discussed.

1 Introduction

Over the past decade, it has become increasingly clear that secreted plant peptides regulate cell fate decisions in a number of plant developmental pathways. The best studied of these belong to the CLE family of peptide ligands which activate signaling cascades to regulate a range of developmental decisions such as the number of stem cells at the shoot and apical meristems and the development of the plant vasculature to produce correct xylem and phloem positioning (see chapters "Peptides Regulating Apical Meristem Development," "Peptides Regulating Root Growth," and "Peptides Regulating Plant Vascular Development"). In this chapter, we review the activity of a different, unrelated, peptide family and describe its role in controlling the development of the specialized cells that are found on the aerial surfaces of plants that are known as guard cells. Guard cell pairs, or

L. Hunt • J.E. Gray (✉)
Department of Molecular Biology and Biotechnology, University of Sheffield, Sheffield S10 2TN, UK
e-mail: j.e.gray@sheffield.ac.uk

H.R. Irving and C. Gehring (eds.), *Plant Signaling Peptides*,
Signaling and Communication in Plants 16, DOI 10.1007/978-3-642-27603-3_6,

stomata, regulate the aperture of microscopic stomatal pores to control plant water loss and gas exchange with the environment. The precise orientation and patterning of stomata on plant surfaces, predominantly leaves, are regulated early during epidermal development by a signal transduction cascade. This is believed to involve an extracellular peptide ligand binding to a putative receptor or receptors on a potentially meristematic cell and regulating the number of its subsequent asymmetric cell divisions. This extracellular peptide family has only recently been characterized in the genetic model species *Arabidopsis thaliana*, and its members are known as the EPIDERMAL PATTERNING FACTORs (or EPFs). The EPF family contains 11 members, which are small secretory peptides characterized by 6 conserved cysteine residues. Several EPFs (including EPF1 and EPF2) have been shown to inhibit stomatal formation, and one has been shown to promote stomatal formation (EPFL9/STOMAGEN). Others do not appear to influence stomata development or are yet to be characterized. Here, we discuss the action of the EPF peptide ligands in controlling stomatal development, how these peptides interact with each other and their potential receptors, and how they may be activated by proteolytic processing.

2 Stomata Are Specialized Cells that Respond to Environment Signals

Stomata are distributed on the surface of leaves so that their pores are optimally positioned to make maximum use of the available CO_2 while minimizing transpirational water loss. Plants are able to adjust their stomatal gas exchange and respond to environmental variables via a reversible short-term response that involves changes in guard cell turgor leading to alterations in pore aperture. This is a well-studied response, and the underlying signaling pathways are relatively well characterized [see review by Kim et al. (2010)]. Plants also operate a less-well-understood longer-term stomatal response to alter gas exchange and adjust to environmental change. This is brought about by altering stomatal density in newly developing leaves. Stomatal densities are known to increase in response to environmental variables including increases in light intensity and in response to reduced atmospheric CO_2 concentrations and to decrease in response to low light intensities and increased atmospheric CO_2 concentrations [see review by Casson and Gray (2008)]. Although stomatal densities vary with environmental conditions, a degree of stomatal spacing is always maintained such that in mature leaves, stomata are always separated from one another by at least one epidermal pavement cell [see review by Bergmann and Sack (2007)]. For optimal pore aperture control and turgor change, it is thought that stomata need to be a least one cell apart, and in recent years, how this is controlled at the molecular level is becoming clearer.

3 Stomatal Development Involves a Series of Asymmetric Divisions

Studies of serial impressions of the developing leaf or cotyledon epidermis, and the use of molecular markers, have revealed the particular cell division patterns that are required to ensure proper stomatal development and spacing [reviewed by Serna and Fenoll (2000)]. The stomatal development pathway is initiated by the stimulation of a protodermal precursor cell to divide asymmetrically. In *A. thaliana*, selected protodermal cells become competent, by an unknown mechanism, to undergo an asymmetric division resulting in the formation of a small triangular cell called a meristemoid, and a larger daughter cell, crucial in ensuring correct spacing between stomata. This asymmetric division marks the entry of these two cells into the stomatal lineage. The larger daughter cell usually exits the stomatal lineage and develops alongside other protodermal cells, into an epidermal pavement cell, but some daughter cells retain the ability to divide asymmetrically and form satellite meristemoids. These asymmetric divisions, known as spacing divisions, are carefully orientated so that satellite meristemoids are not formed adjacent to existing meristemoids or guard cells. Each meristemoid usually divides asymmetrically several times, producing more of these larger daughter cells in what are known as amplifying divisions, as they usually increase the number of pavement cells. When the meristemoid has undergone approximately three asymmetric divisions and is mostly surrounded by its larger sister cells, it undergoes differentiation into an oval-shaped guard mother cell (GMC). This thick-walled cell ultimately divides symmetrically to form the two guard cells of a stoma. The asymmetric divisions of meristemoids are also orientated to ensure that stomata are always separated by at least one pavement cell (Bergmann and Sack 2007). Each of these developmental decision steps, entry division, spacing division, amplifying division, and differentiation, are precisely controlled in response to external and developmental cues.

4 Stomatal Development Is Regulated by a Signaling Pathway Involving Receptors, Peptide Ligands, and Proteases

The application of molecular genetics has been pivotal in revealing components of plant developmental signaling pathways, and among these, the stomatal development pathway is becoming one of the best understood. The identification of *A. thaliana* mutants with altered stomatal patterning from 1995 onward began the process of understanding how, at a molecular level, stomata are oriented and spaced (Yang and Sack 1995). The identification of *sdd1* (*stomatal density and distribution 1*), a mutant with a leaf epidermis that contained an increased density of stomata and also exhibited frequently paired stomata that violated the one-cell-spacing rule, provided the first evidence that an extracellular peptide signaling system could potentially operate in stomatal development (Berger and Altmann 2000;Von Groll et al. 2002). Peptide signaling systems, well characterized in animals, involve cleavage of a propeptide hormone by a protease to generate an active fragment which binds to an extracellular receptor and activates a downstream signaling cascade (see chapters "Plant Peptide Signaling: An Evolutionary Adaptation" and "Processing of Peptides"). Excitingly, the SDD1 gene product turned out to be homologous to the subtilisin protease family of prohormone convertases that, in animals, recognize mono- or dibasic amino acids in a target peptide and cleave at this site to create an active peptide signal. The isolation and characterization of another mutant known as *too many mouths* (or *tmm*), which has a similar epidermal phenotype to *sdd1* and increased stomatal density and stomatal clustering, provided further insights into the putative signaling pathway (Nadeau and Sack 2002). Green fluorescent protein (GFP) fusions suggest that TMM is expressed in the developing epidermis in meristemoids and also in the larger daughter cell of a recent asymmetric division. The TMM gene product is a receptor-like protein with a signal sequence that is believed to target it to the plasma membrane. The predicted protein consists mostly of an extracellular domain which includes ten leucine-rich repeat domains (LRR domains) but also contains paired cysteine residues in the N- and C-terminal non-LRR regions that could potentially mediate interactions with other receptor components and/or a peptide ligand potentially cleaved by SDD1. LRR domains are characteristic of receptor kinases and are known to interact with ligands to transduce signals across the plasma membrane by activation of an intracellular kinase domain. However, TMM does not have a predicted cytoplasmic domain and lacks the cytoplasmic kinase activity required to transduce a signal. Thus, it is believed that TMM may interact with other receptor components, which possess cytoplasmic signaling domains able to transduce a signal that ultimately restricts the progression of some daughter cells of asymmetric divisions through the stomatal linage.

The discovery that the putative receptor-like kinases (or RLKs) ER (ERECTA), ERL1 (ERECTA-LIKE 1), and ERL2 (ERECTA-LIKE 2) (Shpak et al. 2005) are involved in the control of stomatal density prompted the suggestion that these ERECTA family (ERf) receptor-like kinase proteins may act as partners for the TMM protein to mediate a response to activate an intracellular downstream signaling cascade. *ER* shows a broad expression pattern in developing tissues, whereas *ERL1* and *ERL2* expression patterns are specific to stomatal lineage cells of the developing epidermis. Neither *erl1* nor *erl2* mutants show an obvious stomatal development phenotype, but a triple *ererl1erl2* mutant showed a dramatic increase in stomatal density (accompanied by stunted growth and sterility), suggesting these three putative RLKs act redundantly in the control of stomatal density.

Although this chapter focuses on the function of peptide ligands and their receptor, a number of other molecular components that act to signal stomatal development have been characterized (Fig. 1). To summarize these findings briefly,

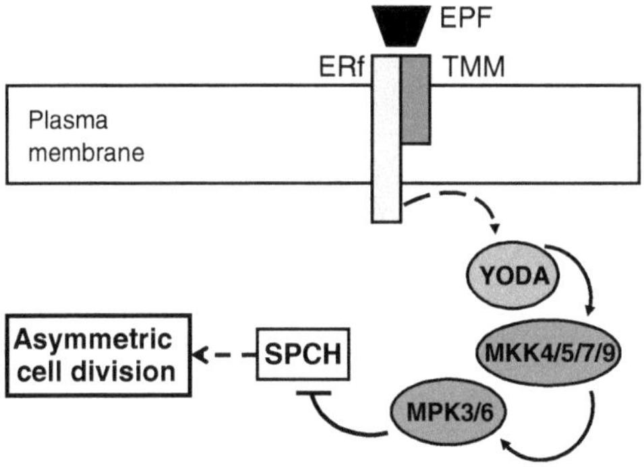

Fig. 1 Binding of EPF peptide to extracellular receptor inhibits asymmetric divisions in the stomatal lineage via a MAPK cascade. The binding of the mature EPF protein to the extracellular portion of the putative TMM/Erf receptor complex activates an intracellular MAPK cascade which ultimately phosphorylates the bHLH transcription factor SPCH. *Curved arrows* represent phosphorylation

stimulation of the receptor kinase complex at the plasma membrane is thought to activate the MAP kinase kinase kinase YODA, which in turn activates a MAP kinase cascade (Bergmann et al. 2004). Basic helix–loop–helix (bHLH) transcription factors, including SPEECHLESS (SPCH), act in the stomatal development pathway, but as promoters rather than inhibitors of meristemoid fate and GMC differentiation. SPCH is destabilized by MAPK phosphorylation, which prevents meristemoid proliferation (Lampard et al. 2008).

In summary, the above results suggested that an extracellular signaling pathway exists in the developing epidermis that involves secreted peptide ligands, in combination with an LRR receptor complex, activating an intracellular MAPK cascade that inhibits excessive stomatal development by restricting the formation or division of meristemoids via phosphorylation of the transcription factor SPCH. However, until 2007, the identity of the secreted peptide signal(s) remained entirely unknown.

5 The Identification of Epidermal Patterning Factor 1 as an Inhibitor of Stomatal Development

The recently discovered secreted peptide signal, which prevents *Arabidopsis* stomata from forming adjacent to one another, is known as Epidermal Patterning Factor 1 (EPF1). This was discovered from a postgenomic screen of plant peptide mutants. Since the completion of the *Arabidopsis* genome sequence and associated EST projects, after improved annotation, the number of encoded short peptides identified has increased. These peptides form several distinct groups such as phytosulfokines (PSKs), CLAVATA 3 (CLV3)/ESR-related peptides (CLEs), and rapid alkalinization factors (RALFs) but also include many other uncharacterized families. EPF1, the first peptide identified as being involved in regulating stomatal development came from a relatively high-throughput screen of plants ectopically overexpressing short,

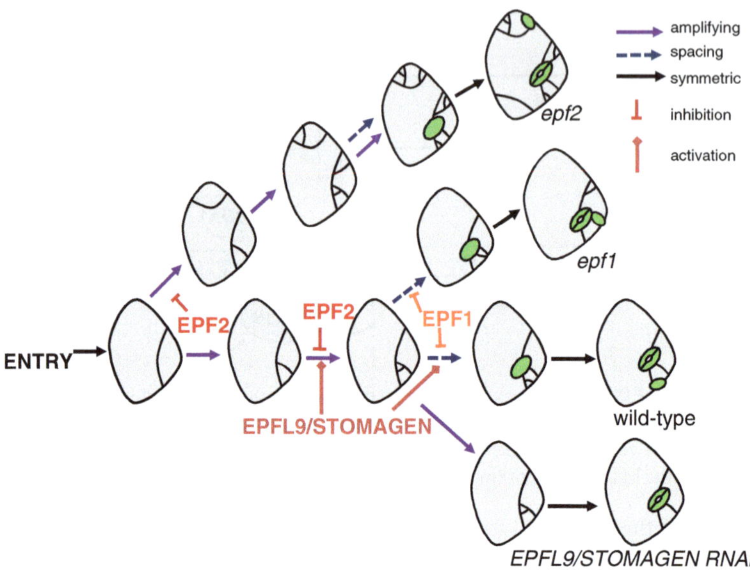

Fig. 2 Stomatal development involves a controlled series of asymmetric divisions followed by a final symmetric division. The initial entry division selects a meristemoid mother cell that undergoes an asymmetric division to form a meristemoid. This can then divide asymmetrically again or divide symmetrically to form the two guard cells that surround the stomatal pore. The phenotype of EPF mutants is illustrated

secreted proteins of less than 150 amino acids in length in an attempt to identify their function (Hara et al. 2007). Each peptide gene was ectopically expressed at a high expression level using the constitutive CaMV 35S RNA promoter and resulting transformants screened for developmental abnormalities. EPF1 overexpression caused a drastic reduction in stomatal numbers, suggesting that it may act as an inhibitory signal for stomatal development. Promoter fusion analysis showed EPF1 expression to be restricted to cells of the stomatal lineage, in particular the meristemoids and GMCs, adding weight to this suggestion. Furthermore, a T-DNA insertion mutant disrupting the *EPF1* gene showed frequent paired stomata in its epidermis, demonstrating a breakdown in the stomatal "one-cell-spacing rule" (Fig. 2). This *epf1* phenotype is reminiscent of that previously described for *sdd1*, and it was therefore expected that this putative secreted peptide signal would prove to be the predicted substrate for the SDD1 protease. However, and somewhat surprisingly, the overexpression of EPF1 peptide in the *sdd1* mutant background was still able to reduce stomatal density, and the *epf1sdd1* double mutant showed an additive effect on stomatal density and clustering. Thus, it appears very unlikely that SDD1 acts as a hormone convertase to process and activate the EPF1 peptide. Although EPF1 does not genetically interact with SDD1, the evidence suggests that it is likely to function as a ligand for the putative TMM ERf receptor complex. Genetic analyses place *EPF1* in the same pathway as *TMM* and *ERf* genes, and mutant crosses with *tmm* or *ererlerl2* show that *epf1* is epistatic to these mutants, as introgression of *epf1* has little or no effect on their phenotypes.

6 EPF Peptide Family Members Are Inhibitors and Activators of Stomatal Development

The EPF peptide family member with the closest amino acid sequence homology to EPF1 is EPF2. Constitutive overexpression of EPF2 results in a phenotypic effect similar to that of EPF1 with a large reduction in stomatal density (Hunt and Gray 2009; Hara et al. 2009), suggesting that these two peptides may mediate similar processes. However, there are differences between the functions of EPF1 and EPF2 in stomatal development. Promoter reporter and microarray experiments indicate that *EPF2* expression occurs earlier in stomatal development than *EPF1* with particularly high expression levels in meristemoids. *epf2* T-DNA disruption mutants have an increase in stomatal density and, in addition, produce excessive numbers of small stomatal lineage cells in their epidermis which fail to differentiate into either guard cells or pavement cells (Fig. 2). Like *epf1*, crosses of *epf2* mutants with *sdd1*show an additive effect, indicating that *SDD1* and *EPF2* operate in independent pathways. Thus, neither EPF1 nor EPF2 appears to be likely substrates for proteolysis by SDD1. Unlike EPF1, EPF2 does not appear to be entirely dependent on the receptor-like protein TMM for its activity. The double mutant *epf1tmm* has a stomatal development phenotype essentially the same as *tmm*, but the phenotype of *epf2tmm* suggests *epf2* is able to partially suppress the *tmm* mutation as the phenotype in mature leaves is intermediate, and in stems, there is an increase in stomata compared to *tmm* (Hunt and Gray 2009). This result indicates EPF2 acts in part via a receptor complex that does not contain TMM. Whether this is an ERf receptor or an unknown receptor has not been determined.

A third member of the *EPF* gene family, the *EPFL6/CHALLAH* gene, was identified as being involved in regulating stomatal development from a screen of a mutagenized population of *tmm1*, which had stomata restored in the hypocotyls and stems (where stomata are normally absent in *tmm* mutants), but had no effect on other tissues (Abrash and Bergmann 2010). EPFL6/CHALLAH is expressed subepidermally in hypocotyls, and overexpression prevents stomatal formation. This is a complex phenotype to interpret. One suggestion is that TMM may be able to sequester EPFL6/CHALLAH peptide and prevent it from binding to the ERECTA family RLK proteins to inhibit stomatal development.

Most recently, a fourth member of the EPF family has been characterized that appears to have essentially the opposite activity to the previously characterized EPFs. The *EPFL9* or *STOMAGEN* gene was selected for study because of its similar expression pattern (from microarray coexpression data) as *EPF1* and *TMM*. In marked contrast to overexpression experiments with the other three characterized EPFs, constitutive overexpression of EPFL9/STOMAGEN increases, rather than decreases, stomatal density (Sugano et al. 2010; Kondo et al. 2010; Hunt et al. 2010). Thus, EPFL9/STOMAGEN is an activator rather than an inhibitor of stomatal development. Surprisingly, promoter GUS fusion and in situ RNA hybridization analysis showed *EPFL9/STOMAGEN* expression to be in mesophyll cells of the developing leaf rather than in epidermal or stomatal cells. This raises the intriguing possibility that the mesophyll cells underlying the epidermis can communicate with other cell layers to determine stomatal density. *EPFL9/STOMAGEN* expression persists through leaf development. It is possible that EPFL9/STOMAGEN could represent the systemic signal that is known to be produced by mature leaves that can adapt stomatal density in developing leaves to environmental conditions (Casson and Gray 2008). However, there is no direct evidence to support this, and as yet, it is not known if EPF proteins can be transported over longer distances.

As described above, EPFL9/STOMAGEN acts to increase stomatal density in early developing leaves, and the purified 45-amino-acid peptide fragment or the chemically synthesized peptide can have the same effect when applied to *Arabidopsis* seedlings (Kondo et al. 2010). Downregulation of EPFL9/STOMAGEN by RNAi expression has the opposite effect, producing plants with reduced stomatal density. Genetic alteration and external application all suggest that EPFL9/STOMAGEN acts through the TMM receptor-like protein. Overexpression of EPFL9/STOMAGEN in an *epf1* and *epf2* or double mutant background suggests that these two "inhibitory peptides" are not necessary for EPFL9/STOMAGEN action (Hunt et al. 2010) which is strengthened by the observation that downregulation of EPFL9/STOMAGEN expression is still able to reduce stomatal density in *epf1*, *epf2*, and *epf1epf2* (Sugano et al. 2010). Reduction of EPFL9/STOMAGEN levels in the *epf2* background reduces stomatal density but does not reduce nonstomatal cell density, suggesting that EPF2 is necessary to prevent excessive meristemoid production, but EPFL9/STOMAGEN acts at a later stage in stomatal development. Overexpression of EPFL9/STOMAGEN in an *sdd1* background suggests SDD1 is not necessary for EPFL9/STOMAGEN action, and external application of active EPFL9/STOMAGEN peptide confirms this. This is not unexpected, though, as the *sdd1* phenotype of increased stomatal density is more likely to result from the inability to activate a negative regulator of stomatal development (such as EPF1) rather than a positive regulator such as EPFL9/STOMAGEN. Thus, although four peptide signals have been shown to be involved in regulating stomatal development, the substrate for SDD1 remains unknown.

Very little is known of the biological function of the other seven EPF family members, although undoubtedly, future work will identify at least some of their biological roles. This may not be a trivial task, however, as the characterization of EPFL6/CHALLAH has shown, as the *chal* stomatal phenotype was only observed when present in a *tmm* or *er* family mutant background. Some information is already available concerning EPFL4 and EPFL5. Overexpression experiments with *EPFL4* and *EPFL5*, like *EPF1*, *EPF2*, and *EPFL6* overexpression experiments, resulted in reduced stomatal densities. On the face of it, this appears to suggest a role for these peptides in stomatal development, but downregulation mutants for either of these genes do not show a stomatal phenotype in wild-type backgrounds (Hara et al. 2009) so their true biological role remains unclear.

Intriguingly, recent experiments with CLE9 indicated a stomatal expression pattern (Jun et al. 2010). Whether CLE9 could be a ligand for ERf or TMM receptor proteins or a substrate for SDD1 proteolysis is currently unknown, but this finding suggests that the EPF family may not be the only peptide signals involved in regulating stomatal development.

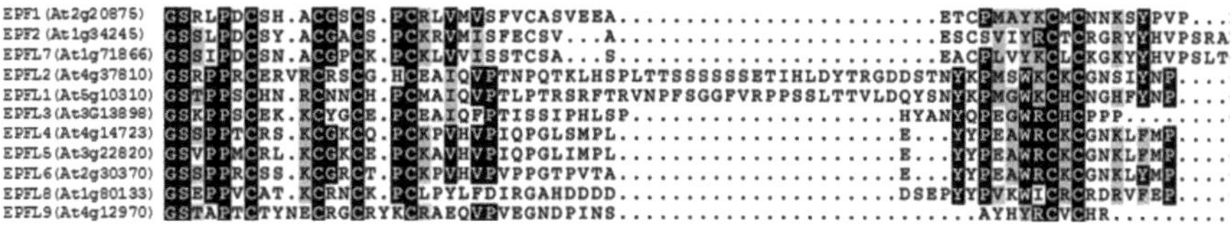

Fig. 3 Alignment of deduced EPF family amino acid sequences. Predicted amino acid sequences, with predicted signal sequences removed, are aligned and conserved residues *shaded black*. TAIR 10 (http://www.arabidopsis.org) gene code is shown in *parentheses*

7 The EPF Family Peptides Share Structural Features

EPF1 does not belong to any of the previously characterized plant peptide families. The EPF family is unique to plants, and homologues are found in a wide range of plant species (Peterson et al. 2010). Analysis of the *Arabidopsis* genome sequence shows EPF1 to be part of an 11-member family [reviewed by Rychel et al. (2010)]. These peptides are named EPF1, EPF2, and EPFL (EPF-like), although CHALLAH (EPFL6) and STOMAGEN (EPFL9) have alternative names. The EPF family peptides are characterized by a signal sequence at their N terminus, suggesting that they are all secreted peptides. After cleavage of their predicted signal sequence, molecular weights of the EPF peptides range from 6.08 to 9.8 kDa. They all have six cysteine residues in conserved positions toward their C terminus (see peptide alignment in Fig. 3). A feature of the EPF family is a variable region between the fourth and fifth conserved cysteine residues, which ranges from as long as 52-amino-acid residues in EPFL1 and EPFL2 to as short as 19-amino-acid residues in EPF2, EPFL7, and EPFL9. Interestingly, EPF1, EPF2, and EPFL7 have an additional pair of cysteines in this variable region (Fig. 3). Although sequence comparisons and alignments allow us to speculate on the important parts of the peptide, only experimental data from peptides purified from plants will reveal the mature active peptide fragment(s). Recently, such experimental evidence has been obtained to show the sequence of the biologically active region of EPFL9/STOMAGEN (Sugano et al. 2010; Kondo et al. 2010). First, affinity chromatography was used to purify the EPFL9/STOMAGEN peptide from *Arabidopsis* plants constitutively overexpressing EPFL9/STOMAGEN. A protein of 5 kDa was purified, corresponding well to the calculated molecular mass of a 45-amino-acid region. N-terminal sequencing suggested that the EPFL9/STOMAGEN peptide was cleaved between residue 57 (methionine) and 58 (isoleucine) before the conserved glycine-serine (GS) motif amino acid residues. Kondo et al. also determined the probable organization of disulphide bonding between cysteine–cysteine pairs. As a result of this work, the peptide structural arrangement for the EPF peptides could be predicted, and this suggested an external loop between the fourth and fifth cysteine residues corresponding to the variable region in the EPF sequences. The positioning of this variable region on the external surface of the EPF peptides indicates that it may be a region that is important for binding the peptide to its receptor or perhaps for participating in other protein interactions. Further evidence that this is an important functional area of the peptide comes from plants harboring the EPFL6/CHALLAH mutation *chal-1*, in which the proline adjacent to the fourth cysteine (a residue conserved in all but EPFL2 and EPFL9/STOMAGEN) is converted to a serine. In these mutants, the *chal* stomatal development phenotype was particularly severe (Abrash and Bergmann 2010).

8 EPFs Are Processed to Generate an Active Peptide Signal

The experiments described above suggest that an active EPF fragment is released by cleavage from the translated peptide. However, although none of the EPFs characterized so far appear to be activated by the SDD1 putative subtilisin protease, sequence analysis reveals that the EPF peptides contain several putative proteolytic sites. Animal prohormone convertases recognize an R/K-Xn-R/K↓ motif (Seidah and Chrétien 1997), where *n* is 0, 2, 4, or 6 and the arrow represents the position of the cleavage site. Among the EPF peptides, this motif is found in all but EPFL7 and EPFL8. In the other nine EPF peptides, the putative protease recognition site is close to the cleavage site experimentally identified in EPFL9/STOMAGEN. Furthermore, ProP analysis (Duckert et al. 2004) identifies a furin recognition motif in EPF1 (RR–RP). In mammals, furin is a member of a family of subtilisin-/kexin-like proprotein convertases which have a more stringent selection of R–X–[R/K]–R↓.

Another characterized group of plant-secreted peptides, the Rapid ALkalinization Factors (RALFs), contains a dibasic RRXL motif that is potentially a cleavage recognition site. Expression of AtRALF23 in the *ats1p* protease mutant background demonstrated that a tagged protein was no longer processed. AtS1P is a membrane-associated subtilisin-like protease (Srivastava et al. 2008, 2009; see chapter "Processing of Peptides"). The RRIL motif in AtRALF23 is also found in the EPFL6/CHALLAH sequence preceding the predicted cleavage site identified experimentally in EPFL9/STOMAGEN. It is not yet known if EPFL6/CHALLAH shows a physical interaction with either ATS1P or SDD1, but genetic interactions suggest that it is not in the same pathway as SDD1 (Abrash and Bergmann 2010). Only one processing step is required to activate AtRALF in contrast to the phytosulfokine, AtPSK4 (Srivastava et al. 2009; see chapter "Processing of Peptides"). The latter is cut by a different subtilisin protease, AtSBT1.1, at the cleavage recognition site RRSLVL↓HTDY. The proteolytic activation of AtPSK4 is likely to require the further trimming of three residues from the N terminus and five residues from the C terminus of the released peptide and, thus, probably requires more than one protease.

So far, experimental evidence suggests the maximum number of amino acids in mature EPFL9 is 45. How this peptide fragment is generated and if there is any further proteolytic processing at the C terminus are not known. Experimental evidence is awaited on whether any of the EPF peptides are a substrate for SDD1 or whether they are processed by a different subtilisin family member or perhaps a different class of protease. The *Arabidopsis* genome sequence encodes numerous putative protease genes and identifying the enzyme(s) that process the EPF peptides has so far proved elusive.

9 Potential Receptors for EPF Family Peptides

The *A. thaliana* genome sequence encodes more than 600 receptor-like kinase (RLK) genes (Shiu and Bleecker 2001). As described above, several of these have so far been shown to play a role in the inhibition of stomatal development: TMM and the three ER family proteins. Current genetic evidence suggests that there could be an interaction between the LRR domain receptor-like protein TMM and the ERf receptor-like kinase proteins although a direct physical interaction remains to be demonstrated. It is possible that EPF peptides can bind directly to either TMM or ERf separately or only when these proteins form a receptor complex, but this is yet to be tested. The EPF stomatal development pathway involving a processed peptide signal, an LRR domain protein, and a receptor protein kinase has obvious similarities to the "CLAVATA" pathway that regulates stem cell number in the shoot apical meristem (see chapter "Peptides Regulating Apical Meristem Development"). The best characterized CLE peptide family member is CLV3, a secreted 15 kDa protein that is processed to a 12- or 13-amino-acid arabinosylated peptide (Fletcher et al. 1999). Until recently, the CLV3 peptide was believed to be perceived by the receptor kinase CLV1, an LRR-RLK structurally similar to ER family proteins (Ogawa et al. 2008), in complex with CLV2, a receptor-like protein, which like TMM lacks a kinase domain (Jeong et al. 1999). However, more recent protein interaction studies indicate that CLV1 and CV2 do not physically interact and that the diversity of possible CLV3 receptor complexes is more numerous than originally believed. CLV1 may form homodimers, or heterodimers by interaction with another RPK called CORYNE (CRN), which can itself form homodimers. CLV2-CRN and CRN-CRN may act together as a tetramer, and furthermore, CLV1 and CLV2-CRN may form a complex (Müller et al. 2008; Zhu et al. 2010; also see chapter "Peptides Regulating Apical Meristem Development"). If EPF perception involves a similar complexity of receptor component interactions as CLE peptides are now believed to use, it seems probable that more EPF receptors are likely to be identified in the future.

As discussed above, the asymmetric divisions of EPF target cells are oriented to prevent stomata forming adjacent to one another. However, we do not know whether the localization of EPF receptors is polarized within the plasma membrane. It is possible that EPF receptors are located homogeneously on the surface of cell, or have lateral polarity, or are predominantly localized on the internal side of the developing epidermal cells in order to perceive a mesophyll-derived signal from EPFL9/STOMAGEN or EPFL6/CHALLAH. However, TMM localization does not appear to be polar in asymmetrically dividing cells (Nadeau and Sack 2002). Recently, the gene product BREAKING OF ASYMMETRY IN THE STOMATAL LINEAGE (BASL) was characterized as having polarized localization in stomatal lineage cells. The BASL protein marks areas of cells that will divide asymmetrically, distal to an

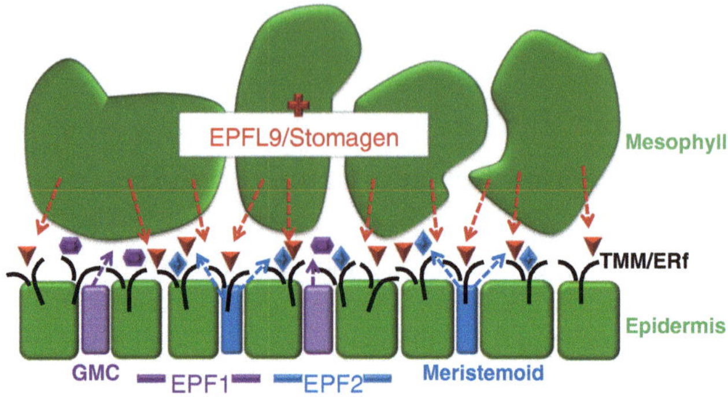

Fig. 4 Stomatal development in the leaf epidermis may be regulated by a balance between positive- and negative-acting EPF family peptides. The positive regular of stomatal development EPFL9/STOMAGEN originating from the developing mesophyll may compete with negative regulators EPF1 and EPF2 derived from meristemoids or guard mother cells (GMC) for binding to the putative TMM/ERf receptor complex

existing meristemoid to ensure correct stomatal spacing. BASL mutants lack asymmetry in stomatal lineage cells, resulting in the formation of paired stomata (Dong et al. 2009), but BASL is unlikely to act in the same pathway as TMM, EPF1, or EPF2 as crosses of mutants in these genes with *basl* show additive phenotypic effects. Thus, it is currently not clear how the EPF peptides and the receptor-like protein TMM orientate cell divisions within the stomatal lineage.

10 Possible Modes of Action of EPF Positive and Negative Regulators of Stomatal Development

The identification of the existence of both positive (EPFL9/STOMAGEN) and negative (EPF1, EPF2, and EPFL6/CHALLAH) regulators of stomatal development within the same peptide signal family has obvious parallels with the CLE family of peptides which has members that either promote or inhibit stem cell differentiation in vascular development (Ito et al. 2006; also see chapter "Peptides Regulating Plant Vascular Development"). The two counteracting pathways in EPF signaling suggest several potential mechanisms of action for the regulation of stomatal development. Firstly, a positive regulator may simply repress expression of a negative regulator. Secondly, the degree of stomatal development may be determined by a competition between negative and positive effectors for the same binding site on a receptor (Fig. 4). In the scenario depicted in Fig. 4, the EPFL/STOMAGEN peptide ligand competes with EPF1 and EPF2 for binding to the TMM ERf receptor complex. When EPF1 or EPF2 are bound, the receptor is activated and the downstream signaling cascade inhibits asymmetric division in target cells and promotes their differentiation into pavement cells. When EPFL9/

STOMAGEN ligand is bound, receptor activation and downstream signaling are blocked, and the target cell retains its meristematic activity. In this model where there is competition for the same ligand binding site on a receptor, the stomatal density may be directly proportional to the ratio of the concentrations of positive to negative acting peptides. A variation on this model would be if the receptor has separate negative and positive ligand binding sites, with binding of one regulator blocking access of another to its binding site.

Thirdly, the ability of a ligand to bind to and activate (or block) the receptor may be affected by its binding to and sequestration by another protein. Such a scenario, whereby the EPFL6/CHALLAH may be sequestered by the receptor-like protein TMM, was outlined above. Alternatively, the peptide ligands may interact with each other in homodimers or heterodimers which could either improve or prevent their binding to and activation of their receptor. Thus, there are many ways that the EPF peptides may interact with each other or their putative receptor components TMM or ERf to promote or inhibit stomatal development, but the true identity of the signaling components that bind to the EPF family peptides and the nature of their interactions are yet to be discovered.

References

Abrash E, Bergmann DC (2010) Regional specification of stomatal production by the putative ligand CHALLAH. Development 137:447–455

Berger D, Altmann T (2000) A subtilisin-like serine protease involved in the regulation of stomatal density and distribution in *Arabidopsis thaliana*. Genes Dev 14:1119–1131

Bergmann DC, Sack F (2007) Stomatal development. Annu Rev Plant Biol 58:163–181

Bergmann DC, Lukowitz W, Somerville CR (2004) Stomatal development and pattern controlled by a MAPKK kinase. Science 304:1494–1497

Casson S, Gray JE (2008) Influence of environmental factors on stomatal development. New Phytol 178:9–23

Dong J, MacAlister CA, Bergmann DC (2009) BASL controls asymmetric cell division in Arabidopsis. Cell 137:1320–1330

Duckert P, Brunak S, Blom N (2004) Prediction of proprotein convertase cleavage sites. Protein Eng Des Sel 17:107–112

Fletcher JC, Brand U, Running MP, Simon R, Meyerowitz EM (1999) Signaling of cell fate decisions by *CLAVATA3* in *Arabidopsis* shoot meristems. Science 283:1911–1914

Hara K, Kajita R, Torii KU, Bergmann DC, Kakimoto T (2007) The secretory peptide gene EPF1 enforces the stomatal one-cell-spacing rule. Genes Dev 21:1720–1725

Hara K, Yokoo T, Kajita R, Onishi T, Yahata S, Peterson KM, Torii KU, Kakimoto T (2009) Epidermal cell density is autoregulated via a secretory peptide, EPIDERMAL PATTERNING FACTOR 2 in Arabidopsis leaves. Plant Cell Physiol 50:1019–1031

Hunt L, Gray JE (2009) The signaling peptide EPF2 controls asymmetric cell divisions during stomatal development. Curr Biol 19:864–869

Hunt L, Bailey KJ, Gray JE (2010) The signaling peptide EPFL9 is a positive regulator of stomatal development. New Phytol 186:609–614

Ito Y, Nakanomyo I, Motose H, Iwamoto K, Sawa S, Dohmae N, Fukuda H (2006) Dodeca-CLE peptides as suppressors of plant stem cell differentiation. Science 313:842–845

Jeong S, Trotochaud AE, Clark SE (1999) The Arabidopsis CLAVATA2 gene encodes a receptor-like protein required for the stability of the CLAVATA1 receptor-like kinase. Plant Cell 11:1925–1934

Jun J, Fiume E, Roeder AH, Meng L, Sharma VK, Osmont KS, Baker C, Ha CM, Meyerowitz EM, Feldman LJ, Fletcher JC (2010) Comprehensive analysis of *CLE* polypeptide signaling gene expression and overexpression activity in Arabidopsis. Plant Physiol 154:1721–1736

Kim TH, Böhmer M, Hu H, Nishimura N, Schroeder JI (2010) Guard cell signal transduction network: advances in understanding abscisic acid, CO_2, and Ca^{2+} signaling. Annu Rev Plant Biol 61:561–591

Kondo T, Kajita R, Miyazaki A, Hokoyama M, Nakamura-Miura T, Mizuno S, Masuda Y, Irie K, Tanaka Y, Takada S, Kakimoto T, Sakagami Y (2010) Stomatal density is controlled by a mesophyll-derived signaling molecule. Plant Cell Physiol 51:1–8

Lampard GR, Macalister CA, Bergmann DC (2008) Arabidopsis stomatal initiation is controlled by MAPK-mediated regulation of the bHLH SPEECHLESS. Science 322:1113–1116

Müller R, Bleckmann A, Simon R (2008) The receptor kinase CORYNE of Arabidopsis transmits the stem cell-limiting signal CLAVATA3 independently of CLAVATA1. Plant Cell 20:934–946

Nadeau JA, Sack FD (2002) Control of stomatal distribution on the Arabidopsis leaf surface. Science 296:1697–1700

Ogawa M, Shinohara H, Sakagami Y, Matsubayashi Y (2008) Arabidopsis CLV3 peptide directly binds CLV1 ectodomain. Science 319:294

Peterson KM, Rychel AL, Torii KU (2010) Out of the mouths of plants: the molecular basis of the evolution and diversity of stomatal development. Plant Cell 222:296–306

Rychel AL, Peterson KM, Torii KU (2010) Plant twitter: ligands under 140 amino acids enforcing stomatal patterning. J Plant Res 123:275–280

Seidah NG, Chrétien M (1997) Eukaryotic protein processing: endoproteolysis of precursor proteins. Curr Opin Biotechnol 8:602–607

Serna L, Fenoll C (2000) Stomatal development in Arabidopsis: how to make a functional pattern. Trends Plant Sci 5:458–460

Shiu SH, Bleecker AB (2001) Plant receptor-like kinase gene family: diversity, function, and signaling. Sci STKE 2001(113):re22

Shpak ED, McAbee JM, Pillitteri LJ, Torii KU (2005) Stomatal patterning and differentiation by synergistic interactions of receptor kinases. Science 309:290–293

Srivastava R, Liu JX, Howell SH (2008) Proteolytic processing of a precursor protein for a growth-promoting peptide by a subtilisin serine protease in Arabidopsis. Plant J 56:219–227

Srivastava R, Liu JX, Guo H, Yin Y, Howell SH (2009) Regulation and processing of a plant peptide hormone, AtRALF23, in Arabidopsis. Plant J 59:930–939

Sugano SS, Shimada T, Imai Y, Okawa K, Tamai A, Mori M, Hara-Nishimura I (2010) Stomagen positively regulates stomatal density in Arabidopsis. Nature 463:241–244

Von Groll U, Berger D, Altmann T (2002) The subtilisin-like serine protease SDD1 mediates cell-to-cell signaling during Arabidopsis stomatal development. Plant Cell 14:1527–1539

Yang M, Sack FD (1995) The too many mouths and four lips mutations affect stomatal production in Arabidopsis. Plant Cell 7:2227–2239

Zhu Y, Wang Y, Li R, Song X, Wang Q, Huang S, Jin JB, Liu CM, Lin J (2010) Analysis of interactions among the CLAVATA3 receptors reveals a direct interaction between CLAVATA2 and CORYNE in Arabidopsis. Plant J 61:223–233

Plant Antimicrobial Peptides

Tatyana Odintsova and Tsezi Egorov

Abstract Antimicrobial peptides (AMPs) are natural antibiotics produced by all living organisms to resist infection by pathogens. They are important effector molecules of the innate immune system both in animals and plants. AMPs are diverse in structure and mode of action and display broad-spectrum antimicrobial activity and thus show promise for engineering pathogen resistance in crops and development of novel pharmaceuticals. A variety of AMP classes have been discriminated, which include defensins, thionins, lipid-transfer proteins, hevein- and knottin-like peptides, and macrocyclic peptides. In this review, the role of AMPs in the plant immune system is briefly discussed, and different families of plant AMPs with respect to their structural peculiarities and biological role are considered. Special emphasis is given to AMPs of wild plants. Defensins belong to the largest AMP family widely distributed throughout the plant and animal kingdoms with a wide range of in vitro biological activities. Current evidence indicates that they interact with specific molecules on the fungal membranes and act on intracellular targets. Thionins show high inhibitory activity against diverse fungi and bacteria including human pathogens and are toxic to some other types of cells, such as mammalian, insect, and plant cells. The antimicrobial effect of thionins is associated with membrane permeabilization through pore formation. Lipid-transfer proteins are not only antimicrobial agents but are also involved in signaling. Hevein-like AMPs comprise peptides with a conserved chitin-binding domain, a variable number of disulfide bridges, and divergent precursor structures. Knottin-like and macrocyclic AMPs form the cystine knot and exhibit insecticidal,

T. Odintsova (✉)
Vavilov Institute of General Genetics, Russian Academy of Sciences, Ul. Gubkina 3, 119991 Moscow, Russian Federation
e-mail: Odintsova2005@rambler.ru

T. Egorov
Shemyakin and Ovchinnikov Institute of Bioorganic Chemistry, Russian Academy of Sciences, Ul. Miklukho-Maklaya 16/10, 117997 Moscow, Russian Federation

H.R. Irving and C. Gehring (eds.), *Plant Signaling Peptides*, Signaling and Communication in Plants 16, DOI 10.1007/978-3-642-27603-3_7, antimicrobial, anti-HIV, hemolytic, and uterotonic activities. Due to exclusive structural and chemical stability, plant cyclotides are regarded as templates into which diverse biological activities may be introduced.

1 Introduction

Plants are constantly exposed to numerous pathogens, such as fungi, oomycetes, bacteria, insects, nematodes, viruses, and viroids; however, diseases seldom occur. To cope with pathogens, plants have evolved a sophisticated multilevel system that enables them to perceive and ward off pathogen attack (Jones and Dangl 2006). By a receptor-mediated mechanism, plants perceive signaling molecules (elicitors) from their own damaged cells (damage-associated molecular patterns, DAMPs) as well as from host and nonhost pathogens (pathogen- or microbe-associated molecular patterns, PAMPs or MAMPs; also see chapter "Peptides as danger signals: MAMPs and DAMPs"). Recognition of plant- or pathogen-derived molecules is an initial step in activation of the innate immunity response. Once recognized, the defense reaction is triggered, culminating in upregulation of defense-related genes. However, successful pathogens may avoid the first-line defense barrier delivering specific effectors (virulence factors) into plant cells. To prevent colonization of tissues by these pathogens, a second class of plant receptors, the nucleotide-binding leucine-rich repeat (NB-LRR) proteins, comes into stage. If an effector is recognized by a corresponding NB-LRR protein, the plant immune cascade is activated. Several low-molecular-weight signaling molecules, such as salicylic and jasmonic acids, methyl jasmonate, ethylene, and nitric oxide, are key regulators of plant immunity pathways. It is generally assumed that plant immune response is complex, involving several parallel defense signaling pathways that interact at key regulatory steps (Pieterse et al. 2009).

The earliest events following recognition of the pathogen by plant receptors are calcium influx, alkalinization of the extracellular space, protein kinase activation, production of reactive oxygen species (ROS), and nitric oxide and transcriptional reprogramming including defense gene activation (Dangl and Jones 2001). Induction of defense-related genes results in a variety of processes restricting spread of a pathogen in plant tissues.

2 Plant Defense Mechanisms

Defense mechanisms deployed by plants to protect themselves from pathogenic microorganisms may be either constitutive or inducible. The latter can be broadly classified into three main categories (1) the production of reactive oxygen and nitrogen species and initiation of the hypersensitive response, (2) the reinforcement of plant cell walls, and (3) the production of antimicrobial compounds that inhibit pathogen growth. The production of ROS (H_2O_2 and O_2^-) resulting in a so-called oxidative burst and catalyzed by NADPH oxidases and cell-wall peroxidases is one of the earliest cellular responses following successful pathogen recognition (Torres et al. 2006). ROS are toxic to microorganisms. In addition, they participate in the reinforcement of plant cell walls inducing cross-linking of glycoproteins and other cell-wall constituents at the primary sites of pathogen penetration. ROS are also signaling molecules mediating defense gene activation and triggering programmed cell death (PCD) directed toward localization of the pathogen inside lesions composed of dead cells, thus limiting the spread of infection through plant tissues. Pathogen invasion initiates reinforcement of plant cell walls. This is achieved by cross-linking of cell-wall components and accumulation of lignin, phenolic compounds, hydroxyproline proteins, and thionins, deposition of novel substances (e.g., callose), and even de novo cell synthesis. Finally, pathogen invasion induces production of antimicrobial compounds.

3 Antimicrobial Compounds

Antimicrobial compounds may be synthesized in plant cells either constitutively in specialized tissues or organs or may be induced by pathogen challenge. According to their chemical nature, they include secondary metabolites, such as constitutively produced phytoanticipins or inducible phytoalexins, and polypeptide-based molecules. Phytoanticipins may be defined as constitutive antifungal compounds present in plants in biologically active state or rapidly produced from inactive precursors following cell damage (Osbourn 1996). They are localized predominantly in the outer cell layers of plant organs or in vacuoles. In contrast to phytoanticipins, phytoalexins are induced on pathogen attack. Both phytoalexins and phytoanticipins are chemically diverse compounds: phenols, phenolic glycosides, saponins, unsaturated lactones, cyanogenic glycosides, and glucosinolates. Their antimicrobial activity is usually more or less unspecific. In accordance with their variant chemical structure, the modes of action of phytoalexins and phytoanticipins differ considerably. For example, saponins form complexes with sterols in membranes of microorganisms inducing pore formation.

Antimicrobial polypeptides may be arbitrarily subdivided into two groups: proteins and peptides. The first group comprises pathogenesis-related (PR-) proteins, including enzyme inhibitors, and ribosome-inactivating proteins (Broekaert et al. 2000). PR proteins were first identified in tobacco plants carrying the *N* gene upon infection with tobacco mosaic virus (Van Loon et al. 2006). Subsequent studies revealed similar proteins in other plant species challenged by pathogens or exposed to other stressful conditions. The role of antimicrobial proteins in plant defense was demonstrated by three independent experimental approaches: in vitro antimicrobial assays, studies of pathogen-induced regulation of gene expression, and analysis of transgenic plants constitutively overexpressing particular proteins. Based on sequence homology, several classes of PR proteins have been identified in plants [reviewed by Van Loon et al. (2006)]. Some families are represented by enzymes (chitinases, glucanases, proteases, oxidases) and are supposed to be involved in a number of diverse functions: disruption of cell walls of fungal or bacterial pathogens by hydrolyzing cell-wall constituents (chitin, glucan, and proteins), establishment of structural barriers in plant cells (peroxidases), and production of ROS (oxalate oxidases). Chitinases may also be involved in defense against nematodes and insects. Other PR protein families comprise enzyme inhibitors. These are suggested to defend plants against herbivorous insects. Some PR proteins do not display any enzymatic activity, but were shown to inhibit growth of fungi and oomycetes by disturbing plasma membrane integrity. Ribosome-inactivating proteins have been implicated in antiviral defense.

4 General Characteristics of Plant Antimicrobial Peptides

Antimicrobial peptides (AMPs) are gene-encoded natural antibiotics that form an ancient and evolutionary conserved defense strategy deployed by all living organisms to combat infection. Due to their high potential to increase pathogen resistance in crops and serve as templates for development of new-generation drugs, this group of defense molecules has been the subject of considerable research in recent years (see reviews Benko-Iseppon et al. 2010; Kido et al. 2010; Padovan et al. 2010; da Rocha Pitta et al. 2010; Ajesh and Sreejith 2009; Farrokhi et al. 2008; Sels et al. 2008; Tavares et al. 2008; Manners 2007; Castro and Fontes 2005; Garcia-Olmedo et al. 1998, 2001; Broekaert et al. 1997). Dedicated databases containing the available information on plant AMPs encompassing taxonomic, microbiological, and physicochemical data have been created (Hammani et al. 2009). Plant AMPs are effector molecules considered to be an important defense barrier to pathogens and pests. This now widely accepted view follows from in vitro biological assays and is strongly supported by three lines of evidence, such as observations of increased resistance of transgenic plants constitutively expressing AMPs, upregulation of AMP genes in response to stress, and prevalent localization in peripheral cells layers and in tissues especially sensitive to infection. Numerous AMPs have been isolated from different plant species. It is now evident that plant species produce unique arrays of AMPs to prevent pathogen invasion of their tissues. Plant AMPs share similar properties, such as small size (30–90 residues), positive charge, and a high portion of hydrophobic residues (about 30%) that allow them to fold into an amphiphilic structure with distinct patches of hydrophobic and positively charged amino acids (Hancock and Sahl 2006). These structural features ensure effective interaction with plasma membranes of pathogenic microorganisms assumed to be their primary target. Plant AMPs differ in amino acid sequences and adopt diverse conformations. A vast majority of them belong to cysteine-rich polypeptides, their structure being stabilized by intrachain disulfide bonds (2–5), providing the molecules with high structural stability. Examples of plant AMP three-dimensional structures are presented in Fig. 1. Although the number of disulfide bonds in peptides may be identical, the molecular scaffold may be radically different. Plant AMP classification is therefore based on the homology of amino acid sequences, the so-called cysteine motifs (signatures), and the three-dimensional structures. A variety of AMP classes have been discriminated (Table 1). They include defensins, thionins, lipid-transfer proteins, hevein- and

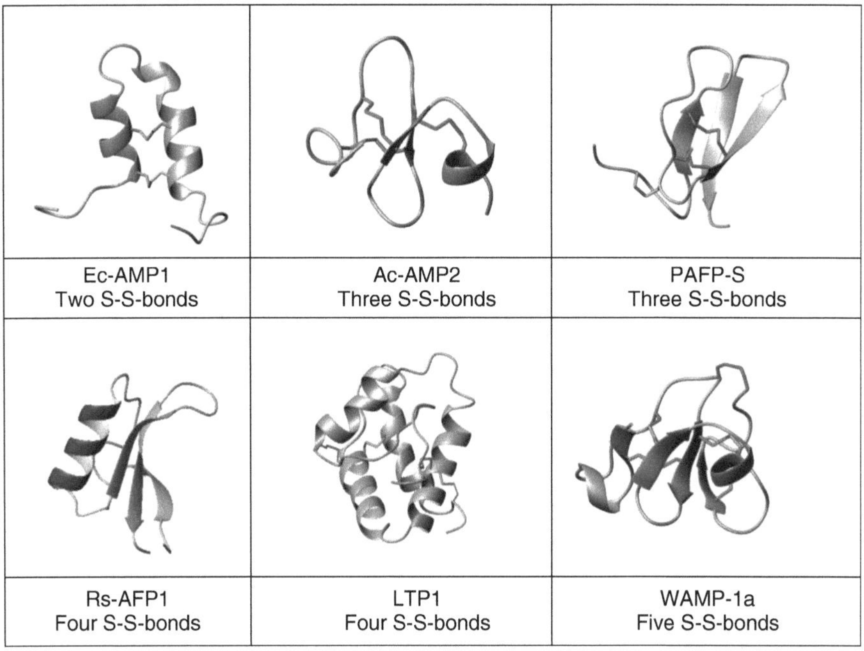

Fig. 1 Three-dimensional structures of selected plant AMPs deposited in the protein database. The antimicrobial peptide Ec-AMP1 from *Echinochloa crusgalli* (2L2R); the hevein-like peptide Ac-AMP2 from *Amaranthus caudatus* (1MMC); the knottin-like peptide PAFP from *Phytolacca americana* (1DKC); the defensin Rs-AFP1 from *Raphanus sativus* (1AYJ); the lipid transfer protein LTP1 from *Triticum aestivum* (1LPT); the hevein-like peptide WAMP-1a from *Triticum kiharae* (2lb7)

Table 1 Plant AMP families

Family	Number of amino acid residues	Number of disulfide bonds
nsLipid-transfer proteins	90–95 (class 1)	4
	70 (class 2)	
Defensins	45–54	4 or 5 (flower defensins)
Thionins	45–47	3, 4
Hevein-like	43	3–5
Knottin-like	36–37	3
Macrocyclic	29–31	3

knottin-like peptides, and macrocyclic peptides. Other molecular types that do not fall into the above-mentioned classes and are specific to some plant species have been described. These include snakins, shepherdins, and 4-Cys AMPs from *Impatiens balsamina* and *Zea mays* (Garcia-Olmedo 2001). Despite structural diversity, a common conserved structural motif, the so-called γ-core, has been identified in all AMPs studied. It is supposed to be the ancient AMP type, which was modified during evolution by the addition of diverse structural elements (α-helices and/or β-strands) (Yount and Yeaman 2004). Although the molecular types of plant AMPs described above have been known for more than two decades, reports of novel structural types are appearing in the literature, and it now becomes evident that some unique types earlier supposed to be only present in single plant species have a wider distribution across the plant kingdom than previously suggested. Thus, the diversity of AMPs in plants still needs to be investigated. Defensins belong to the most well-studied AMP family. It is the most ancient AMP class: homologous peptides have been identified in diverse eukaryotic organisms, from fungi to plants and higher vertebrates. However, plants also possess AMPs that are not found outside the plant kingdom, such as thionins, lipid-transfer proteins, and snakins. AMPs of different families differ in the processing and subcellular targeting. While some of them are produced as rather simple precursor proteins targeted to the apoplast or to the vacuole, others are derived from complex multidomain precursors subjected to proteolytic processing leading to the production of multiple AMPs. Plant AMPs have a broad spectrum of in vitro activity against phytopathogens, the vast majority of them are active against fungi and some are inhibitory to bacteria and herbivorous insects.

The mode of action of plant AMPs remains largely unknown and is only beginning to be unraveled. The microbial plasma membrane is supposed to be the primary target for most of them. Recent advances in our understanding of the mode of action of plant AMPs are restricted to a limited number of peptides, mainly belonging to the thionin and defensin families. Several mechanisms for the nonspecific interaction between plant AMPs and plasma membranes of microorganisms have been suggested. The cationic and amphiphilic nature of AMPs ensures their direct interaction with anionic cell surfaces of microbial pathogens: lipopolysaccharides in Gram-negative and teichoic acids in Gram-positive bacteria. Subsequently, AMPs come in contact with membrane phospholipids. After the insertion into the membrane bilayer, they act either by disruption of the membrane integrity through its thinning, formation of pores and/or interference with the barrier function, or by affecting intracellular targets. Several models describe the interactions of AMPs with plasma membranes resulting in the formation of barrel-stave or toroidal channels, and the disruption of the membrane integrity by a "carpet" of AMPs on the membrane surface (Hancock and Sahl 2006). For some AMP families (defensins), a receptor-mediated mechanism of fungal growth inhibition was postulated. In addition to membranes of pathogens, intracellular targets for some AMPs have also been identified. AMPs have attracted considerable interest for engineering pathogen resistance in crops and as promising leads for drug design. The great yield losses of crops caused by pathogenic microorganisms and pests stimulate the search for new genes conferring pathogen resistance. AMPs exhibiting broad-spectrum resistance are good candidates for genetic transformation of plants. Some progress has already been achieved in this field. Among numerous plant species, wild plants and weeds represent a valuable, still poorly explored source of various antimicrobials including AMPs. To compete with crops, weeds have developed a highly sophisticated multifactorial adaptation system, which makes them successful as invasive species. The molecular basis of this system is formed by diverse phytotoxins, which are released into the environments and suppress growth of other plants, and effective pathogen resistance mechanisms providing better competitiveness and adaptation to stressful conditions. In our studies of weeds, high broad-spectrum antimicrobial activity of AMPs was shown (Rogozhin et al. 2011; Slavokhotova et al. 2011; Odintsova et al. 2008). AMPs also bear great potential as models for the creation of new anti-infective drugs, both bactericidal and antimycotic, since in the last decades, the incidence of microbial infections is rising, together with the number of pathogenic microorganisms resistant to conventional antibiotics (da Rocha Pitta et al. 2010).

In recent years, isolation of AMPs from different plant species has been an active area of research. With the advent of genome-scale sequencing technologies and bioinformatics tools for mining AMPs in plant genomes, an amazing abundance of AMP-like genes in the sequenced genomes was discovered. Silverstein et al. (2007), using sequence motifs for plant AMPs, searched for related sequences in the complete genomes of *Arabidopsis thaliana* and *Oryza sativa* and identified 825 genes encoding cysteine-rich AMPs in the *Arabidopsis* genome and 598 sequences in the rice genome. The identified AMP genes include thionins, defensins, lipid-transfer proteins, snakins, protease inhibitors, and pollen allergens and account for ~2–3% of the gene repertoire of each model species, and 59% of the discovered sequences were shown to be expressed. The functional roles of multiple AMP genes in plant genomes probably involve protection against rapidly evolving microorganisms that acquire tolerance to particular AMP molecules. It has been suggested that abundant AMP genes ensure functional diversity and provide adaptation of the plant immune system to more evolutionary flexible pathogen populations (Manners 2007).

5 Families of AMPs

5.1 Defensins

Defensins belong to the largest and best-studied AMP family (see reviews Carvalho and Gomes 2009; Stotz et al. 2009b; Lay and Anderson 2005; Pellegrini and Franco 2005; Thomma et al. 2002; Broekaert et al. 1995). Plant defensins are small (45–54 amino acids) cationic peptides with a characteristic motif of eight cysteine residues forming four disulfide bridges with a single exception of flower defensins, which possess five disulfide bonds. Defensins were first discovered in wheat and barley seeds and initially assigned to a novel group of thionins named γ-thionins (Colilla et al. 1990; Mendez et al. 1990). However, subsequent studies showed that they differ structurally from thionins but display similarity to mammalian and insect defensins, so they were renamed "plant defensins." Homologous peptides were then isolated from sorghum seeds. They were shown to inhibit insect α-amylases (Bloch and Richardson 1991). Defensins represent a unique AMP group widely distributed throughout the plant and animal kingdoms and discovered in invertebrates, birds, mammals, and plants. Recently, defensin-like peptides have been identified in fungi and myxobacteria (Carvalho and Gomes 2009). More than 300 defensin-like sequences were identified in the completely sequenced *Arabidopsis* genome (Silverstein et al. 2005). Except for the conserved cysteine residues, defensin amino acid sequences are highly variable among plant species (Carvalho and Gomes 2009). On the whole, more conservation is observed within plant families, as shown for the Brassicaceae (Terras et al. 1993) and Poaceae defensins (Odintsova et al. 2007). However, in some instances, a striking similarity of defensins from unrelated families was reported (Rogozhin et al. 2011). Thus, *Nigella sativa* (Caryophyllaceae family) defensins show high sequence similarity (73%) with the defensins Rs-AFP1 and Rs-AFP2 from *Raphanus sativus* seeds belonging to the Brassicaceae family, thus providing evidence for the conservation of the defensin structure among evolutionarily diverged plant species and pointing to the common origin of all plant defensins. The three-dimensional structure of several plant defensins has been determined by NMR spectroscopy. Despite considerable sequence variation, the fold of the molecule is similar and involves a single α-helix and a triple-stranded β-sheet. The main structural element of the defensin molecule is the so-called cysteine-stabilized α-helix β-sheet motif (CSαβ), in which two cysteine residues separated by one turn of the α-helix are connected to two cysteines which are located a single amino acid apart in the third β-strand (Thomma et al. 2002). In some defensins, unique structural elements have been revealed such as 3_{10} helix in VrD1 and an additional disulfide bridge connecting the loop between β1 and α-helix with α-helix in a floral defensin PhD1. Plant defensins share both structural and functional similarity with defensins of insects and mammals (Aerts et al. 2008). The structural similarity indicates common origin of all defensins. The discovery of defensins in fungi suggests that they appeared in the evolution before the divergence of fungi and insects (Carvalho and Gomes 2009). Defensins have been identified in a vast majority of monocot and dicot plants including the families Poaceae, Brassicaceae, Fabaceae, Chenopodiaceae, Asteraceae, Solanaceae, Liliaceae, Hippocastanaceae, Ginkgoaceae, Rosaceae, Amaranthaceae, Cucurbitaceae (reviewed in Carvalho and Gomes 2009), Ranunculaceae (Rogozhin et al. 2011), and Cariophyllaceae (Slavokhotova et al. 2011) and even in gymnosperms (Sharma and Lönneborg 1996). Most plant defensins were isolated from seeds; however, they were also found in other plant organs, such as leaves, flowers, tubers, seedpods, and fruits. Some members of the family are induced upon pathogen attack, while others are constitutively expressed in particular tissues or organs (Lay and Anderson 2005).

In vitro studies have revealed a wide range of activities of plant defensins that imply their involvement in biotic and abiotic stress response. The protective role of defensins in vivo is supported by enhanced tolerance to infections of transgenic plants with constitutive expression of the corresponding genes (Stotz et al. 2009b; Lay and Anderson 2005). Most plant defensins display antifungal activity (da Rocha Pitta et al. 2010). It deserves special attention that some members of the defensin family are active not only against phytopathogenic fungi but against human pathogens as well, such as *Candida albicans* (Stotz et al. 2009b). According to their effect on fungi, plant defensins are separated into two groups: the morphogenetic, causing morphological changes in fungi such as hyperbranching, and the nonmorphogenetic, inducing only growth inhibition (Thomma et al. 2002). Some defensins are active against bacteria (da Rocha Pitta et al. 2010). Antimicrobial activity depends both on the type of defensin and the pathogen species. Inhibition of protein synthesis in vitro (Mendez et al. 1990, 1996), inhibition of proteases (Melo et al. 2002; Wijaya et al. 2000) and α-amylases (Pelegrini et al. 2008; Bloch and Richardson 1991), blockage of L-type Ca^{2+} (Spelbrink et al. 2004) and sodium channels (Kushmerick et al. 1998), cytotoxic activity on human tumor cell lines (da Rocha Pitta et al. 2010; Anaya-Lopez et al. 2006; Pellegrini and Franco 2005) and even on particular plant cells (de Zélicourt et al. 2007) has also been reported.

Defensins of cultivated cereals (wheat, barley, sorghum, rice, and maize) studied so far represent a unique group in the sense that they lack substantial antifungal activity. Thus, Osborn et al. (1995) compared the antifungal and antibacterial activities of defensins from plants belonging to the families Asteraceae, Fabaceae, Hippocastanaceae, and Saxifragaceae with those of γ-1 purothionin from wheat and three defensins from *Sorghum bicolor*. They showed that of the five fungi tested, only SIαI from *S. bicolor* possessed any antifungal activity at concentrations higher than 50 μg/ml and only on three of the fungi. Sorghum defensins were reported to inhibit insect gut α-amylases (Bloch and Richardson 1991). Barley defensins were demonstrated to inhibit protein synthesis in both eukaryotic and prokaryotic cell-free systems (Mendez et al. 1996). Gamma-hordothionin also inhibited the α-amylase activity from human saliva (Mendez et al. 1990). Maize defensins (γ-zeathionins) were shown to block sodium channels in animal cells (Kushmerick et al. 1998). Our data showed that defensins from *Triticum kiharae*, a synthetic allopolyploid produced by crossing *Triticum timopheevii* and *Aegilops squarrosa* and *T. aestivum*, a cultivated wheat species, failed to inhibit growth of the filamentous fungi tested (Odintsova et al. 2008). However, defensins from a cereal weed species *Echinochloa crusgalli*, in contrast to cultivated cereals studied so far, possess rather potent antifungal activity (Odintsova et al. 2008). This finding represents a spectacular example of preservation of antifungal functions in defensins to achieve fitness, competitiveness, and adaptation in agroecosystems, which were largely lost during cultivation of cereals. Cultivation is aimed at maximizing the organ, which serves as food for humans. This breeding selection is achieved at the expense of other adaptive traits originally present in the wild forms including pathogen resistance; thus, cultivated plants are generally very sensitive to infection. In contrast, *E. crusgalli*, which is one of the world's worst weeds occurring in all agricultural regions, retained this useful property. We suppose that one of the possible mechanisms underlying successful competition of this weed with crops could be the preservation of the antifungal activity of defensins resulting in increased resistance. The latter property is likely to contribute to high adaptation of this weed species to constantly changing environments.

In a limited number of studies, defensins were shown to be involved in abiotic stress response. For example, Koike et al. (2002) observed expression activation of the defensin-like gene Tad1 during cold acclimation of winter wheat, suggesting its role in low-temperature-induced resistance to pathogens during winter hardening. In another study, screening of the cDNA library of the zinc hyperaccumulating plant *Arabidopsis halleri* in the yeast *Saccharomyces cerevisiae* revealed four defensin genes (AhPDFs), which confer zinc tolerance to yeast cells, providing evidence for the role of defensins in metal physiology (Mirouze et al. 2006). In another study, a defensin gene PDF1.2 was shown to be constitutively expressed at a high level in the halophyte *Thellungiella halophila* (Taji et al. 2004). Using SuperSAGE (Matsumura et al. 2003)—an optimized Serial Analysis of Gene Expression (SAGE) technology for transcriptome profiling—a search for AMP transcripts in different tissues of Brazilian plant crops, cowpea (*Vigna unguiculata*), soybean (*Glycine max*), and sugarcane hybrids (*Saccharum spp.*) was carried out, and tags annotated as defensins were demonstrated to be differentially expressed in response to abiotic stress (Kido et al. 2010). The expression of defensin genes in response to biotic and abiotic factors illustrates cross talk between different signal transduction pathways and gene expression programs.

Structure–function relationships remain the most intriguing issue in defensin research, providing clues for identification of the determinants of diverse biological activities displayed by different members of the defensin family. Several approaches have been used to address the problem (1) comparison of amino acid sequences of highly homologous peptides contrasting in biological (e.g., antifungal) activity, (2) site-directed mutagenesis, (3) generation of chimeric molecules combining different portions of highly homologous peptides that differ in biological activity (strong or weak). The pioneering work exploiting the first two approaches was performed by the Belgian group using radish defensins as a model (De Samblanx et al. 1997). Two adjacent sites important for the antifungal activity were identified. The first site comprises the loop connecting β-strands 2 and 3, and the second site is composed of the loop connecting β-strand 1 and the α-helix with residues on the α-helix and β-strand 3 (De Samblanx et al. 1997). An interesting conclusion followed from the comparison of amino acid sequences of the two highly homologous Poaceae defensins: the *E. crusgalli* D1 defensin (Odintsova et al. 2008) and the *T. kiharae* defensin Tk-AMP-D1 (Odintsova et al. 2007). The former exhibited potent antifungal activity against the fungal pathogens, while the latter was only very weakly active. Sequence comparison showed that the main variation in the amino acid sequences between D1 of *E. crusgalli* and Tk-AMP-D1 of *T. kiharae* resides in the loop connecting the α-helix and β-strand 2, in β-strand 2, and in the loop connecting β-strand 2 and β-strand 3. From these observations, it was hypothesized that just these regions are responsible for the higher antifungal activity of *E. crusgalli* D1 defensin compared to *T. kiharae* Tk-AMP-D1 defensin. Similar conclusions were drawn from the comparison of two structurally related *Medicago sativa* (MsDef1) and *Medicago truncatula* (MtDef2) defensins (Spelbrink et al. 2004), which share 65% amino acid sequence homology but differ considerably in antifungal activity. Testing of the antifungal activity of the chimeric molecules containing portions of both peptides in different combinations showed that the major determinants of the antifungal activity are located in the C-terminal part of the defensin molecule and that the amino acid residue arginine in MsDef1 is crucial for its antifungal activity.

In addition to the role in defense, certain plant defensins were implicated in other cellular functions. The role of plant defensins in growth and development was demonstrated for *A. thaliana* defensins (Allan et al. 2008). Dual function in defense and development has been recently suggested for tomato defensins (Stotz et al. 2009a). Several defensin-like genes were shown to be involved in signaling between male and female gametophytes (Okuda et al. 2009).

The exact mechanisms of action of most plant defensins are poorly understood. The mode of action of antifungal defensins from radish RsAFP2 and dahlia DmAMP1 has been studied in more detail (Thevissen et al. 2004). Evidence was provided that they interact with specific targets on the fungal membrane, since high-affinity binding sites on fungal cells and membrane fractions of yeasts composed of specific sphingolipids or glucosylceramides for Dm-AMP1 and Rs-AFP2, respectively, were identified. The radish defensin RsAFP2 was shown to induce ROS production in *C. albicans* leading to membrane permeabilization (Aerts et al. 2007). For some members of the defensin family, penetration through the membrane and interaction with intracellular targets have been demonstrated (van der Weerden et al. 2008; Lobo et al. 2007). Thus, the antifungal activity of the plant defensin NaD1 was shown to involve specific interaction with the fungal cell wall, followed by permeation of the plasma membrane and entry of NaD1 into the cytoplasm, resulting in its granulation and subsequent cell death (van der Weerden et al. 2010). The defensin Psd1 from *Pisum sativum* interacts with an intracellular target, cyclin F related to regulation of the cell cycle in *Neurospora crassa*, thus interfering with its normal progression (Lobo et al. 2007). Functional studies of the MsDef1 defensin from *Medicago sativa* demonstrated that it blocked the L-type Ca^{2+} channel in mammalian cells, it was further suggested that it acts similarly in fungal cells (Spelbrink et al. 2004). Since Ca^{2+} is a key regulator of vital processes in fungi including hyphal elongation, the blockage of Ca^{2+} channels is likely to lead to abnormal hyphal growth (hyperbranching).

Most plant defensins are synthesized as precursor proteins, which consist of a signal peptide and the mature peptide domain. The signal peptide targets the defensin molecule to the extracellular space. Some flower defensins additionally possess a C-terminal prodomain. In most defensins, the signal peptide is acidic. By analogy with mammalian defensins, it has been suggested that it neutralizes the toxic effect of the mature peptide.

More than 300 putative defensin-like genes were discovered in the *Arabidopsis* genome (Silverstein et al. 2005), most of them form clusters suggesting common origin by successive gene duplication events. Each gene consists of two exons and a single intron in the signal peptide domain. In a single case, two introns in the defensin gene were reported (Carvalho and Gomes 2009). It is unclear whether all of the identified genes are expressed and under which conditions. Gene expression studies show that defensin genes usually demonstrate complex developmentally regulated and tissue-/organ-specific expression patterns (Thomma et al. 2002). Each organ expresses at least one defensin gene (Stotz et al. 2009b). There is considerable variation among species. Thus, in the wild species *E. crusgalli*, *N. sativa*, and *Stellaria media*, two defensins were found in dormant seeds (Rogozhin et al. 2011; Slavokhotova et al. 2011; Odintsova et al. 2008), while in the polyploid wheat species *T. kiharae* and *T. aestivum*, 13 defensins were discovered (Odintsova et al. 2007; Egorov et al. 2005). They form three groups (I–III) of structurally related polypeptides. Group III defensins revealed striking similarity with *Hordeum vulgare* defensins, suggesting their ancient origin before the divergence of the *Triticum* and *Hordeum* genera. Analysis of defensins of diploid *Triticum* and *Aegilops* species, the putative genome donors to polyploid wheat, showed that they possessed defensins identical to those found in polyploid wheat species. These data point to high conservation of the defensin structure in the evolution of the Poaceae family and suggest a vital function of these molecules. However, the diploid species have a smaller number of defensin polypeptides than polyploid wheats. This finding indicates that the diversity of defensins revealed in *T. kiharae* and *T. aestivum* is partially due to genome complexity of polyploid wheat species arising from a combination of three related genomes.

Defensin genes were used for genetic transformation of plants to increase pathogen resistance (Table 2). Diverse transgenes as well as different recipient plants were employed (Stotz et al. 2009b; Lay and Anderson 2005). In a number of cases, elevated resistance to particular pathogens was achieved in transgenic plants overexpressing a defensin gene. These data confirm the protective role of defensins in vivo. However, in some instances, morphological changes in transgenic plants were noted, as well as the adverse effect of a constitutively expressed transgene on the transgenic plant physiology (cell growth, fertility, regeneration efficiency) (Stotz et al. 2009b).

5.2 Thionins

Thionins are small cysteine-rich AMPs, which were first discovered in wheat flour more than half a century ago. Later, they were discovered to be ubiquitous among the plant kingdom. About 100 sequences of thionins from about 15 plant species have been reported (Stec 2006). They were found in seeds, roots, stems, and leaves. According to the number of cysteine residues, thionins are separated into two classes: those with 6 and 8 cysteine residues (Broekaert et al. 1997). Five classes of thionins are distinguished based on amino acid sequences and disulfide bond patterns with high sequence homology between members of one class. Thionins of class I were discovered in cereal seeds. They belong to the 8-Cys thionins, consist of 45 amino acid residues, and are highly positively charged. Class II thionins were isolated from the parasitic plant *Pyrularia pubera* and from *H. vulgare* leaves. They also belong to

Table 2 Examples of AMPs used for plant transformation resulting in increased pathogen resistance

Name	Family	Donor plant	Recipient plant
Rs-AFP2	Defensin	Radish	Tobacco, tomato, oil rape
Alf-AFP	Defensin	Alfalfa	Potato
DRR230-a	Defensin	Pea	Tobacco, canola
Dm-AMP1	Defensin	Dahlia	Eggplant
BSD1	Defensin	Chinese cabbage	Tobacco
Spi1	Defensin	Norway spruce	Tobacco
WT1	Defensin	Wasabi	Rice
Pn-AMP2	Hevein-like	Morning glory	Tobacco, tomato
Hordothionin	Thionin	Barley	Tobacco
Alpha-thionin	Thionin	Barley	Tobacco
Thionin	Thionin	Arabidopsis	Arabidopsis
Thionin	Thionin	Oat	Rice
LTP	Lipid-transfer protein	Barley	Arabidopsis, tobacco
Ace-AMP1+chitinase	Lipid-transfer protein	Onion	Carrot
LTP	Lipid-transfer protein	Motherwort	Tobacco

the 8-Cys thionins and contain 46–47 amino acid residues; although their net charge is positive, it is lower than that of class I thionins. Thionins of class III were found in different mistletoe species, they belong to the 6-Cys thionins, consist of 45–46 amino acid residues, and the charge of their molecule is similar to that of class II thionins. Thionins of class IV were isolated from *Crambe abyssinica* seeds, and they possess 46 amino acids, belong to the 6-Cys peptides, and are neutral. Thionins of class V are represented by a wheat thionin which lacks the C-terminal region present in class I and II thionins and, in contrast to other thionins, is nontoxic.

By X-ray analysis, the crystal structures of 6-Cys crambin from *Cr. abyssinica*, 8-Cys thionins of wheat and barley, and viscotoxin A3 were determined (Stec 2006). Several thionin solution structures were resolved by NMR spectroscopy (Stec 2006). In all instances, the three-dimensional structure of thionins was shown to resemble the letter L, in which the long arm is formed by two antiparallel α-helices, and the short arm, by two β-strands. A groove between these two regions plays a functional role (Stec et al. 2004). Comparison of amino acid sequences of different thionins revealed several conserved amino acid residues besides the cysteine residues. To elucidate the role of these residues in thionin functionality, a peptide corresponding to positions 7–15 of the thionin with a substitution of cysteine at position 12 with a serine was synthesized. It exhibited less tight binding to membrane phospholipid, phosphatidylserine, than the native thionin. Lys1 and Arg10 constitute a phosphate-binding site, while Ser2 and Tyr13 comprise a glycerol-binding site. The position 8 is important for antimicrobial activity. The synthesis of the central antiparallel double helix of the class II *Py. pubera* thionin resulted in a molecule that was fully active against most tested pathogens; thus, this structural element is likely to play a key role in the thionin antimicrobial activity (Vila-Perello et al. 2005).

As early as in 1942, Stuart and Harris (1942) demonstrated that thionins inhibited growth of fungi and bacteria in vitro. Similar effects on phytopathogenic bacteria were later shown providing evidence for the role of these peptides in antimicrobial defense in planta. Subsequent studies confirmed high antimicrobial activity of thionins both against Gram-negative and Gram-positive bacteria, phytopathogenic fungi, and oomycetes; the IC_{50} (concentration of a peptide that causes 50% inhibition of pathogen growth) ranges from 1 to 15 μg/ml depending on the thionin–pathogen combination (da Rocha Pitta et al. 2010; Stec 2006). The antimicrobial activity of thionins is enhanced in the presence of other proteins, such as 2S albumins, lipid-transfer proteins, protease inhibitors, and wheat puroindolines. Some Gram-negative bacteria are insensitive to thionins. In has been suggested that this may be due to the protective role of cell-wall lipopolysaccharides (Titarenko et al. 1997). Some thionins were shown to inhibit human pathogens, for example, a thionin Thi2.1 from *A. thaliana* expressed in bovine epithelial cells was active against *C. albicans* (Loeza-Ángeles et al. 2008). Wheat thionins exhibited leishmanicidal activity most likely associated with permeabilization of the plasma membrane leading to the immediate death of the parasite (da Rocha Pitta et al. 2010). Thionins were also shown to be toxic to some other types of cells, such as mammalian, insect, and plant cells (da Rocha Pitta et al. 2010; Loeza-Ángeles et al. 2008; Stec 2006). High antimicrobial activity of thionins makes them promising templates for designing highly active chemical analogues of a more simple structure (Vila-Perello et al. 2006).

The protective role of thionins in vivo is supported by their upregulation by pathogens, localization in tissues and cells primarily exposed to pathogens, and enhanced pathogen resistance of transgenic plants constitutively expressing thionin genes. For example, high-level expression of a barley thionin gene in transgenic tobacco conferred resistance to *Pseudomonas syringae* (Carmona et al. 1993). Transgenic rice plants overexpressing an oat thionin gene became more resistant to bacterial infection (Iwai et al. 2002).

The antimicrobial effect of thionins is associated with membrane permeabilization as was convincingly shown for cereal thionins (Thevissen et al. 1996). Several models for direct interaction of thionins with membranes have been suggested (Stec et al. 2004). For viscotoxin A3, the formation of pores in membranes was shown. In addition to affect the integrity of membranes, thionins probably have intracellular targets. They were shown to activate endogenous phospholipase A2 and adenylate cyclase. Wheat β-purothionin inhibits protein kinase C and binds to calmodulin. Based on similarities between the helix-turn-helix (HTH) motifs of thionins and the HTH DNA-binding proteins, direct interaction of thionins with DNA and RNA was hypothesized (Li et al. 2002). Purothionins inhibit ribonucleotide reductase and β-glucuronidase. In addition to direct antimicrobial function, thionins display an immunomodulatory activity (Carvalho and Gomes 2007).

Thionins are synthesized as precursor proteins containing a signal peptide, the mature peptide domain and a C-terminal prodomain. Cleavage of the signal peptide results in thionin activation. The C-terminal domain neutralizes the toxicity of the mature peptide domain (Padovan et al. 2010).

5.3 *Plant Nonspecific LTPs*

Plant nonspecific lipid-transfer proteins (nsLTPs) are basic proteins (7–10 kDa) that increase the intermembrane exchange and(or) transfer of polar lipids, fatty acids, or acyl-CoA in vitro (Yeats and Rose 2008). Since these proteins display no selectivity with respect to polar lipids, they were called nonspecific lipid-transfer proteins. In vitro bioassays showed that some nsLTPs display antimicrobial activity (da Rocha Pitta et al. 2010), while others inhibit animal α-amylases (Zottich et al. 2011). Diverse other functions have been suggested for these proteins, and they include synthesis of cutin, beta-oxidation, somatic embryogenesis, and plant signaling (Carvalho and Gomes 2007).

Based on structural properties, plant LTPs are subdivided into two families. LTPs of family 1 are larger (molecular weight of about 9 kDa) than family 2 LTPs (about 7 kDa). Amino acid sequence homology between the two LTP families is rather low (about 30% sequence identity), although the cysteine signature is similar, the disulfide bond pattern is different: in family 1 LTPs, Cys5 is connected with Cys8, and Cys6 with Cys1, while in family 2 LTPs, Cys5 is linked to Cys1, and Cys6 to Cys8. Several conserved amino acid residues have been identified in the LTP1 family, Asp43, Arg44, and Lys52 are among them. The three-dimensional structure of several LTP1s was determined by X-ray crystallography or NMR spectroscopy, and compact globular structure of their molecules stabilized by four disulfide bridges was revealed (Fig. 1). The tertiary structure of LTPs belonging to family 2 is much less well studied and was determined for rice and wheat proteins. The similarity between the protein scaffold for both LTP1 and LTP2 proteins was demonstrated. A common feature of plant LTPs belonging to both families is the presence of a hydrophobic tunnel-like cavity formed by four α-helices held together by four disulfide bridges (Yeats and Rose 2008). In LTP1 family, this cavity is covered by a lateral chain of amino acid residues such as Ala, Arg, Ile, Leu, Lys, Pro, Ser, Thr, Tyr, and Val and has two entrances (Carvalho and Gomes 2007). Two charged amino acid residues Arg44 and Lys35 are located at the larger entrance and are supposed to interact with lipids. In LTP2 proteins, instead of the hydrophobic tunnel, there is a triangular hollow box. This box is more flexible than the cavity in LTP1 proteins. The ability of the cavity to accommodate hydrophobic molecules is crucial for the lipid-transfer activity of LTPs. The LTP1 from *Allium cepa* seeds differs from other members of the family in that its cavity is filled with bulky side chains of aromatic residues, and it is therefore unable to bind and transport free lipids (Yeats and Rose 2008). Cereal LTPs can accommodate one or two lipid molecules, while *Nicotiana tabacum* LTP1 binds only one. The LTP2 proteins transfer lipids more efficiently than those of the LTP1 family likely due to a more flexible hydrophobic cavity. LTPs differ considerably in antimicrobial activity. Thus, a mung bean LTP was active against a broad spectrum of pathogens including *Fusarium solani*, *Fusarium oxysporum*, *Pythium aphanidermatum*, *Sclerotium rolfsii*, and *Staphylococcus aureus* (Wang et al. 2004). High antimicrobial activity was shown for the onion LTP Ace-AMP1 (IC_{50} from 1 to 6 μg/ml), while the radish LTP was only moderately active (IC_{50} from 7 to 100 μg/ml). The wheat and maize LTPs displayed rather low antifungal activity against the pathogens tested (Rogozhin et al. 2009). In addition to antifungal activity, some LTPs also inhibit growth of pathogenic bacteria (da Rocha Pitta et al. 2010). The wheat LTP increases antifungal activity of thionins. It is of particular interest that Ace-AMP1 has no effect on the antimicrobial activity of thionins that implies that both proteins differ in the mode of action. Antimicrobial activity of family 2 LTPs was not demonstrated. The reason for the considerable variation in antimicrobial activity of plant LTPs remains largely unknown. In the rice LTP1, ten specific residues important for the antimicrobial activity have been identified by site-directed mutagenesis (Tassin et al. 1998).

It is generally accepted that LTPs target microbial phospholipid membranes. For wheat LTPs, interaction with phospholipid monolayers was demonstrated; however, in contrast to the barley LTP, these proteins failed to penetrate through the phospholipid bilayers (Tassin et al. 1998) that correlated with higher antifungal activity of the barley protein. The protein fraction enriched in the chilli pepper LTP was shown to permeabilize yeast plasma membranes as was demonstrated by SYTOX Green assay (Diz et al. 2006). The LTP1 from *Helianthus annuus* produced the same effect on *F. solani* spores, resulting in growth inhibition and decreased spore viability (Regente et al. 2005).

Data on the role of LTPs in signaling is accumulating (Yeats and Rose 2008), and *T. aestivum* LTP was shown to bind with high affinity to the tobacco membrane at the same site as elicitins, signaling polypeptides secreted by oomycetes, which induce hypersensitive response and SAR in tobacco plants. A recombinant LTP1 from *N. tabacum* was demonstrated to bind jasmonic acid, and the complex produced was able to bind the elicitin receptor.

Most LTPs are located extracellularly; however, the LTP1 from *Ricinus communis* was shown to be located in the glyoxysome and possibly involved in beta-oxidation. In *T. aestivum*, LTP1s were found in aleurone granules, and in *V. unguiculata*, they were discovered in protein storage vacuoles (Carvalho and Gomes 2007). In *B. oleracea*, LTPs were associated with the waxy surface of leaves and were presumably involved in transport of cutin monomers.

Differential expression of LTP genes has been shown for a number of plant species and plant tissues at different developmental stages and physiological conditions including stress responses (Jung et al. 2005). Expression of LTP genes was studied in cowpea (*V. unguiculata*) in different tissues of adult plants under normal and stress conditions (Jung et al. 2003). Three LTP mRNAs obtained from pepper (*Capsicum annuum*) were differentially expressed in leaves, stems, and fruit tissues in response to different pathogens. Two of them were also induced by abiotic stress factors, such as drought, high salinity, low temperature, and wounding. Several regulatory elements in the promoter region of the *C. annuum* LTP1 gene have been identified (Jung et al. 2005). Among them are the low-temperature-responsive element, the drought-responsive element, the pathogen-responsive element, and the ethylene-responsive element, suggesting the role of LTPs in plant stress response.

Transformation of plants with LTP genes was performed in a number of plant species, which resulted in increased resistance to pathogens (Kido et al. 2010). For example, transformation of tobacco plants with a barley LTP2 gene enhanced resistance to *Ps. syringae* as evidenced by a decreased number of necrotic lesions in transgenic plants and their reduced size. Transgenic wheat plants, overexpressing Ace-AMP1, showed enhanced resistance to *Blumeria graminis* f. sp. *tritici* and *Neovossia indica* (Carvalho and Gomes 2007), which correlated with an induction of several defense-related genes (Phenylalanine Ammonia Lyase, PR-2 and PR-3).

In completely sequenced genomes of *A. thaliana*, rice, and poplar, about a dozen LTP1-encoding genes have been identified. Homologous sequences were found in gymnosperms but were not discovered in mosses. LTPs of family 2 are, in all probability, restricted to angiosperms (Yeats and Rose 2008). LTPs are produced as precursor proteins with a signal peptide, suggesting targeting to the apoplast, and only Ace-AMP1 from *Al. cepa* has an additional C-terminal propeptide (Carvalho and Gomes 2007).

5.4 Hevein-Like AMPs

Hevein-type AMPs show structural similarity to hevein, the 43-amino-acid residue chitin-binding peptide isolated from the rubber tree *Hevea brasiliensis* L. (Van Parijs et al. 1991). They comprise a single-hevein-domain subfamily in a large group of chitin-binding proteins (Beintema 1994; Raikhel and Lee 1993), which share a common property, the ability to bind chitin, a β-1,4-linked polymer of *N*-acetylglucosamine (GlcNAc), and related glycoconjugates containing GlcNAc or *N*-acetyl-D-neuraminic acid (NeuNAc). Since chitin does not occur in higher plants but instead is a component of cell walls of fungi and exoskeleton of invertebrates, such as insects and nematodes, it has been hypothesized that chitin-binding proteins are involved in plant defense against microorganisms and pests. All known chitin-binding proteins contain a common structural motif of 30–43 amino acid residues with several cysteine and glycine residues at conserved positions named the chitin-binding domain, which is responsible for chitin binding. Besides hevein, the members of this family are lectins, chitinases, some wound-induced proteins, and a number of AMPs (Raikhel and Lee 1993), later classified as "hevein-type." Although hevein-type AMPs share some sequence homology, they differ in the number of disulfide bonds. Most of them possess 8 cysteine residues forming 4 disulfide bonds and, in this respect, are close to the chitin-binding domains of class I/IV chitinases (Beintema 1994; Raikhel and Lee 1993). Two hevein-like peptides Pn-AMP1 and Pn-AMP2 isolated from *Pharbitis nil* (Koo et al. 1998) and oat avesin (Li and Claeson 2003) belong to 8-Cys peptides. Truncated variants with only 6 cysteine residues also occur, among them the hevein-like peptides from the *Amaranthus* genus: Ac-AMP1 and Ac-AMP2 from *Amaranthus caudatus* seeds (Broekaert et al. 1992), Ay-AMP from *Amaranthus hypochondriacus* (Rivillas-Acevedo and Soriano-Garcia 2007), and Ar-AMP from *Amaranthus retroflexus* (Lipkin et al. 2005), as well as the IWF4 from sugar beet leaves (Nielsen et al. 1997). In the 6-Cys hevein-like peptides, the C-terminal region of the canonical hevein-like domain is missing. Three 10-Cys hevein-like AMPs that differ from each other in the cysteine motifs have been described to date (Odintsova et al. 2009; Huang et al. 2002; Van den Bergh et al. 2002). Two of them were isolated from the bark of trees (*Euonymus europaeus* L. and *Eucommia ulmoides*) (Huang et al. 2002; Van den Bergh et al. 2002). Another structural variant of 10-Cys AMPs represented by two nearly identical peptides WAMP-1a and WAMP-1b (from "wheat AMP") was isolated from seeds of *T. kiharae* (Odintsova et al. 2009). The characteristic feature of the molecular scaffold of the 10-Cys hevein-like AMPs is the presence of an additional fifth disulfide bond (Fig. 1), although its location is radically different in these peptides. Wheat peptides and Ee-CBP from *Eucommia ulmoides* show striking similarity to the hevein-like domains of cereal class I chitinases and *Eucommia ulmoides* chitinase, respectively, both in terms of amino acid sequence and cysteine patterns (Odintsova et al. 2009; Van den Bergh et al. 2004). The functional significance of the effect of additional cysteine residues on the structure and functions of hevein-type AMPs remains to be elucidated. In WAMPs, the fifth disulfide bond brings together the N- and C-terminal regions of the polypeptide chain, enriched in basic amino acids (Arg3, Arg9, Lys12, Arg42, and Arg45 in WAMP-1b) resulting in a cluster of basic residues. The second structural peculiarity of WAMPs that discriminates these peptides from all known chitin-binding polypeptides is the unique structure of the chitin-binding site in which a conserved serine residue at position 20 is substituted for glycine, although three aromatic residues (Tyr22, Phe24, and Tyr31) are well conserved. Analysis of chitin-binding properties of WAMPs in vitro showed that both peptides bind chitin, demonstrating that the serine/glycine substitution is not crucial for binding, although its precise role in the efficiency of binding remains to be explored.

For several hevein-like AMPs (Ac-AMP1, Ar-AMP, Pn-AMP, and IWF4), it was shown that they are produced as precursor proteins of a similar structure and consist of a signal peptide, the mature peptide, and the C-terminal prodomain of about 30 residues long. The 10-Cys peptide Ee-CBP from *Euonymus europaeus* is synthesized as a chitinase-like precursor; instead of a short C-terminal prodomain, it contains a long chitinase domain, which is linked to the hevein domain by a hinge region and cleaved during proteolytic processing (Van den Bergh et al. 2004). In contrast to Ee-CBP, WAMPs are synthesized as preproproteins containing a putative signal peptide of 34 amino acids, the mature peptide domain (45 residues), and a short C-terminal prodomain (37 amino acids) (our unpublished data). However, the WAMP cDNA sequences are also structurally related to portions of chitinase genes, as was shown by alignments of translated cDNA sequences with protein database disclosing "the remnants" of the chitinase gene. The homologous regions covered the hevein-like domain and the C-terminal regions (~25 C-terminal amino acid residues) of the catalytic domains of the enzymes. This may suggest a mechanism by which WAMP genes arose. One might speculate that they originated from class I/IV chitinase genes by a similar mechanism that includes a frame-shift deletion of a considerable portion (~700 bp) of the catalytic domain (our unpublished data). It is of interest that a WAMP homologue termed LAMP was discovered in a wild cereal species *Leymus arenarius*, providing evidence that 10-Cys hevein-like AMPs have a wider distribution in the Poaceae family than previously expected (Utkina et al. 2010).

The three-dimensional structures of several chitin-binding proteins have been determined, and these include hevein, the Ac-AMP2 from *Amaranthus caudatus*, wheat hemagglutinin containing four chitin-binding domains, *Urtica dioica* agglutinin composed of two hevein domains, the 10-Cys peptide from *Euc. ulmoides* (Huang et al. 2004; Xiang et al. 2004), and WAMP-1a (Fig. 1). The fold of the hevein domain is similar in all chitin-binding proteins. Analysis of the three-dimensional structure of hevein complexes with oligosaccharides revealed the amino acid residues of the peptide involved in specific hydrophobic interactions with oligosaccharides (Trp21, Trp2, and Trp30). These interactions are stabilized by a hydrogen bond with a Ser19. Homologous conserved residues were identified in other chitin-binding proteins, except for Ser19 in WAMPs from wheat seeds. 1H-NMR was used to determine the 3D-structure of WAMP-1a (Dubovskii, Balashova, Vassilevski, Odintsova, Slavokhoptova, Grishin, Egorov, Arseniev, unpublished data). The peptide exhibits a stable and compact fold due to the presence of an array of disulfide and hydrogen bonds, and, likely, several salt bridges, and forms four antiparallel β-strands and two helical turns. Similar fold and position of disulfide bridges are found in chitin-bonding domains of cereal class I chitinases. The three-dimensional structures of the 10-Cys peptides EAFP1 and EAFP2 from *Euc. ulmoides* were determined by X-ray analysis and NMR spectroscopy. The amphipathic topology of the peptides was revealed, disclosing a hydrophobic cluster containing a chitin-binding site and a cluster of positively charged residues (Huang et al. 2004; Xiang et al. 2004). The occurrence of these clusters probably accounts for the dual antimicrobial activity of this peptide both against chitin-containing and nonchitin pathogens.

The antimicrobial activity of chitin-binding proteins was first reported for *U. dioica* lectin. Subsequent studies confirmed antimicrobial activity of hevein-like AMPs, although hevein displays rather low inhibitory activity (Van Parijs et al. 1991). Some chitin-binding proteins are inactive against microorganisms but exhibit insecticidal activity (Raikhel and Lee 1993); the wheat lectin WGA is a spectacular example. The range of fungi sensitive to hevein-like peptides is rather broad and includes both filamentous fungi and yeasts. Some hevein-like AMPs are active against bacteria as well. The Ac-AMP1 and Ac-AMP2 inhibit growth of Gram-positive bacteria but have no effect on Gram-negative bacteria (Broekaert et al. 1992). The WAMP-1a peptide is active both against Gram-positive and Gram-negative bacteria (Odintsova et al. 2009). The antifungal activity of Ee-CBP is increased in the presence of endogenous chitinase (Van den Bergh et al. 2004).

The mode of action of hevein-like chitin-binding AMPs is poorly understood. It is supposed that they bind cell-wall chitin in the growing fungal hyphae, thus disturbing normal cell wall morphogenesis. However, this does not explain their activity against chitin-free (e.g., oomycetes and bacteria) pathogens. For the hevein-like peptide Pn-AMP1 from *Pharbitis nil*, Pn-AMP1-induced actin depolarization was shown to account for growth arrest in yeasts *Saccaromyces cerevisiae* and *Candida albicans*, and this process is mediated by cell-wall mannoproteins rather than chitin (Koo et al. 2004).

Transgenic tomato and tobacco plants overexpressing the hevein-like peptide Pn-AMP2 were obtained and shown to be resistant to a number of chitin-containing and chitin-free pathogens (Lee et al. 2003; Koo et al. 2002).

5.5 *Knottin-Like and Macrocyclic Peptides (Plant Cyclotides)*

Knottin-like peptides have six disulfide-linked cysteine residues forming a specific structure, the so-called cystine knot, in which a ring formed by two disulfide bridges is penetrated by a third disulfide bond. It is interesting that the cystine knot is present in functionally diverse proteins from a wide range of organisms, such as microorganisms, insects, and plants. Knottin-like peptides named Mj-AMP1 and Mj-AMP2 were first isolated from seeds of the plant *Mirabilis jalapa* (Cammue et al. 1992). The peptides are highly homologous and differ by only four amino acid residues. They inhibited in vitro a wide range of phytopathogenic fungi (*Botrytis cinerea*, *Colletotrichum lindemuthianum*, and *Venturia inaequalis*) and Gram-positive bacteria. However, transgenic plants constitutively expressing these peptides did not show enhanced resistance to *B. cinerea* and *Alternaria longipes*. Another knottin-like peptide PAFP-s was isolated from *Phytolacca americana* (Shao et al. 1999). It displayed antifungal activity against *F. oxysporum* and *Trichoderma viride* (Shao et al. 1999). The α-amylase inhibitor from amaranth also belongs to the family of knottin-like peptides (Chagolla-Lopez et al. 1994).

Plant cyclic peptides (cyclotides) were discovered over a decade ago (Schöpke et al. 1993). They represent small molecules of about 30 amino acid residues with a head-to-tail cyclized backbone and six highly conserved cysteine residues (Craik 2001). Similarly to knottin-like AMPs, the crucial core of cyclotides is a cystine knot. The consensus three-dimensional structure of plant cyclotides includes a small beta-sheet, which is tightly associated with the cystine knot. Several residues that are important for the compact structure of cyclotides are conserved in different members of the family, such as a Glu residue in loop 1 and a hydroxyl-containing Thr or Ser immediately following this Glu residue. Due to the peculiarities of the protein scaffold including the cyclic backbone, cyclotides display exceptional stability to proteolytic degradation and chemical and thermal treatments. Cyclotides are widely distributed among the Rubiaceae and Violaceae families, and recently, homologous sequences have been discovered in the Poaceae and Apocynaceae families, suggesting their wider distribution throughout the plant kingdom. Cyclotides are synthesized as larger precursors containing a signal peptide, a short hydrophobic C-terminal region, and one to three mature cyclotide domains that are each preceded by an N-terminal repeat domain. Posttranslational processing includes enzymatic cleavage, disulfide bond formation, and cyclization. Cyclotides exhibit a number of functions such as insecticidal, antimicrobial, anti-HIV, hemolytic, and uterotonic activities (Craik et al. 2007). All these activities are likely to be associated with membrane disruption. Neither the precise mode of action nor their roles in plants have been elucidated, although significant insecticidal activity of cyclotides points to their role in plant defense. Expression of cyclotide-like genes was analyzed in different tissues of barley and rice, and tissue-specific expression pattern was demonstrated. In barley, the highest expression level was observed in fully expanded tissues; in rice, low transcription levels were recorded in all tissues, except for mature root.

Due to exclusive structural and chemical stability, plant cyclotides are regarded as templates into which diverse biological activities may be inserted (Craik et al. 2007; Craik 2001).

6 Therapeutical and Biotechnological Applications

The increasing resistance of pathogenic microorganisms to conventional antibiotics has become a growing threat for humans for nearly as long as people have been using them. Drug resistance is an especially difficult problem for immunocompromised patients. Heavy use of antibiotics in these patients increases selection pressure for mutations conferring drug resistance. According to the data collected by the Centers for Disease Control and Prevention in the United States, more than 70% of the bacteria that cause hospital-acquired infections are resistant to at least one of the antibiotics most commonly used to treat them (http://www.cdc.gov/drugresistance/index.htm). Antimicrobial resistance is driving up health-care costs and increases the severity of disease and death rates. The necessity to combat the problem of antimicrobial resistance stimulated a search for new antimicrobial agents active against resistant pathogen strains. Naturally occurring AMPs including those of plant origin have prospects as new antibiotics due to their rapid action and broad-spectrum antimicrobial activity (da Rocha Pitta et al. 2010; Hancock and Sahl 2006). An obvious advantage of AMPs is that resistance is clearly more difficult to achieve than for conventional antibiotics. The reason for this is that they act on membranes of microorganisms by disrupting their integrity. However, to date, only four cationic peptides based on animal AMPs have advanced into phase 3 clinical-efficacy trials (Hancock and Sahl 2006). Among the limitations of cationic AMPs for clinical use are their possible insufficiently investigated toxicity for humans, potential susceptibility for proteases, and the high cost of production. Several recombinant DNA methods have been utilized for manufacturing AMPs in bacteria, yeasts, plants, and animal cells. However, none of them proved commercially feasible. A promising approach consists in the use of shorter broadly active peptides. The potential strategies for the use of AMPs include their employment as single anti-infective agents or in combination with conventional antibiotics, or as immunomodulatory agents to enhance innate immunity responses (da Rocha Pitta et al. 2010). The latter application is mostly connected with AMPs from higher vertebrates, such as mammalian defensins and cathelicidins (Zaiou 2007; Hancock and Sahl 2006). Another important use of AMPs may be in the development of novel anticancer drugs, since several plant AMPs were shown to exhibit antiproliferative activity on human tumor cell lines (da Rocha Pitta et al. 2010).

Diseases induced by pathogenic microorganisms and pests cause considerable yield losses (up to 45%) in cultivated crops and decrease quality and safety of agricultural products. The use of chemically aggressive fungicides and pesticides reduces the losses only by 7%. The impressive achievements of genetic engineering provide new strategies for crop protection and production of highly productive, pathogen- and pest-resistant and environmentally friendly crops. AMPs with high broad-spectrum activity against phytopathogens represent a source of valuable genes for improvement of crops via genetic transformation. As single genes or in combination of two or more genes, they may be directly incorporated into the plant genome to enhance resistance to pathogens. Increased resistance to certain pathogens has been already achieved for several plant AMPs under laboratory conditions (Table 2). Another approach to enhance resistance consists in modulating expression of plant defense genes. Although quite intriguing, it is still in its infancy. Improved understanding of the pathways used to induce AMP production should provide additional tools for engineering resistance.

References

Aerts AM, Francois IE, Meert EMK, Li QT, Cammue BPA, Thevissen K (2007) The antifungal activity of RsAFP2, a plant defensin from *Raphanus sativus*, involves the induction of reactive oxygen species in *Candida albicans*. J Mol Microbiol Biotechnol 13:243–247

Aerts AM, Francois IEJA, Cammue BPA, Thevissen K (2008) The mode of antifungal action of plant, insect and human defensins. Cell Mol Life Sci 65:2069–2079

Ajesh K, Sreejith K (2009) Peptide antibiotics: an alternative and antimicrobial strategy to circumvent fungal infections. Peptides 30:999–1006

Allan A, Snyder AK, Preuss M, Nielsen EE, Shah DM, Smith TJ (2008) Plant defensins and virally encoded fungal toxin KP4 inhibit plant root growth. Planta 227:331–339

Anaya-Lopez JL, Lopez-Meza JE, Baizabal-Aguirre VM, Cano-Camacho H, Ochoa-Zarzosa A (2006) Fungicidal and cytotoxic activity of *Capsicum chinense* defensin expressed by endothelial cells. Biotechnol Lett 28:1101–1108

Beintema JJ (1994) Structural features of plant chitinases and chitin-binding proteins. FEBS Lett 350:159–163

Benko-Iseppon AM, Galdino SL, Calsa T Jr, Kido EA, Tossi A, Belarmino LC, Crovella S (2010) Overview on plant antimicrobial peptides. Curr Protein Pept Sci 11:181–188

Bloch CJ, Richardson M (1991) A new family of small (5 kDa) protein inhibitors of insect α-amylases from seeds of sorghum (*Sorghum bicolor* (L.) Moebch.) have sequence homologies with wheat γ-purothionins. FEBS J 279:101–104

Broekaert WF, Mariën W, Terras FRG, De Bolle MFC, Proost P, Van Damme J, Dillen L, Claeys M, Rees SB, Vanderleyden J, Cammue BPA (1992) Antimicrobial peptides from *Amaranthus caudatus* seeds with sequence homology to the cysteine-glycine-rich domain of chitin-binding proteins. Biochemistry 31:4308–4314

Broekaert WF, Terras FRG, Cammue BPA, Osborn RW (1995) Plant defensins: novel antimicrobial peptides as components of the host defense system. Plant Physiol 108:1353–1358

Broekaert WF, Cammue BPA, De Bolle MFC, Thevissen K, De Samblanx GW, Osborn RW (1997) Antimicrobial peptides from plants. Crit Rev Plant Sci 16:297–323

Broekaert WF, Terras FRG, Cammue BPA (2000) Induced and preformed antimicrobial proteins. In: Slusarenko AJ, Fraser RSS, van Loon LC (eds) Mechanisms of resistance to plant diseases. Kluwer, The Netherlands, pp 371–479

Cammue BPA, De Bolle MFC, Terras FRG, Proost P, Van Damme J, Rees SB, Vanderleyden J, Broekaert WF (1992) Isolation and characterization of a novel class of plant antimicrobial peptides from *Mirabilis jalapa* L. seeds. J Biol Chem 267:2228–2233

Carmona MJ, Molina A, Fernandez JA, Lopez-Fando JJ, Garcia-Olmedo F (1993) Expression of the α-thionin gene from barley in tobacco confers enhanced resistance to bacterial pathogens. Plant J 3:457–462

Carvalho AO, Gomes VM (2007) Role of plant lipid transfer proteins in plant cell physiology–A concise review. Peptides 28:1144–1153

Carvalho AO, Gomes VM (2009) Plant defensins–prospects for the biological functions and biotechnological properties. Peptides 30:1007–1020

Castro MS, Fontes W (2005) Plant defense and antimicrobial peptides. Protein Pept Lett 12:13–18

Chagolla-Lopez A, Blanco-Labra A, Patthy A, Sanchez R, Pongor S (1994) A novel α-amylase inhibitor from amaranth (*Amaranthus hypochondriacus*) seeds. J Biol Chem 269:23475–23680

Colilla FJ, Rocher A, Mendez E (1990) Gamma-purothionins: amino acid sequence of two polypeptides of a new family of thionins from wheat endosperm. FEBS Lett 270:191–194

Craik DJ (2001) Plant cyclotides: circular, knotted peptide toxins. Toxicon 39:1809–1813

Craik DJ, Clark RJ, Daly NL (2007) Potential therapeutic applications of the cyclotides and related cystine knot mini-proteins. Expert Opin Investig Drugs 16:595–604

da Rocha Pitta MG, da Rocha Pitta MG, Galdino SL (2010) Development of novel therapeutic drugs in humans from plant antimicrobial peptides. Curr Protein Pept Sci 11:236–247

Dangl JL, Jones JDG (2001) Plant pathogens and integrated defence responses to infection. Nature 411:826–833

De Samblanx GW, Goderis IJ, Thevissen K, Raemaekers R, Fant F, Borremans F, Acland DP, Osborn RW, Patel S, Broekaert WF (1997) Mutational analysis of a plant defensin from radish (*Raphanus sativus* L.) reveals two adjacent sites important for antifungal activity. J Biol Chem 272:1171–1179

de Zélicourt A, Letousey P, Thoiron S, Campion C, Simoneau P, Elmorjani K, Marion D, Simier P, Delavault P (2007) Ha-DEF1, a sunflower defensin, induces cell death in *Orobanche* parasitic plants. Planta 226:591–600

Diz MSS, Carvalho AO, Rodrigues R, Neves-Ferreira AGC, Da Cunha M, Alves EW, Okorokova-Façanha AL, Oliveira MA, Perales J, Machado OL, Gomes VM (2006) Antimicrobial peptides from chilli pepper seeds causes yeast plasma membrane permeabilization and inhibits the acidification of the medium by yeasts. Biochim Biophys Acta 1760:1323–1332

Egorov TA, Odintsova TI, Pukhalsky VA, Grishin EV (2005) Diversity of wheat antimicrobial peptides. Peptides 26:2064–2073

Farrokhi N, Whitelegge JP, Brusslan JA (2008) Plant peptides and peptidomics. Plant Biotechnol J 6:105–134

Garcia-Olmedo F, Molina A, Alamillo JM, Rodriguez-Palenzuela P (1998) Plant defense peptides. Biopolymers (Pept Sci) 47:479–491

Garcia-Olmedo F, Rodriguez-Palenzuela P, Molina A, Alamillo JM, Lopez-Solanilla E, Berrocal-Lobo M, Poza-Carrion C (2001) Antibiotic activities of peptides, hydrogen peroxide and peroxynitrite in plant defence. FEBS Lett 498:219–222

Hammani R, Hamida JB, Vergoten G, Fliss I (2009) PhytAMP: a database dedicated to antimicrobial plant peptides. Nucl Acids Res 963(Database Issue):8–15

Hancock REW, Sahl H-G (2006) Antimicrobial and host-defense peptides as new anti-infective therapeutic strategies. Nat Biotechnol 24:1551–1557

Huang R-H, Xiang Y, Liu X-Z, Zhang Y, Hu Z, Wang D-C (2002) Two novel antifungal peptides distinct with a five-disulfide motif from the bark of *Eucommia ulmoides* Oliv. FEBS Lett 521:87–90

Huang R-H, Xiang Y, Tu G-Z, Zhang Y, Wang D-C (2004) Solution structure of *Eucommia* antifungal peptide: a novel structural model distinct with a five-disulfide motif. Biochemistry 43:6005–6012

Iwai T, Kaku H, Honkura R, Nakamura S, Ochiai H, Sasaki T, Ohashi Y (2002) Enhanced resistance to seed-transmitted bacterial diseases in transgenic rice plants overproducing an oat cell-wall-bound thionin. Mol Plant Microbe Interact 15:515–521

Jones JDG, Dangl JL (2006) The plant immune system. Nature 444:323–329

Jung HW, Kim KD, Hwang BK (2003) Three pathogen-inducible genes encoding lipid-transfer proteins from pepper are differentially activated by pathogens, abiotic and environmental stresses. Plant Cell Environ 26:915–928

Jung HW, Kim KD, Hwang BK (2005) Identification of pathogen-responsive regions in the promoter of a pepper lipid transfer protein gene (CALTP1) and the enhanced resistance of the CALTP1 transgenic *Arabidopsis* against pathogen and environmental stresses. Planta 221:361–373

Kido EA, Pandolfi V, Houllou-Kido LM, Andrade PP, Marcelino FC, Nepomuceno AL, Abdelnoor RV, Burnquist WL, Benko-Iseppon AM (2010) Plant antimicrobial peptides: an overview of SuperSAGE transcriptional profile and a functional review. Curr Protein Pept Sci 11:220–230

Koike M, Okamoto T, Tsuda S, Imai R (2002) A novel plant defensin-like gene of winter wheat is specifically induced during cold acclimation. Biochem Biophys Res Commun 298:46–53

Koo JC, Lee SY, Chun HJ, Cheong YH, Choi JS, Kawabata S-I, Miyagi M, Tsunasawa S, Ha KS, Bae DW, Han C-D, Lee BL, Cho MJ (1998) Two hevein homologs isolated from the seed of *Pharbitis nil* L. exhibit potent antifungal activity. Biochim Biophys Acta 1382:80–90

Koo JC, Chun HJ, Park HC, Kim MC, Koo YD, Koo SC, Ok HM, Park SJ, Lee S-H, Yun D-J, Lim CO, Bahk JD, Lee SY, Cho MJ (2002) Over-expression of a seed hevein-like antimicrobial peptide from *Pharbitis nil* enhances resistance to a fungal pathogen in transgenic tobacco plants. Plant Mol Biol 50:441–452

Koo JC, Lee B, Young ME, Koo SC, Cooper JA, Baek D, Lim CO, Lee SY, Yun D-J, Cho MJ (2004) Pn-AMP1, a plant defense protein, induces actin depolarization in yeasts. Plant Cell Physiol 45:1669–1680

Kushmerick C, de Souza CM, Santos Cruz J, Bloch C, Beirao PS (1998) Functional and structural features of gamma-zeathionins, a new class of sodium channel blockers. FEBS Lett 440:302–306

Lay FT, Anderson MA (2005) Defensins–components of the innate immune system in plants. Curr Protein Pept Sci 6:85–101

Lee OS, Lee B, Park N, Koo JC, Kim YH, Prasad D, Karigar C, Chun HJ, Jeong BR, Kim DH, Nam J, Yun J-G, Kwak S-S, Cho MJ, Yun D-J (2003) Pn-AMPs, the hevein-like proteins from *Pharbitis nil* confer disease resistance against phytopathogenic fungi in tomato, *Lycopersicum esculentum*. Phytochemistry 62:1073–1079

Li S-S, Claeson P (2003) Cys/Gly-rich proteins with a putative single chitin-binding domain from oat (*Avena sativa*) seeds. Phytochemistry 63:249–255

Li SS, Gullbo J, Lindholm P, Larsson R, Thunberg E, Samuelsson G, Bohlin L, Claeson P (2002) Ligatoxin B, a new cytotoxic protein with a novel helix-turn-helix DNA-binding domain from the mistletoe *Phoradendron liga*. Biochem J 361:405–413

Lipkin A, Anisimova V, Nikonorova A, Babakov A, Krause A, Bienert M, Grishin E, Egorov T (2005) An antimicrobial peptide Ar-AMP from amaranth (*Amaranthus retroflexus* L.) seeds. Phytochemistry 66:2426–2431

Lobo DS, Pereira IB, Fragel-Madeira L, Medeiros LN, Cabral LM, Faria J, Bellio M, Campos RC, Linden R, Kurtenbach E (2007) Antifungal *Pisum sativum* defensin 1 interacts with *Neurospora crassa* cyclin F related to the cell cycle. Biochemistry 46:987–996

Loeza-Ángeles H, Sagrero-Cisneros E, Lara-Zárate L, Villagómes-Gómez E, López-Meza JE, Ochoa-Zarzosa A (2008) Thionin Thi2.1 from *Arabidopsis thaliana* expressed in endothelial cells shows antibacterial, antifungal and cytotoxic activity. Biotechnol Lett 10:1713–1719

Manners JM (2007) Hidden weapons of microbial destruction in plant genomes. Genome Biol 8:225–234

Matsumura H, Reich S, Ito A, Saitoh H, Kamoun S, Winter P, Kahl G, Reuter M, Kruger DH, Terauchi R (2003) Gene expression analysis of plant host-pathogen interactions by SuperSAGE. Proc Natl Acad Sci USA 100:15718–15723
Melo FR, Rigden DJ, Franco OL, Mello LV, Ary MB, Grossi-de-Sa MF, Bloch C (2002) Inhibition of trypsin by cowpea thionin: characterization, molecular modeling and docking. Proteins 48:311–319
Mendez E, Moreno A, Colilla FJ, Pelaez F, Limas GG, Mendez R, Soriano F, Salinas M, de Haro C (1990) Primary structure and inhibition of protein synthesis in eukaryotic cell–free system of a novel thionin, gamma-hordothionin, from barley endosperm. J Biochem 194:533–539
Mendez E, Rocher A, Calero M, Girbes T, Citores L, Soriano F (1996) Primary structure of omega-hordothionin, a member of a novel family of thionins from barley endosperm, and its inhibition of protein synthesis in eukaryotic and prokaryotic cell-free systems. Eur J Biochem 239:67–73
Mirouze M, Sels J, Richard O, Czernic P, Loubet S, Jacquier A, Francois IEJA, Cammue BPA, Lebrun M, Berthomieu P, Marques L (2006) A putative novel role for plant defensins: a defensin from zinc hyper-accumulating plant, *Arabidopsis halleri*, confers zinc tolerance. Plant J 47:329–342
Nielsen KK, Nielsen JE, Madrid SM, Mikkelsen JD (1997) Characterization of a new antifungal chitin-binding peptide from sugar beet leaves. Plant Physiol 113:83–91
Odintsova TI, TsA E, Musolyamov AKh, Odintsova MS, Pukhalsky VA, Grishin EV (2007) Seed defensins from *T. kiharae* and related species: genome localization of defensin-encoding genes. Biochimie 89:605–612
Odintsova TI, Rogozhin EA, Baranov Y, Musolyamov AK, Yalpani N, Egorov TA, Grishin EV (2008) Seed defensins of barnyard grass *Echinochloa crusgalli* (L.) Beauv. Biochimie 90:1667–1673
Odintsova TI, Vassilevski AA, Slavokhotova AA, Musolyamov AK, Finkina EI, Khadeeva NV, Rogozhin EA, Korostyleva TV, Pukhalsky VA, Grishin EV, Egorov TA (2009) A novel antifungal hevein-type peptide from *Triticum kiharae* seeds with a unique 10-cysteine motif. FEBS J 276:4266–4275
Okuda S, Tsutsui H, Shiina K, Sprunck S, Takeuchi H, Yui R, Kasahara RD, Hamamura Y, Mizukami A, Susaki D, Kawano N, Sakakibara T, Namiki S, Itoh K, Otsuka K, Matsuzaki M, Nozaki H, Kuroiwa T, Nakano A, Kanaoka MM, Dresselhaus T, Sasaki N, Higashiyama T (2009) Defensin-like polypeptide LUREs are pollen tube attractants secreted from synergid cells. Nature 458:357–361
Osborn RW, De Samblanx GW, Thevissen K, Goderis I, Torrekens S, Van Leuven F (1995) Isolation and characterization of plant defensins from seeds of Asteraceae, Fabaceae, Hippocastanaceae and Saxifragaceae. FEBS Lett 368:257–262
Osbourn AE (1996) Preformed antimicrobial compounds and plant defense against fungal attack. Plant Cell 8:1821–1831
Padovan L, Scocchi M, Tossi A (2010) Structural aspects of plant antimicrobial peptides. Curr Protein Pept Sci 11:210–219
Pelegrini PB, Lay FT, Murad AM, Anderson MA, Franco OL (2008) Novel insights on the mechanism of action of alpha-amylase inhibitors from the plant defensin family. Proteins 73:719–729
Pellegrini PB, Franco OL (2005) Plant gamma-thionins: novel insights on the mechanism of action of a multi-functional class of defense proteins. Int J Biochem Cell Biol 37:2239–2253
Pieterse CMJ, Leon-Reyes A, Van der Ent S, Van Wees SCM (2009) Networking by small-molecule hormones in plant immunity. Nat Chem Biol 5:308–316
Raikhel NV, Lee H-I (1993) Structure and function of chitin-binding proteins. Annu Rev Plant Physiol Plant Mol Biol 44:591–615
Regente MC, Giudici AM, Villalain J, de la Canal L (2005) The cytotoxic properties of a plant lipid transfer protein involve membrane permeabilization of target cells. Lett Appl Microbiol 40:183–189

Rivillas-Acevedo LA, Soriano-Garcia M (2007) Isolation and biochemical characterization of an antifungal peptide from *Amaranthus hypochondriacus* seeds. J Agric Food Chem 55:10156–10161

Rogozhin EA, Odintsova TI, Musoliamov AKh, Smirnov AN, Babakov AV, TsA E, Grishin EV (2009) The purification and characterization of a novel lipid transfer protein from caryopsis of barnyard grass (*Echinochloa crusgalli*). Prikl Biokhim Mikrobiol 45:403–409

Rogozhin EA, Oshchepkova YI, Odintsova TI, Khadeeva NV, Veshkurova ON, Egorov TA, Grishin EV, Salikhov SI (2011) Novel antifungal defensins from *Nigella sativa* L. seeds. Plant Physiol Biochem 49:131–137

Schöpke T, Hasan Agha MI, Kraft R, Otto A, Hiller K (1993) Hämolytisch aktive komponenten aus *Viola tricolor* L. und *Viola arvensis* Murray. Sci Pharm 61:145–153

Sels J, Mathys J, De Coninck BM, Cammue BP, De Bolle MF (2008) Plant pathogenesis-related (PR) proteins: a focus on PR peptides. Plant Physiol Biochem 46:941–950

Shao F, Hu Z, Xiong Y, Huang Q, Wang C, Zhu R, Wang D (1999) A new antifungal peptide from the seeds of *Phytolacca americana*: characterization, amino acid sequence and cDNA cloning. Biochem Biophys Acta 1430:262–268

Sharma P, Lönneborg A (1996) Isolation and characterization of a cDNA encoding a plant defensin-like protein from roots of Norway spruce. Plant Mol Biol 31:707–712

Silverstein KA, Graham MA, Paape TD, VandenBosch KA (2005) Genome organization of more than 300 defensin-like genes in *Arabidopsis*. Plant Physiol 138:600–610

Silverstein KA, Moskal WA Jr, Wu HC, Underwood BA, Graham MA, Town CD, VandenBosch KA (2007) Small cysteine-rich peptides resembling antimicrobial peptides have been under-predicted in plants. Plant J 51:262–280

Slavokhotova AA, Odintsova TI, Rogozhin EA, Musolyamov AK, Andreev YA, Grishin EV, Egorov TA (2011) Isolation, molecular cloning and antimicrobial activity of novel defensins from common chickweed (*Stellaria media* L.) seeds. Biochimie 93:450–456

Spelbrink RG, Dilmac N, Allen A, Smith TJ, Shah DM, Hockerman GH (2004) Differential antifungal and calcium channel-blocking activity among structurally related plant defensins. Plant Physiol 135:2055–2067

Stec B (2006) Plant thionins – the structural perspective. Cell Mol Life Sci 63:1370–1385

Stec B, Markman O, Rao U, Heffron G, Henderson S, Vernon LP, Brumfeld V, Teeter MM (2004) Proposal for molecular mechanism of thionins deduced from physico-chemical studies of plant toxins. J Pept Res 64:210–224

Stotz HU, Spence B, Wang Y (2009a) A defensin from tomato with dual function in defense and development. Plant Mol Biol 71:131–143

Stotz HU, Thomson JG, Wang Y (2009b) Plant defensins: defense, development and application. Plant Signal Behav 4:1010–1012

Stuart LS, Harris TH (1942) Bactericidal and fungicidal properties of a crystalline protein from unbleached wheat flour. Cereal Chem 19:288–300

Taji Y, Seki M, Satou M, Sakurai T, Kobayashi M, Ishiyama K, Narusaka Y, Narusaka M, Zhu JK, Shinozaki K (2004) Comparative genomics in salt tolerance between Arabidopsis and Arabidopsis–related halophyte salt cress using Arabidopsis microarray. Plant Physiol 135:1697–1709

Tassin S, Broekaert WF, Marion D, Acland DP, Ptak M, Vovelle F, Sodano P (1998) Solution structure of Ace-AMP1, a potent antimicrobial protein extracted from onion seeds. Structural analogies with plant non-specific lipid transfer proteins. Biochemistry 37:3623–3637

Tavares LS, de Santos M, Viccini LF, Mereira JS, Miller RNG, Franco OL (2008) Biotechnological potential of antimicrobial peptides from flowers. Peptides 29:1842–1851

Terras FRG, Torrekens S, Van Leuven F, Osborn RW, Vanderleyden J, Cammue BPA, Broekaert WF (1993) A new family of basic cysteine-rich plant antifungal proteins from Brassicaceae species. FEBS Lett 316:233–240

Thevissen K, Ghazi A, De Samblanx GW, Brownlee C, Osborn RW, Broekaert WF (1996) Fungal membrane responses induced by plant defensins and thionins. J Biol Chem 271:15018–15025

Thevissen K, Warnecke DC, Francois IEJA, Leipelt M, Heinz E, Ott C (2004) Defensins from insects and plants interact with fungal glucosylceramides. J Biol Chem 279:3900–3905
Thomma BP, Cammue BP, Thevissen K (2002) Plant defensins. Planta 216:193–202
Titarenko E, Lopez-Solanilla E, Garcia-Olmedo F, Rodriguez-Palenzuela P (1997) Mutants of *Ralstonia (Pseudomonas solanacearum)* sensitive to antimicrobial peptides are altered in their LPS structure and are avirulent. J Bacteriol 179:6699–6704
Torres MA, Jones JDG, Dangl JL (2006) Reactive oxygen species signaling in response to pathogens. Plant Physiol 141:373–378
Utkina LL, Zhabon EO, Slavokhotova AA, Rogozhin EA, Shiyan AN, Grishin EV, Egorov TA, Odintsova TI, Pukhalskii VA (2010) Heterologous expression in *Escherichia coli* cells of a synthetic gene encoding a novel hevein-like peptide of *Leymus arenarius*. Russ J Genet 46:1–7
Van den Bergh KPB, Proost P, Van Damme J, Coosemans J, Van Damme EJM, Peumans WJ (2002) Five disulfide bridges stabilize a hevein-type antimicrobial peptide from the bark of spindle tree (*Euonymus europaeus* L.). FEBS Lett 530:181–185
Van den Bergh KPB, Rougé P, Proost P, Coosemans J, Krouglova T, Engelborghs Y, Peumans WJ, Van Damme EJM (2004) Synergistic antifungal activity of two chitin-binding proteins from spindle tree (*Euonymus europaeus* L.). Planta 219:221–232
Van der Weerden NL, Lay FT, Anderson MA (2008) The plant defensin, NaD1, enters the cytoplasm of *Fusarium oxysporum* hyphae. J Biol Chem 283:14445–14452
Van der Weerden NL, Hancock REW, Anderson MA (2010) Permeabilization of fungal hyphae by the plant defensin NaD1 occurs through a cell wall dependent process. J Biol Chem 285:37513–37520
Van Loon LC, Rep M, Pieterse CM (2006) Significance of inducible defense-related proteins in infected plants. Annu Rev Phytopathol 44:135–162
Van Parijs J, Broekaert WF, Goldstein IJ, Peumans WJ (1991) Hevein: an antifungal protein from rubber-tree (*Hevea brasiliensis*) latex. Planta 183:258–264
Vila-Perello M, Sanchez-Vallet A, Garcia-Olmedo F, Molina A, Andreu D (2005) Structural dissection of a highly knotted peptide reveals minimal motif with antimicrobial activity. J Biol Chem 280:1661–1668
Vila-Perello M, Tognon S, Sanchez-Vallet A, Garcia-Olmedo F, Molina A, Andreu D (2006) A minimalist design approach to antimicrobial agents based on a thionin template. J Med Chem 49:448–451
Wang SY, Wu JH, Ng TB, Ye XY, Rao PF (2004) A non-specific lipid transfer protein with antifungal and antibacterial activities from the mung bean. Peptides 25:1235–1242
Wijaya R, Neumann GM, Condron R, Hughes AB, Polya GM (2000) Defense proteins from seed of *Cassia fistula* include a lipid transfer protein homologue and a protease inhibitory plant defensin. Plant Sci 159:243–255
Xiang Y, Huang R-H, Liu X-Z, Zhang Y, Wang D-C (2004) Crystal structure of a novel antifungal protein distinct with five disulfide bridges from *Eucommia ulmoides* Oliver at an atomic resolution. J Struct Biol 148:86–97
Yeats TH, Rose JKC (2008) The biochemistry and biology of extracellular plant lipid-transfer proteins (LTPs). Protein Sci 17:191–198
Yount NY, Yeaman MR (2004) Multidimensional signatures in antimicrobial peptides. Proc Natl Acad Sci USA 101:7363–7368
Zaiou M (2007) Multifunctional antimicrobial peptides: therapeutic targets in several human diseases. J Mol Med 85:317–329
Zottich U, Da Cunha M, Carvalho AO, Dias GB, Silva NC, Santos IS, do Nacimento VV, Miguel EC, Machado OL, Gomes VM (2011) Purification, biochemical characterization and antifungal activity of a new lipid transfer protein (LTP) from *Coffea canephora* seeds with α-amylase inhibitor properties. Biochim Biophys Acta 1810:375–383

The Role of Plant Peptides in Symbiotic Interactions

Virginie Mortier, Ulrike Mathesius, and Sofie Goormachtig

Abstract Plants form symbioses with nitrogen-fixing rhizobia that cause the formation of root nodules. Plant peptides of the CLAVATA3/embryo-surrounding region (CLE), early nodulin 40 (ENOD40), rapid alkalinization factor (RALF), *devil1* (*DVL1*)/*rotundifolia4* (*ROT4*), and nodule-specific cysteine-rich (NCR) families have been implicated in all stages of nodulation. While CLE peptides have roles in controlling the proliferation of nodule cell divisions locally and systemically, ENOD40 peptides act locally in nodule initiation. RALF and DLV1-related peptides play a role in bacterial infection thread formation, while the large family of NCR proteins is nodule specific and controls bacteroid differentiation in mature nodules. Interestingly, parasitic nematodes, which induce feeding structures in host roots that involve similar induction of cell division and differentiation as nodules, are the only organisms known to encode CLE peptides outside the plant kingdom. While rhizobia induce the expression of CLE peptides by the host, parasitic nematodes directly inject CLE peptides into developing feeding cells, mimicking the host peptides.

1 Nodulation: A Macroscopic View

The symbiotic interaction of legume plants with rhizobia results in the development of novel root organs, nodules, the inner cells of which are colonized by bacteria that fix nitrogen for the host. In return, the microsymbionts receive carbon sources and

V. Mortier • S. Goormachtig
Department of Plant Systems Biology, VIB, 9052 Ghent, Belgium

Department of Plant Biotechnology and Genetics, Ghent University, 9052 Ghent, Belgium

U. Mathesius (✉)
Department of Plant Science, Research School of Biology, Australian National University, Linnaeus Way, Canberra, ACT 0200, Australia
e-mail: Ulrike.Mathesius@anu.edu.au

H.R. Irving and C. Gehring (eds.), *Plant Signaling Peptides*,
Signaling and Communication in Plants 16, DOI 10.1007/978-3-642-27603-3_8,

a protective niche. Nodulation is a sophisticated program that requires strictly coordinated bacterial infection and de novo organ formation. How this is achieved varies between different legume species, and here we mainly focus on the Nod factor-dependent root hair infection strategy and nodule development as observed in the model legume *Medicago truncatula*.

The symbiosis is initiated by a specific chemical signal exchange and recognition. When the rhizobia sense a specific array of flavonoid compounds secreted by the host, the expression of nodulation (nod) genes is switched on, resulting in the production and secretion of lipochitooligosaccharide signals, the Nod factors (D'Haeze and Holsters 2002). The specific structure of the Nod factors is sensed by LysM receptor-like kinases located in the plasma membrane of susceptible root hairs and activates the nodulation process. The first responses observed take place within seconds to minutes after Nod factor application and consist of plasma membrane depolarization and extracellular alkalinization (Felle et al. 1999). A few minutes later, Ca^{2+} oscillations are observed in and around the nucleus, and it is currently thought that this event is central in the activation of the epidermal and cortical responses and the coordination between both by triggering the expression of nodulin genes, classified as early nodulin (ENOD) and late nodulin (LNOD), depending whether the induction of their expression precedes or accompanies nitrogen fixation (Crespi and Galvez 2000; Miwa et al. 2006; Schultze and Kondorosi 1998; Sun et al. 2007; Talukdar et al. 2009). Morphological changes are an arrest of root hair growth, followed by swelling and deformation of the root hair tip (Gage 2004). As a result, the root hair will curl and entrap a rhizobial colony. This event is commonly referred to as root hair curling (RHC). Subsequently, the plant cell wall is locally degraded within the curl, followed by invagination of the plasma membrane. Cell wall material is deposited around the invaginated membrane ensuing the formation of a tubular structure, the infection thread, through which the bacteria are guided towards deeper root cell layers (Gage 2004). The cortical cells respond to Nod factors by dedifferentiation and reiteration of the cell cycle: while the inner cortical cells progress through the cell cycle and develop into a nodule primordium, the outer cortical cells arrest in G2 for preinfection thread formation via cytoplasmic bridges through which infection threads will pass (Timmers 2008; van Spronsen et al. 2001; Yang et al. 1994). Once the infection threads reach the cells of the nodule primordium, the bacteria are released inside the cytoplasm via infection droplets, surrounded by a plant-derived peribacteroid membrane. Within these new organelles, the symbiosomes, bacteria develop into bacteroids to fix nitrogen (Brewin 2004; Kouchi et al. 2010).

Two types of nodules have been intensively studied: determinate and indeterminate ones. While determinate nodules develop from division of outer cortical cells and in mature nodules all meristematic tissue has been consumed, indeterminate nodules arise from the inner cortex and are characterized by a persistent apical meristem (Crespi and Galvez 2000; Ferguson et al. 2010). Hence, mature indeterminate nodules are zonated with a meristem, an infection zone, a fixation zone, and a senescence zone. The nodule parenchyma surrounds the central zone and contains the vascular tissue, via which nutrients are exchanged between both symbiotic partners (Van de Velde et al. 2006).

2 Involvement of Classical Plant Hormones in Nodulation

There is ample evidence that the classical hormones cytokinin, auxin, abscisic acid (ABA), gibberellins, ethylene, and jasmonic acid are involved in the initiation and coordination of the nodulation process (Boot et al. 1999; Frugier et al. 2008; Gonzalez-Rizzo et al. 2006; Grunewald et al. 2009; Heidstra et al. 1997; Mathesius et al. 1998; Murray et al. 2007; Pacios-Bras et al. 2003; Oldroyd and Downie 2008; Smit et al. 1995; Tirichine et al. 2007; van Noorden et al. 2006; van Spronsen et al. 2001; Wasson et al. 2006). A central regulator of nodule organogenesis is cytokinin. Most convincingly, the *Lotus japonicus* knockout mutant for the cytokinin receptor gene *LOTUS HISTIDINE KINASE 1 (LHK1)*, *M. truncatula* transgenic plants with suppressed expression of *CYTOKININ RECEPTOR 1* (*CRE1*), and stable mutants in the same gene are defective in nodule primordia formation (Gonzalez-Rizzo et al. 2006; Murray et al. 2007; Plet et al. 2011). Similarly, the ectopic expression of a cytokinin-degrading enzyme in *L. japonicus* resulted in reduced nodulation (Lohar et al. 2004). Additionally, an *L. japonicus* gain-of-function mutant for the LHK1 receptor provoked spontaneous nodules, indicating that cytokinin signaling is both necessary and sufficient for nodule formation (Tirichine et al. 2007). In addition to cytokinin, auxin is also required for nodule formation (Ding and Oldroyd 2009; Oldroyd and Downie 2008). In white clover (*Trifolium repens*), the auxin flow within the root vasculature is transiently inhibited at the site of infection, leading to auxin accumulation in the cortical region where the nodule primordia form (Mathesius et al. 1998). A reduction in auxin flow has been confirmed by radioactive auxin tracer experiments for *M. truncatula* and vetch (*Vicia faba*) but not *L. japonicus* (Boot et al. 1999; Pacios-Bras et al. 2003; van Noorden et al. 2006; Wasson et al. 2006). The involvement of auxin during early nodulation is also supported by the fact that application of synthetic auxin transport inhibitors, such as 1-naphthylphthalamic acid (NPA), leads to spontaneous nodule formation (Hirsch and Fang 1994). Moreover, proteome studies in *M. truncatula* revealed a high overlap in protein changes in the roots 24 h after auxin or *Sinorhizobium meliloti* treatment (van Noorden et al. 2007). A positive role for gibberellins during nodulation was suggested, by the reduced nodule number observed on a gibberellin biosynthetic mutant. Moreover, exogenous application of gibberellins rescued the phenotype by restoring the nodule number (Kawaguchi et al. 1996; Ferguson et al. 2005). A role of gibberellin in the intercellular invasion process of lateral root base nodulation in *Sesbania rostrata* has also been suggested (Lievens et al. 2005). ABA is believed to be involved in nodulation at the level of nodule initiation, where it would interfere with NF signaling, leading to a reduction in nodule number, and at the point of nodule development, where it can suppress cytokinin-induced nodulation processes (Ding et al. 2008; Ferguson and Mathesius 2003). It has also been suggested that ABA plays a role at the latest stages of nodulation because an increase in ABA levels is observed at the onset of nodule senescence (Ferguson and Mathesius 2003).

High concentrations of ethylene inhibit nodulation, by exerting a negative effect at multiple levels during nodulation. First, ethylene suppresses the Nod factor signaling pathway, either at or during Nod factor-induced calcium spiking, leading to inhibition of root hair deformation, shortening of the calcium spike period, blocking of bacterial infection, and suppression of nodulin genes expression (Oldroyd et al. 2001). In addition, ethylene has been shown to regulate infection thread formation and to dictate the positioning of nodules between the protoxylem poles (Penmetsa and Cook 1997; Heidstra et al. 1997). Indeed, 1-aminocyclopropane-1-carboxylic acid (ACC) oxidase involved in the last step of ethylene biosynthesis is highly expressed at the protophloem poles (Heidstra et al. 1997). Presumably, an ethylene gradient is made with the lowest concentration located at the protoxylem poles, allowing nodulation at that place. Moreover, the addition of inhibitors of ACC synthase (AVG) and antagonists of ethylene action (Ag^+ ions) had a positive effect on nodule number (Nukui et al. 2000; Oldroyd et al. 2001). Analysis of the ethylene-insensitive *sickle* (*skl*) mutant confirmed the negative effect of ethylene on nodulation, as this mutant has a hyperinfection phenotype (Penmetsa and Cook 1997; Penmetsa et al. 2008). Similarly, transgenic *L. japonicus* plants expressing a mutated ethylene receptor gene produce an elevated number of infection threads and nodule primordia (Nukui et al. 2004). Like ethylene, jasmonic acid inhibits nodulation by suppressing calcium spiking and Nod factor-induced gene expression (Sun et al. 2006). Moreover, several lines of evidence indicate that jasmonic acid signaling in the leaves is involved in the autoregulation of nodulation (AON) signal transduction pathway, a process by which the plant controls the number of nodules (Kinkema and Gresshoff 2008).

3 Peptide Signals and Nodulation

Peptide hormones have been extensively studied in animals where they act as the most common signal molecules to regulate and coordinate spatial and temporal expression of genes for the development of a multicellular organism. Peptide hormones consist of short stretches of amino acids (an informal dividing line is mostly set at 50 amino acids in length) that are mostly posttranslationally modified, for example, by glycosylation, arabinosylation, and hydroxyprolination. Animal peptides are highly diverse in sequence and length, but members of the same family are highly conserved. While some peptides are derived from small ORFs and supposedly act within the cytoplasm, most peptides are derived from the C-terminal end of preproproteins, often by cleavage at the level of dibasic amino acid residues (also see chapters "Plant Peptide Signaling: An Evolutionary Adaptation" and "Processing of Peptides"). In addition, the presence of signal peptides at the N-terminal end of the preproproteins causes those peptides to be secreted in the extracellular medium where they act non-cell-autonomously as intercellular signaling molecules.

A smaller number of plant peptides have been identified until now mostly by chance, and many are expected to be discovered by conducting more direct searches (see chapters "Plant Antimicrobial Peptides" and "Methods to Identify New Partners of Plant Signaling Peptides"). The identified peptides play a role in various aspects of plant development, growth, and defense.

Systemin, the first plant peptide hormone to be described, is involved in systemic defenses triggered by insect attack or mechanical wounding (McGurl et al. 1992; Pearce et al. 1991). Systemin signaling has been thoroughly investigated, and currently, it is thought that the peptide acts as an enhancer of jasmonic acid signaling in solanaceous species (Degenhardt et al. 2010).

Phytosulfokines (PSKs), 5-amino-acid-long peptides with sulfated tyrosine residues, were the next peptides to be discovered. They regulate cellular dedifferentiation, proliferation, and differentiation in suspension cultures of carrot, rice, and *Asparagus officinalis* (Hanai et al. 2000; Matsubayashi and Sakagami 1996; Yang et al. 2000a, b). Moreover, PSKs trigger tracheary element differentiation of cultured *Zinnia* mesophyll cells at nanomolar concentrations (Matsubayashi et al. 1999). PSKs also promote various stages of plant growth, including adventitious bud and root formation, pollen germination, and somatic embryogenesis (Chen et al. 2000; Hanai et al. 2000; Igasaki et al. 2003; Kobayashi et al. 1999; Yamakawa et al. 1998; also see chapters "Peptides Regulating Root Growth" and "Peptides Regulating Plant Vascular Development").

The S-locus cysteine-rich proteins (SCRs) and S-locus protein 11 (SP11) are involved in pollen self-incompatibility in *Brassicaceae* (Kachroo et al. 2002; Schopfer et al. 1999; also see chapter "The S-LOCUS CYSTEINE RICH PROTEIN (SCR): A Small Peptide with a High Impact on the Evolution of Flowering Plants"). POLARIS plays a role in root vascular patterning by maintaining responsiveness to exogenous auxin and cytokinin (Casson et al. 2002). "Inflorescence deficient in abscission" (IDA) controls floral organ abscission (Butenko et al. 2003; also see chapter "Methods to Identify New Partners of Plant Signaling Peptides").

Finally, peptides from the "CLAVATA3/embryo-surrounding region" (CLE), early nodulin 40 (ENOD40), rapid alkalinization factor (RALF), *devil1* (*DVL1*)/ *rotundifolia4* (*ROT4*), and nodule-specific cysteine-rich (NCR) families carry out a range of different functions but have in common that at least one of their members is involved in the process of nodulation. Here we will discuss these nodule-related peptides and how they control the nodulation process.

3.1 CLE Peptides and Nodulation

CLE peptides have been identified in the genomes of mono- and dicotyledonous plants, in peptide families of up to 40 members (Cock and McCormick 2001;

Mitchum et al. 2008; Oelkers et al. 2008). They were also found in the moss *Physcomitrella patens*, the green alga *Chlamydomonas reinhardtii*, and the pteridophyte *Selaginella moellendorffii*, indicating that they occurred already at the early beginning of plant evolution (Oelkers et al. 2008; also see chapter "Plant Peptide Signaling: An Evolutionary Adaptation").

CLE peptides (<15 kDa) are 12–13 amino acids long and are cleaved from a 14-amino-acid-long conserved domain, located at or close to the C terminus of larger CLE proteins (Ito et al. 2006; Kondo et al. 2006; Ni and Clark 2006; also see chapters "Peptides Regulating Plant Vascular Development" and "Processing of Peptides"). Posttranslational modifications such as hydroxyprolination and arabinosylation have been reported (Ito et al. 2006; Kondo et al. 2006; Ohyama et al. 2009). A hydrophobic signal peptide can be distinguished at the N-terminal site of most CLE proteins suggesting that the proteins are targeted to the secretory pathway and that the peptides act as intercellular signals (Sawa et al. 2006). The region between the signal peptide and the CLE domain is highly variable, and its sequence can be substituted by the sequence of other, unrelated proteins, without disrupting the function of the peptide (Ni and Clark 2006). Despite variability in the core structure of the peptides, all members of the CLE family share certain structural parameters such as charge, length, and hydrophilicity. The genomes of *Oryza sativa*, *M. truncatula*, *Tritium aestivum*, and *S. moellendorffii* encode CLE members with multiple CLE domains (Mortier et al. 2010; Oelkers et al. 2008). Processing of several CLE peptides from one proprotein might be a mechanism to provide an amplification effect. These multiple CLE domains are probably the result of duplication events that occurred during evolution.

CLE peptides are involved in balancing cell division and differentiation during several aspects of plant development. They control shoot and root meristem activity, apical dominance, organ size, vascular differentiation, and nodulation (Brand et al. 2000; Hirakawa et al. 2008; Hobe et al. 2003; Mortier et al. 2010; Okamoto et al. 2009; Reid et al. 2011; also see chapters "Peptides Regulating Apical Meristem Development," "Peptides Regulating Root Growth," and "Peptides Regulating Plant Vascular Development"). The only CLE peptides identified outside the plant kingdom are encoded by plant parasitic nematodes, as discussed in Sect. 4.

Due to a high level of functional redundancy observed between the different CLE members, the gene-specific functions of only a small number of CLE peptides (CLV3, TDIF, and CLE40) have so far been identified. CLV3, the best studied peptide of the CLE family, is involved in shoot apical meristem (SAM) homeostasis in *Arabidopsis*. *Clv3* mutants have an expanded SAM, fasciation of stems and leaves and supernumerary floral organs, caused by the accumulation of stem cells in the shoot apical and flower meristems (Clark et al. 1995; see chapter "Peptides Regulating Apical Meristem Development"). Ectopic expression of *CLV3*, on the other hand, results in plants in which the SAM is not maintained, and organogenesis from the shoot tip is arrested (Brand et al. 2000). Hence, *CLV3* plays a key role in the suppression of stem cell proliferation in the SAM. Further studies have shown that *CLV3* is part of a cell nonautonomous negative feedback regulation loop, comprising among others the homeodomain transcription factor *WUSCHEL* (*WUS*) (Brand et al. 2000; Schoof et al. 2000). *WUS*, which is expressed in the cells of the organizing center of the SAM, induces expression of *CLV3* in the overlying neighboring cells and confers their stem cell identity (Mayer et al. 1998; Rojo et al. 2002), while *CLV3* downregulates *WUS* expression and thereby confines the size of the organizing center (Lenhard and Laux 2003; Schoof et al. 2000). Recruitment of cells from the peripheral zone to the central zone and vice versa were observed as well, suggesting that *CLV3* is also involved in setting the boundary between both zones (Müller et al. 2006; Reddy and Meyerowitz 2005).

The CLV3 peptide is perceived by a complex interplay of membrane receptors. The peptide is directly bound by the leucine-rich repeat receptor-like kinase (LRR-RLK) CLV1, which can form homodimers or heterodimers in combination with their close homologs BARELY ANY MERISTEM1 (BAM1) or BAM2 (Guo and Clark 2010; Ogawa et al. 2008; Trotochaud et al. 1999, 2000). In addition, genetic and physiological studies have recently demonstrated that the LRR receptor-like protein CLV2 interacts with a receptor-like protein kinase lacking an extracellular domain, named CORYNE (CRN) or suppressor of LLP1 2 (SOL2), to mediate CLV3-dependent SAM homeostasis in parallel to CLV1 (Bleckmann et al. 2010; Guo et al. 2010; Guo and Clark 2010; Miwa et al. 2008; Müller et al. 2008; Zhu et al. 2010a, b). However, different regulatory mechanisms might be observed throughout the plant kingdom as an evolutionary analysis suggested that CLV2 and CRN are restricted to vascular plants, while CLV1 is highly conserved (Miwa et al. 2009). Still more receptors and putative CLE peptides might contribute to SAM homeostasis, such as the recently identified CLV1-unrelated LRR-RLK RECEPTOR-LIKE PROTEIN KINASE2/TOADSTOOL2 (RPK2/TOAD2) (Kinoshita et al. 2010; Sawa and Tabata 2011). Other components of the CLV-WUS pathways are a kinase-associated protein phosphatase (KAPP) and POLTERGEIST (POL) type 2C phosphatase which negatively regulate CLV signaling through interaction with CLV1 (Trotochaud et al. 1999; Yu et al. 2003). Next to their expression in the center of the SAM, BAM receptors are also expressed at the flanks of the SAM and are proposed to sequester exterior CLV3-like CLE peptides thereby preventing their entrance in the core region of the SAM (DeYoung and Clark 2008).

The signaling components for stem cell maintenance in the shoot and in the root apical meristems are largely conserved (see chapters "Peptides Regulating Apical Meristem Development" and "Peptides Regulating Root Growth"), hence it is not surprising that CLE peptides were also shown to control root apical meristem homeostasis. Loss of function of the *Arabidopsis CLE40* caused roots to exhibit a slightly waving pattern, a reduced length, and strong slant indicative of a defect in gravity response (Hobe et al. 2003). A reduction in root length was also reported in experiments in which CLE peptides were supplemented to the growth medium of *Arabidopsis* or *M. truncatula* seedlings or in which CLE genes were ectopically expressed (Fiers et al. 2005; Ito et al. 2006; Mortier et al. 2010; Oelkers et al. 2008; Strabala et al. 2006; Wang et al. 2011; Whitford et al. 2008). In these experiments, a high level of redundancy was observed between members with related sequences. Probably, the spatial and temporal expression patterns regulate the range of action of related CLE peptides. Indeed, expression analysis of *Arabidopsis*, soybean, and *M. truncatula* CLE genes indicated different expression patterns for different *CLE* genes, often restricted to well-defined tissues (Mortier et al. 2010, 2011; Sharma et al. 2003).

Another well-characterized CLE peptide, tracheary element differentiation inhibitory factor (TDIF), isolated from *Zinnia elegans* cell cultures during in vitro experiments, suppresses tracheary element differentiation and controls the rate and orientation of vascular cell division (Etchells and Turner 2010; Ito et al. 2006; also see chapter "Peptides Regulating Plant Vascular Development"). The sequence of the TDIF peptide is identical to CLE41 and CLE44 and highly homologous to CLE42 and CLE46. In vitro experiments performed with these *Arabidopsis* peptides resulted in a suppression of xylem development, while the other CLE peptides tested did not exhibit TDIF activity (Ito et al. 2006). Applying mature CLV3 peptides to the *Z. elegans* cell culture resulted in the opposite effect, the promotion of tracheary element cell formation (Ito et al. 2006). Hence, it was proposed that two counteracting pathways are involved in CLE signaling, one promoting and one inhibiting cell differentiation. Nevertheless, Whitford et al. (2008) reported synergistic actions between CLV3-like and TDIF-like peptides. Ectopic expression or exogenous application of TDIF-like peptides, but not of CLV3-like peptides, resulted in differentiation and proliferation of vascular cells. However, when both types of peptides were simultaneously overexpressed, or ectopically applied onto *A. thaliana* seedlings, the effect of the TDIF gain-of-function was enhanced, resulting in a massive proliferation of vascular cells. Additional experiments revealed that this cell proliferation was auxin mediated (Whitford et al. 2008). Recently, a novel role for TDIF-related CLE peptides was demonstrated in promoting axillary bud formation (Yaginuma et al. 2011). Both during root apical meristem homeostasis and vascular patterning, CLE peptides are perceived by membrane-bound receptor-like kinases. *Arabidopsis* CRINKLY4 (ACR4) was suggested as the receptor for CLE40 in roots and to be involved in the suppression of the WUS homolog, *WUS-RELATED HOMEOBOX5* (*WOX5*), to maintain distal root stem cells (Stahl and Simon 2009; Stahl et al. 2009). In accordance, TDIF was shown to bind in vitro the PHLOEM INTERCALATED WITH XYLEM/TDIF RECEPTOR (PXY/TDR) receptor expressed in the procambial cells to inhibit vascular differentiation (Hirakawa et al. 2008).

There are two reasons to propose that CLE peptides might control nodule formation. First, it is because of the similarity between the CLE peptide receptors CLV1, BAM 1-3, and TDR, and the *M. truncatula* receptor SUPER NUMERIC NODULES (SUNN) and its orthologs of *L. japonicus*, pea, and soybean [HYPERNODULATION ABERRANT ROOT FORMATION 1 (HAR1), SYMBIOSIS29 (SYM29), and NODULE AUTOREGULATION RECEPTOR KINASE (NARK)], all belonging to the subclass XI of the LRR-RLKs (Krusell et al. 2002; Nishimura et al. 2002; Schnabel et al. 2005; Searle et al. 2003). These legume receptors, although most similar to CLV1, seem not to control SAM organization but rather are involved in AON, a systemic mechanism that involves root to shoot signaling to control nodule numbers (Kosslak and Bohlool 1984; Carroll et al.

1985a, b; Delves et al. 1986; Magori and Kawaguchi 2009; Nutman 1952; Pierce and Bauer 1983). The similarity to CLV1 suggested that SUNN/HAR-1/SYM29/NARK might perceive CLE peptides (Krusell et al. 2002; Nishimura et al. 2002; Schnabel et al. 2005; Searle et al. 2003). Second, CLE peptides might be important to balance cell division and differentiation of nodule primordia and nodule meristems.

Indeed, in *L. japonicus*, soybean, as well as in *M. truncatula*, *CLE* genes were identified with upregulated expression during nodulation (Mortier et al. 2010, 2011; Okamoto et al. 2009; Reid et al. 2011). Another CLE gene, GmNIC1, was inducible by nitrate but not by rhizobia (Reid et al. 2011). Tissue-specific expression analysis in *M. truncatula* revealed that the expression of two *CLE* genes, *MtCLE12* and *MtCLE13*, was linked to proliferation and differentiation during nodulation (Mortier et al. 2010). *MtCLE13* expression was upregulated early in nodule development, and a high-to-low expression gradient, radiating from the inner towards the outer cortical cell layers, was observed in a root cortical region defining the incipient nodule. At later stages, *MtCLE12* and *MtCLE13* were expressed in the meristem and early infection zone of mature, elongated nodules.

In *L. japonicus*, *M. truncatula*, and soybean, overexpression of nodule-related CLE proteins, *LjCLE-RS1*, *LjCLE-RS2*, *MtCLE12*, *MtCLE13*, *GmRIC1*, *GmRIC2*, and *GmNIC1*, resulted in the abolishment or strong reduction of nodulation (Mortier et al. 2010; Okamoto et al. 2009; Reid et al. 2011; Lim et al. 2011). This effect involved long-distance signaling and was dependent on SUNN/HAR-1/NARK, suggesting that these CLE peptides might be involved in AON (Mortier et al. 2010; Okamoto et al. 2009; Reid et al. 2011; Saur et al. 2011). In *M. truncatula*, this inhibition was shown to occur at the stage of Nod factor signal transduction (Mortier et al. 2010; Saur et al. 2011).

Whether these CLE peptides directly bind SUNN/HAR-1/NARK is currently not known. Many studies have shown that SUNN/HAR-1/NARK act in the shoot to provoke AON although the genes are also expressed in the root (Delves et al. 1986; Francisco and Harper 1995; Jiang and Gresshoff 2002; Krusell et al. 2002; Nishimura et al. 2002, Schnabel et al. 2005; Searle et al. 2003). Hence, because the peptide genes are expressed in the nodule, direct binding would indicate a long-distance transport of the CLE peptides from the root to the shoot. This would be a novel feature, as so far all characterized CLE peptides act within short distances from the cells in which they are produced (Fukuda et al. 2007; Hirakawa et al. 2008; Miwa et al. 2009; Stahl et al. 2009; Whitford et al. 2008).

In *L. japonicus*, another shoot-controlled LRR-RLK, KLAVIER (KLV), as well as a CLV2 ortholog are involved in AON (Krusell et al. 2010; Oka-Kira et al. 2005). This suggests complex receptor interactions in the shoot to control AON, similar to the CLV1, CLV2, and CRN interactions observed to sustain SAM homeostasis and the root CLE peptides that are suggested to interact with a CLV2/CRN receptor complex (Meng and Feldman 2010; Meng et al. 2010). Although the *sunn* mutant did not show any SAM phenotypes, *Lotus* and pea *clv2* and *klavier* mutants have a shoot phenotype similar to the *Arabidopsis clv1* and *clv2* phenotypes (Krusell et al. 2010; Miyazawa et al. 2010; Oka-Kira et al. 2005). Hence, the same receptor complex might rule AON and SAM/root apical meristem homeostasis, but different plants might use different contributions of each receptor to induce different effects. The ACR4 and TDR/PXY receptors acting in the root apical meristem and vascular tissue, respectively, might interact with currently unidentified receptors.

3.2 The ENOD40 Peptide, a Never Ending Story?

The ENOD40 gene was initially isolated from soybean as an early nodulation-induced gene encoding a short mRNA (±0.7 kb) lacking a long open reading frame (Yang et al. 1993). The gene is widely conserved across the plant kingdom and was found in several other legumes such as *M. truncatula*, *L. japonicus*, and *S. rostrata* but also in nonlegumes including monocots, such as rice, barley, rye grass, maize and sorghum, and dicots, such as tobacco, tomato, and citrus (Compaan et al. 2003; Corich et al. 1998; Gultyaev and Roussis 2007; Kouchi et al. 1999; Larsen 2003; Matvienko et al. 1996; Vleghels et al. 2003). ENOD40 sequences carry two conserved nucleotide regions and several short small open reading frames. The 10- to 13-amino-acid-long translation product of open reading frame A was detected in soybean (Röhrig et al. 2002). During in vitro experiments with soybean, a second translation product (open reading frame B) was detected, which is 24–27 amino acids long and partially overlaps with open reading frame A (Röhrig et al. 2002). In addition to a function for *ENOD40* as a peptide(s), *ENOD40* might play a role as an untranslated RNA molecule (Crespi et al. 1994; Bardou et al. 2011). This is supported by the highly conserved secondary structure of the transcript and the isolation of an *ENOD40* RNA-binding protein (RBP1) in *M. truncatula* (Campalans et al. 2004; Crespi et al. 1994; Gultyaev and Roussis 2007; Charon et al. 2010). Possibly, the *ENOD40* mRNA requires the ENOD40 peptide(s) for its stability and full activity.

The expression of *ENOD40* has been studied in many legumes. It is expressed early after rhizobial infection in the root pericycle opposite the protoxylem poles and in the dividing cortical cells of the nodule primordium (Crespi et al. 1994; Kouchi and Hata 1993; Papadopoulou et al. 1996; Takeda et al. 2005; Yang et al. 1993). At later stages of nodulation, *ENOD40* is expressed in the preinfection zone, in uninfected cells of the central tissue in mature nodules, as well as in the pericycle of the nodule vascular bundles (Corich et al. 1998; Papadopoulou et al. 1996; Takeda et al. 2005; Yang et al. 1993). These nodulation-related expression patterns were reported for bean, soybean, *M. truncatula*, *M. sativa*, *L. japonicus*, and *S. rostrata* (Compaan et al. 2003; Corich et al. 1998; Grønlund et al. 2005; Papadopoulou et al. 1996; Takeda et al. 2005; Wan et al. 2007; Yang et al. 1993). The two *ENOD40* genes identified in *L. japonicus* are differentially regulated. Both are specifically upregulated during nodule organogenesis but show different expression patterns upon rhizobial infection (Takeda et al. 2005). Colocalization of different *ENOD40* genes within one species was also reported for *M. sativa*,

M. truncatula, *L. japonicus*, and *T. repens* (Fang and Hirsch 1998; Flemetakis et al. 2000; Varkonyi-Gasic and White 2002; Wan et al. 2007).

Like CLE peptides, ENOD40 is involved in the control of cell division and differentiation for nodule primordium formation. Ectopic expression of *ENOD40* in *M. truncatula* induced extensive cortical cell divisions and accelerated nodulation (Charon et al. 1997, 1999; Crespi et al. 1994). The latter phenotype was the result of an increase in nodule primordia and a proliferation response of the region close to the root tip (Charon et al. 1999). It is thought that ENOD40 is not an inducer of cell division per se, but rather that it relies on other factors for cell cycle activation and subsequent nodule primordium formation (Mathesius et al. 2000; Minami et al. 1996; Sousa et al. 2001). In *M. truncatula*, downregulation of *ENOD40*, via cosuppression or RNAi resulted in a reduction in the number of nodules and nodule development, was arrested (Charon et al. 1999; Crespi et al. 1994; Wan et al. 2007). The same phenotype was reported for RNAi knockdown lines of *ENOD40* in *L. japonicus* (Kumagai et al. 2006). Bacterial infection of root hairs was not affected, illustrating that *ENOD40* is required for nodule initiation and organogenesis, but not for the early infection events (Kumagai et al. 2006). In *M. truncatula*, impaired bacteroid development was also observed in the RNAi lines (Wan et al. 2007). This could be a consequence of an imbalance between cell division and differentiation rather than that *ENOD40* being essential for bacteroid development per se. The expression analysis together with the functional analysis thus demonstrates that *ENOD40* is involved in the establishment and differentiation of the nodule primordium and meristem. The fact that *ENOD40* is expressed in the pericycle opposite protoxylem poles prior to cortical cell divisions suggests that *ENOD40* might counteract the effects of ethylene to promote mitotic activation of root cortical cells for nodule formation (Heidstra et al. 1997). Indeed, the *ENOD40* expression pattern is complementary to the expression pattern of ACC oxidase, a key enzyme in the synthesis of ethylene (Dey et al. 2004; Heidstra et al. 1997).

Perhaps interlinked with a role in controlling primordium development, *ENOD40* might function in nutrient mobilization. The ENOD40 peptides A and B from soybean were shown to interact in vitro with the 93-kDa subunit of sucrose synthase (Röhrig et al. 2002), and the covalent binding of peptide A to cysteine 264 of sucrose synthase increased the enzyme cleavage activity (Röhrig et al. 2004). These findings suggest a role for *ENOD40* in nutrient supply by controlling the breakdown of sucrose for nodule development (Gordon et al. 1999).

ENOD40 transcripts were also detected outside developing nodules, and gene homologs were identified in non-legumes, suggesting a more general biological function. More specifically, *ENOD40* transcripts were found in stems, in the parenchyma cells surrounding the protoxylem, suggesting a role in vascular development (Corich et al. 1998; Kouchi et al. 1999). In addition, transcripts were found in developing flowers and lateral roots (Corich et al. 1998; Papadopoulou et al. 1996; Varkonyi-Gasic and White 2002; Vleghels et al. 2003). Since all these tissues are linked with cell division and differentiation, *ENOD40* may have a common underlying role in organ and tissue development. Indeed, ectopic expression of *ENOD40* in alfalfa affected the formation of somatic embryos under in vitro culture conditions (Crespi et al. 1994), while its ectopic expression in *A. thaliana* caused a reduction in cell size (Guzzo et al. 2005). This function is probably interlinked with phytohormone signaling pathways because ectopic expression of *ENOD40* in tobacco Bright Yellow-2 cells resulted in elongation growth by alteration of ethylene biosynthesis kinetics and changes in phytohormonal signaling were reported for *M. truncatula* and rice plants ectopically expressing *MtENOD40* (Dey et al. 2004; Ruttink et al. 2006). Despite these studies, a detailed insight in *ENOD40* action is still not obtained and awaits further investigation.

3.3 RALF Peptides Involved in Nodule Organogenesis and Infection

RALF is a 49-amino-acid-long peptide, processed from the C terminus of its preproprotein, probably by cleavage at a dibasic amino acid motif (Bedinger et al. 2010). At the N terminus of the preproprotein, the presence of a signal peptide suggested extracellular secretion (Pearce et al. 2001). Highly conserved homologs are present in various species throughout the plant kingdom, with 39 members in *A. thaliana* and 40 members in *M. truncatula* (Combier et al. 2008; Olsen et al. 2002; Pearce et al. 2001; also see chapter "Plant Peptide Signaling: An Evolutionary Adaptation"). RALF peptide genes are expressed in various tissues including shoots, roots, flowers, leaves, and nodules (Combier et al. 2008; Germain et al. 2005; Pearce et al. 2001). RALF was originally discovered during the purification of tobacco systemin as a component that induces rapid alkalinization of the medium, as well as rapid activation of MAPK activity in tobacco cells (Pearce et al. 2001). In contrast to systemin, no defense responses were observed in tobacco cell cultures when supplied at nanomolar concentrations (Pearce et al. 2001). RALF has been suggested to be involved in the regulation of root growth and development, as synthetic tomato RALF peptides, when supplied to tomato and *Arabidopsis* seedling, induced a small enlargement of meristematic cells and an immediate arrest of root growth (Pearce et al. 2001; also see chapter "Peptides Regulating Root Growth"). Knockdown of RALF expression in *Nicotiana attenuata* resulted in plants with longer roots and abnormal root hair development (Wu et al. 2007). According to Covey et al. (2010), RALF is involved in pollen tube elongation. Finally, Combier et al. (2008) showed evidence for the involvement of an *M. truncatula RALF*, named *MtRALF1*, during nodulation.

Microarray and expression analysis confirmed that *MtRALF1* is transiently upregulated by Nod factor treatment (Combier et al. 2008). A reduced number of nodules were observed on roots ectopically expressing *MtRALF1*, while the number of aborted infection threads, with a bulbous and saclike phenotype, was strongly increased (Combier et al. 2008). The few nodules that could form on the roots ectopically expressing *MtRALF1* had an abnormal development and remained small, with poor rhizobial colonization. These results suggest that *MtRALF1* might be involved in nodule development and rhizobial infection, more specifically during the initial progression of infection threads through the outer cortical cell layers. As both inner and outer cortical cells initiate the cell cycle upon inoculation, RALF peptides might be involved in a common step in cell cycle activation. Until now, two proteins were identified as possible components of a RALF receptor complex (Scheer and Ryan 1999). Further investigation of these proteins, and the identification of new RALF interacting partners, is needed to obtain a clear insight in the role of this peptide.

3.4 DVL1/ROT4

DVL1/ROT4 are peptides that are also derived from the C terminus of larger preproproteins. In *A. thaliana*, a gene family of 23 members was identified and the peptides are derived from a conserved 29-amino-acid-long domain (Narita et al. 2004; Wen et al. 2004). The peptide genes were identified during screens for leaf shape and fruit mutants of *A. thaliana*. A dominant mutation of ROT4 (*rot4-1D*) resulted in plants with short leaves and floral organs, which was mainly caused by a decrease in cell proliferation (Narita et al. 2004). *Dvl1-1D* plants exhibited shortened petioles and siliques, rounder leaves, and moderately horned fruits (Wen et al. 2004). Hence, ROT4 and DVL1 affect plant growth and development by regulating polar cell proliferation. The functions of the DVL1/ROT4 family members are expected to be redundant, as ectopic expression caused similar pleiotropic effects on plant development (Wen et al. 2004). However, divergent expression patterns were reported for the family members: *ROT4* is expressed in young leaves and in the SAM; the expression of *DVL1* has been localized to leaves; and the other family members are mostly expressed in flowers and roots (Narita et al. 2004; Wen et al. 2004). In *M. truncatula*, ten members of the *DVL1/ROT4* family were identified, and expression analysis revealed that one of them, *MtDVL1*, is transiently upregulated after Nod factor treatments during initial phases of nodulation (Combier et al. 2008). Ectopic expression of *MtDVL1* resulted in a reduction in nodule number and in a strong increase in the number of aborted infection threads in outer cortical cell layers, identical to what was observed for *MtRALF1* (Combier et al. 2008). Nevertheless, in contrast to what was observed for *MtRALF1*, nodule development and nitrogen fixation were as normal. Based on all these phenotypes, *MtDVL1* is supposed to act during rhizobial infection at the initial progression of infection threads through the outer cortical cell layers. Because *DVL1* family members are involved in cell proliferation, Combier et al. (2008) suggest a role for *MtDVL1* during cellular events leading to the differentiation for preinfection thread formation. While DVL peptides control polar cell proliferation in lateral organs, they might also play a role in the polar growth of root hairs and as such be indirectly linked to nodulation (Narita et al. 2004; Wen et al. 2004). Evidence for this hypothesis is still lacking, but molecular analysis of root hair growth might reveal the answer.

3.5 *Long Peptides or Small Proteins: The Role of NCRs in Symbiosome Development*

All the above-mentioned peptides seem to affect cell division and differentiation in a certain way. Another group of peptides or small proteins, the NCR proteins, play an important role in controlling the microsymbiont during nodulation. NCRs are 30–50 amino acids long of which 4 or 6 amino acids are cysteine residues located at conserved positions in the C terminus. The other amino acid residues are extremely divergent. A signal peptide at the N terminus targets the proteins to the secretory pathway (Mergaert et al. 2003). More than 300 NCR members have been discovered up to now, and they all occur in the galegoid legumes such as *Medicago* spp. (Alunni et al. 2007; Fedorova et al. 2002; Graham et al. 2004; Györgyey et al. 2000, Mergaert et al. 2003), *Pisum sativum* (Kardailsky et al. 1993; Kato et al. 2002; Scheres et al. 1990), *Trifolium repens* (Crockard et al. 2002), *Astragalus sinicus* (Chou et al. 2006), *Galega orientalis* (Kaijalainen et al. 2002), and *Vicia faba* (Frühling et al. 2000).

NCRs are exclusively nodule specific, except for two members which were also expressed in mycorrhizal roots (Mergaert et al. 2003). Different spatial and temporal expression patterns were observed for different NCR genes suggesting that NCRs play a role in different nodule tissues and cell types and at several stages of nodule organogenesis (Mergaert et al. 2003). Many NCRs might work synergistically, as overlapping expression patterns were also observed. From the results of macroarray experiments, NCRs seemed to be coregulated with a calmodulin (CaM)-like protein and two signal peptide peptidases (Mergaert et al. 2003). Mergaert and associates (2003) therefore suggest that these genes might be involved in similar functions or pathways (Young 2000). The exact function of the NCRs is not known, but Mergaert et al. (2003) and Alunni et al. (2007) suggested that they act as antimicrobial peptides especially recruited for bacteroid differentiation (Brogden 2005). Indeed, NCRs were recently indicated to be involved in the terminal and irreversible differentiation of rhizobia to bacteroids in the *S. meliloti–M. truncatula* symbiosis (Van de Velde et al. 2010).

4 CLE Peptide Signaling During Nematode–Plant Interactions

The formation of nodules as feeding structures for rhizobia resembles the development of cyst- or gall-formed feeding structures of plant parasitic nematodes. While nodulation is a mutualistic process, nematode infection causes major damage to plants. Cyst and root knot nematodes are obligate sedentary endoparasites that complete their life cycle in the plant. Intercellular invasion close to the root tip is followed by selection of target cells usually closely behind the root tip (Bird 2004). Nematodes produce a complex mixture of effectors in esophageal glands which they then inject into target plant cells using a stylet. This triggers cell division and differentiation, similar to reactivation of the cell cycle during nodulation (Goverse et al. 2000), and involves activation of similar genes in the host (Favery et al. 2002; Koltai et al. 2001; Mathesius 2003; Weerasinghe et al. 2005). Cyst nematode secretions stimulate the formation of syncytia by progressive dissolution of neighboring cell walls. The syncytial protoplasm shows typical features of a plant meristem, including dense cytoplasm, small vacuoles, prominent nuclei, and a high metabolic rate. Root knot nematodes cause endoreduplication of target cells (typically vascular parenchyma cells), leading to the formation of giant cells, which are surrounded by dividing pericycle, endodermis, and cortical cells, which together form a root gall. There is strong evidence that plant parasitic nematodes use CLE peptides to initiate feeding structures in the host root.

An esophageal gland cell cDNA library yielded two CLE-related proteins from the soybean cyst nematode *Heterodera glycines* (Gao et al. 2003; Wang et al. 2005). Five further genes, which contained multiple CLE motifs, were identified from the potato cyst nematode *Globodera rostochiensis* (Lu et al. 2009) and two from the beet cyst nematode *H. schachtii* (Wang et al. 2011). Expression of cyst nematode CLE genes in the *clv3* mutant of *Arabidopsis* restored the wild-type phenotype, suggesting that the nematode peptides are functional in the plant (Wang et al. 2005; Lu et al. 2009). Silencing of cyst nematode CLE genes by soaking with RNA or by ingestion of host-derived RNA interference constructs was shown to reduce nematode infection of the host (Bhakhetia et al. 2007; Patel et al. 2008). An antibody localized the HgCLE 1 and 2 propeptides in the nematode as well as in the syncytium cytoplasm, where it might be posttranslationally modified before targeting to the apoplast (Wang et al. 2010). Deletion studies showed that while the CLE motif is necessary for function of the HgCLE peptides, the signal peptide is dispensable. Instead, it is likely that the variable domains of the cyst nematode CLE peptides are necessary for targeting of the peptide to the apoplast (Wang et al. 2010). Cyst nematode CLE peptides are likely to be perceived by the same or very similar receptors to those perceiving plant CLE peptides. For example, the CLV2 and CORYNE receptors, which usually perceive the CLV3 peptide, are expressed in feeding sites (Replogle et al. 2011). *Crn-1/clv2-1* double mutants showed partial resistance to *H. schachtii*, suggesting that these receptors are necessary for infection but likely are not the only receptors for nematode CLE peptides (Replogle et al. 2011). Subsequently, Guo et al. (2011) showed that *G. rostochiensis* CLE peptides are processed by host proteases and bind directly to the CLV2, BAM1, and BAM2 receptors.

A root knot nematode-specific CLE gene, 16D10, was identified from *Meloidogyne incognita* (Huang et al. 2006b). Again, this protein is produced in esophageal glands and has a CLE motif as well as a secretion signal. It also includes a phosphorylation site and is likely to interact with transcription factors of the GRAS family. While its expression in plants caused root elongation and branching, it was not able to restore the *clv3* mutant phenotype (Huang et al. 2006b). Silencing of 16D10 in the nematode by expression of an RNA interference construct in the host plant significantly reduced infection rate from root knot nematodes, strongly

Table 1 Summary of peptides involved in plant biotic interactions

Peptide family	Expression patterns	Mutant or knockdown phenotypes	Overexpression phenotype	Putative functions in symbiosis
CLE peptides	Proliferating nodule primordia; meristem and infection zone of mature nodules	N/A	Reduced nodule numbers	Control of nodule numbers via systemic signaling
	Injected into feeding cells by nematodes	Reduced nematode infection	*WUS*-like phenotypes, reduced root length	Initiation of nematode feeding sites
ENOD40	Early nodule primordia Uninfected cells in nodules Nodule vascular bundle	Reduced numbers of nodule primordia	Increased numbers of cortical cell divisions and nodules	Cell division and differentiation during nodule development; sucrose supply in nodules
RALF peptides	Induced by Nod factor treatment	N/A	Abortion of infection threads, reduced numbers of nodules	Infection thread formation and nodule development
DLV1/ROT4	Induced by Nod factor treatment	N/A	Abortion of infection threads	Infection thread formation
NCRs	Nodules only	Defects in bacteroid differentiation	N/A	Antimicrobial peptides Bacteroid differentiation

suggesting that this peptide is involved in feeding site formation (Huang et al. 2006a). As CLE peptides have otherwise been found exclusively in plants, a future question remains whether CLE peptide-encoding genes have been acquired by nematodes through horizontal gene transfer or by convergent evolution (Mitchum et al. 2008).

5 Conclusions

Rhizobia have evolved the ability to induce specific plant peptides in the host that play a role in the local and systemic regulation of nodule development (Table 1, Fig. 1). All except the NCR family of peptides are very closely related or identical

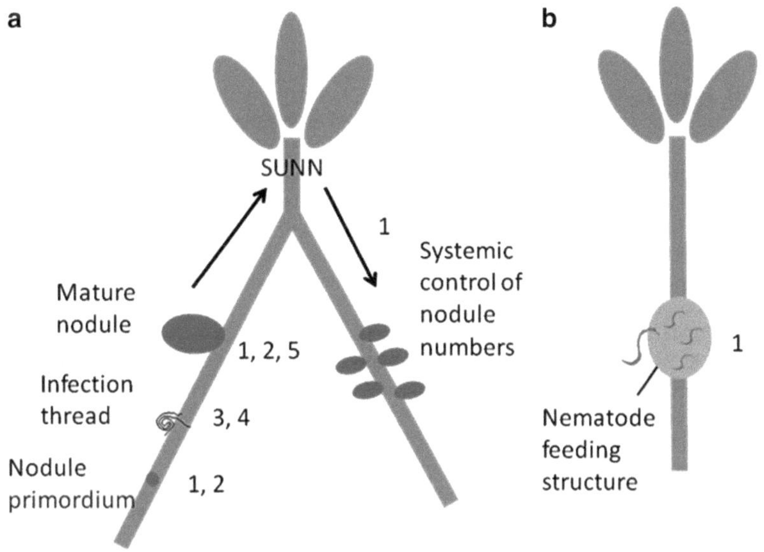

Fig. 1 Schematic diagram of symbiotic processes controlled by peptides. (**a**) Nodulation. Peptides of the CLE (1), ENOD40 (2), RALF (3), DLV1/ROT4 (4), and NCR (5) families have roles at different stages of nodulation, including nodule primordium formation, infection thread development, nodule and bacteroid differentiation, as well as the systemic control of nodule numbers involving the receptor-like kinase SUNN/NARK/HAR1/SYM29 acting in the shoot. (**b**) Plant–parasitic nematode interactions. CLE peptides are encoded by parasitic cyst and root knot nematodes and injected into developing feeding cells into the host via the nematode stylet. CLE peptides are required for successful development of nematode feeding structures including root galls and syncytia. Numbers refer to groups of peptides as indicated in legend to *panel* (**a**)

to peptides involved in nonsymbiotic aspects of plant development. It therefore appears that rhizobia have modified the action of these peptides to achieve the development of a new root organ. In contrast, parasitic nematodes that also induce cell divisions inside host roots and invade roots have acquired genes encoding CLE peptides, which they inject into host cells to manipulate their development. With new methodologies to detect peptides both by bioinformatics and through mass spectrometry (also see chapters "Methods to Isolate and Identify New Plant Signaling Peptides" and "Methods to Identify New Partners of Plant Signaling Peptides"), it is likely that many more peptides will be discovered that play a role in symbiotic interactions of plants with microorganisms. It would be of particular interest to find out about the long-distance mobility, exact processing, and perception of these peptides in the plant and whether new receptors have evolved to perceive symbiosis-related peptides compared to similar or identical peptides that play a role during other developmental processes.

References

Alunni B, Kevei Z, Redondo-Nieto M, Kondorosi A, Mergaert P, Kondorosi E (2007) Genomic organization and evolutionary insights on GRP and NCR genes, two large nodule-specific gene families in *Medicago truncatula*. Mol Plant Microbe Interact 20:1138–1148

Bardou F, Merchan F, Ariel F, Crespi M (2011) Dual RNAs in plants. Biochimie 93:1950–1954

Bedinger PA, Pearce G, Covey PA (2010) RALFs: peptide regulators of plant growth. Plant Signal Behav 5:1342–1346

Bhakhetia M, Urwin PE, Atkinson HJ (2007) qPCR analysis and RNAi define pharyngeal gland cell-expressed genes of *Heterodera glycines* required for initial interactions with the host. Mol Plant Microbe Interact 20:306–312

Bird DM (2004) Signaling between nematodes and plants. Curr Opin Plant Biol 7:372–376

Bleckmann A, Weidtkamp-Peters S, Seidel C, Simon R (2010) Stem cell signalling in Arabidopsis requires CRN to localize CLV2 to the plasma membrane. Plant Physiol 152:166–176

Boot KJM, van Brussel AAN, Tak T, Spaink HP, Kijne JW (1999) Lipochitin oligosaccharides from *Rhizobium leguminosarum* bv. *viciae* reduce auxin transport capacity in *Vicia sativa* subsp *nigra* roots. Mol Plant Microbe Interact 12:839–844

Brand U, Fletcher JC, Hobe M, Meyerowitz EM, Simon R (2000) Dependence of stem cell fate in Arabidopsis in a feedback loop regulated by CLV3 activity. Science 289:617–619

Brewin NJ (2004) Plant cell wall remodeling in the *Rhizobium*-legume symbiosis. Crit Rev Plant Sci 23:293–316

Brogden KA (2005) Antimicrobial peptides: pore formers or metabolic inhibitors in bacteria? Nat Rev Microbiol 3:238–250

Butenko MA, Patterson SE, Grini, PE, Stenvik, GE, Amundsen, SS , Mandal A, Aalen RB (2003) INFLORESCENCE DEFICIENT IN ABSCISSION controls floral organ abscission in Arabidopsis and identifies a novel family of putative ligands in plants. Plant Cell 15: 2296–2307

Campalans A, Kondorosi A, Crespi M (2004) *Enod40*, a short open reading frame-containing mRNA, induces cytoplasmic localization of a nuclear RNA binding protein in *Medicago truncatula*. Plant Cell 16:1047–1059

Carroll BJ, McNeil DL, Gresshoff PM (1985a) A supernodulation and nitrate-tolerant symbiotic (nts) soybean mutant. Plant Physiol 78:34–40

Carroll BJ, McNeil DL, Gresshoff PM (1985b) Isolation and properties of soybean [*Glycine max* (L.) Merr.] mutants that nodulate in the presence of high nitrate concentrations. Proc Natl Acad Sci USA 82:4162–4166

Casson SA, Chilley PM, Topping JF, Evans IM, Souter MA, Lindsey K (2002) The POLARIS gene of *Arabidopsis* encodes a predicted peptide required for correct root growth and leaf vascular patterning. Plant Cell 14:1705–1721

Charon C, Johansson C, Kondorosi E, Kondorosi A, Crespi M (1997) *enod40* induces dedifferentiation and division of root cortical cells in legumes. Proc Natl Acad Sci USA 94:8901–8906

Charon C, Sousa C, Crespi M, Kondorosi A (1999) Alteration of *enod40* expression modifies *Medicago truncatula* root nodule development induced by *Sinorhizobium meliloti*. Plant Cell 11:1953–1966

Charon C, Moreno AB, Bardou F, Crespi M (2010) Non-protein-coding RNAs and their interacting RNA-binding proteins in the plant cell nucleus. Mol Plant 3:729–739

Chen YF, Matsubayashi Y, Sakagami Y (2000) Peptide growth factor phytosulfokine-alpha contributes to the pollen population effect. Planta 211:752–755

Chou MX, Wei XY, Chen DS, Zhou JC (2006) Thirteen nodule-specific or nodule-enhanced genes encoding products homologous to cysteine cluster proteins or plant lipid transfer proteins are identified in *Astragalus sinicus* L. by suppressive subtractive hybridization. J Exp Bot 57: 2673–2685

Clark SE, Running MP, Meyerowitz EM (1995) Clavata3 is a specific regulator of shoot and foral meristem development affecting the same processes as Clavata1. Development 121: 2057–2067

Cock JM, McCormick S (2001) A large family of genes that share homology with CLAVATA3. Plant Physiol 126:939–942

Combier JP, Kuster H, Journet EP, Hohnjec N, Gamas P, Niebel A (2008) Evidence for the involvement in nodulation of the two small putative regulatory peptide-encoding genes MtRALFL1 and MtDVL1. Mol Plant Microbe Interact 21:1118–1127

Compaan B, Ruttink T, Albrecht C, Meeley R, Bisseling T, Franssen H (2003) Identification and characterization of a *Zea mays* line carrying a transposon-tagged ENOD40. Biochim Biophys Acta 1629:84–91

Corich V, Goormachtig S, Lievens S, Van Montagu M, Holsters M (1998) Patterns of *ENOD40* gene expression in stem-borne nodules of *Sesbania rostrata*. Plant Mol Biol 37:67–76

Covey PA, Subbaiah CC, Parsons RL, Pearce G, Lay FT, Anderson MA, Ryan CA, Bedinger PA (2010) A pollen-specific RALF from tomato that regulates pollen tube elongation. Plant Physiol 153:703–715

Crespi M, Galvez S (2000) Molecular mechanisms in root nodule development. J Plant Growth Regul 19:155–166

Crespi MD, Jurkevitch E, Poiret M, d'Aubenton-Carafa Y, Petrovics G, Kondorosi E, Kondorosi A (1994) *enod40*, a gene expressed during nodule organogenesis, codes for a non-translatable RNA involved in plant growth. EMBO J 13:5099–5112

Crockard A, Bjourson J, Dazzo B, Cooper JE (2002) A white clover nodulin gene, *dd23b*, encoding a cysteine cluster protein, is expressed in roots during the very early stages of interaction with *Rhizobium leguminosarum* biovar *trifolii* and after treatment with chitolipooligosaccharide Nod factors. J Plant Res 115:439–447

D'Haeze W, Holsters M (2002) Nod factor structures, responses, and perception during initiation of nodule development. Glycobiology 12:79R–105R

Degenhardt DC, Refi-Hind S, Stratmann JW, Lincoln DE (2010) Systemin and jasmonic acid regulate constitutive and herbivore-induced systemic volatile emissions in tomato, *Solanum lycopersicum*. Phytochemistry 71:2024–2037

Delves AC, Mathews A, Day DA, Carter AS, Carroll BJ, Gresshoff PM (1986) Regulation of the soybean-*Rhizobium* nodule symbiosis by shoot and root factors. Plant Physiol 82:588–590

Dey M, Complainville A, Charon C, Torrizo L, Kondorosi A, Crespi M, Datta S (2004) Phytohormonal responses in *enod40*-overexpressing plants of *Medicago truncatula* and rice. Physiol Plant 120:132–139

DeYoung BJ, Clark SE (2008) BAM receptors regulate stem cell specification and organ development through complex interactions with CLAVATA signaling. Genetics 180:895–904

Ding Y, Oldroyd GE (2009) Positioning the nodule, the hormone dictum. Plant Signal Behav 4:89–93

Ding Y, Kalo P, Yendrek C, Sun J, Liang Y, Marsh JF, Harris JM, Oldroyd GE (2008) Abscisic acid coordinates nod factor and cytokinin signaling during the regulation of nodulation in *Medicago truncatula*. Plant Cell 20:2681–2695

Etchells JP, Turner SR (2010) The PXY-CLE41 receptor ligand pair defines a multifunctional pathway that controls the rate and orientation of vascular cell division. Development 137: 767–774

Fang Y, Hirsch AM (1998) Studying early nodulin gene *ENOD40* expression and induction by nodulation factor and cytokinin in transgenic alfalfa. Plant Physiol 116:53–68

Favery B, Complainville A, Vinardell JM, Lecomte P, Vaubert D, Mergaert P, Kondorosi A, Kondorosi E, Crespi M, Abad P (2002) The endosymbiosis-induced genes *ENOD40* and *CCS52a* are involved in endoparasitic-nematode interactions in *Medicago truncatula*. Mol Plant Microbe Interact 15:1008–1013

Fedorova M, van de Mortel J, Matsumoto PA, Cho J, Town CD, VandenBosch KA, Gantt JS, Vance CP (2002) Genome-wide identification of nodule-specific transcripts in the model legume *Medicago truncatula*. Plant Physiol 130:519–537

Felle HH, Kondorosi E, Kondorosi A, Schultze M (1999) Elevation of the cytosolic free [Ca2+] is indispensable for the transduction of the Nod factor signal in alfalfa. Plant Physiol 121: 273–280

Ferguson BJ, Mathesius U (2003) Signaling interactions during nodule development. J Plant Growth Regul 22:47–72

Ferguson BJ, Ross JJ, Reid JB (2005) Nodulation phenotypes of gibberellin and brassinosteroid mutants of pea. Plant Physiol 138:2396–2405

Ferguson BJ, Indrasumunar A, Hayashi S, Lin MH, Lin YH, Reid DE, Gresshoff PM (2010) Molecular analysis of legume nodule development and autoregulation. J Integr Plant Biol 52:61–76

Fiers M, Golemiec E, Xu J, van der Geest L, Heidstra R, Stiekema W, Liu CM (2005) The 14-amino acid CLV3, CLE19, and CLE40 peptides trigger consumption of the root meristem in *Arabidopsis* through a CLAVATA2-dependent pathway. Plant Cell 17:2542–2553

Flemetakis E, Kavroulakis N, Quaedvlieg NE, Spaink HP, Dimou M, Roussis A, Katinakis P (2000) *Lotus japonicus* contains two distinct *ENOD40* genes that are expressed in symbiotic, nonsymbiotic, and embryonic tissues. Mol Plant Microbe Interact 13:987–994

Francisco PB, Harper JE (1995) Translocatable leaf signal autoregulates soybean nodulation. Plant Sci 107:167–176

Frugier F, Kosuta S, Murray JD, Crespi M, Szczyglowski K (2008) Cytokinin: secret agent of symbiosis. Trends Plant Sci 13:115–120

Frühling M, Albus U, Hohnjec N, Geise G, Pühler A, Perlick AM (2000) A small family of broad bean codes for late nodulins containing conserved cysteine clusters. Plant Sci 152:67–77

Fukuda H, Hirakawa Y, Sawa S (2007) Peptide signaling in vascular development. Curr Opin Plant Biol 10:477–482

Gage DJ (2004) Infection and invasion of roots by symbiotic, nitrogen-fixing rhizobia during nodulation of temperate legumes. Microbiol Mol Biol Rev 68:280–300

Gao B, Allen R, Maier T, Davis EL, Baum TJ, Hussey RS (2003) The parasitome of the phytonematode *Heterodera glycines*. Mol Plant Microbe Interact 16:720–726

Germain H, Chevalier E, Caron S, Matton DP (2005) Characterization of five RALF-like genes from *Solanum chacoense* provides support for a developmental role in plants. Planta 220: 447–454

Gonzalez-Rizzo S, Crespi M, Frugier F (2006) The *Medicago truncatula* CRE1 cytokinin receptor regulates lateral root development and early symbiotic interaction with *Sinorhizobium meliloti*. Plant Cell 18:2680–2693

Gordon AJ, Minchin FR, James CL, Komina O (1999) Sucrose synthase in legume nodules is essential for nitrogen fixation. Plant Physiol 120:867–878

Goverse A, de Almeida EJ, Verhees J, van der Krol S, Helder J, Gheysen G (2000) Cell cycle activation by plant parasitic nematodes. Plant Mol Biol 43:747–761

Graham MA, Silverstein KA, Cannon SB, VandenBosch KA (2004) Computational identification and characterization of novel genes from legumes. Plant Physiol 135:1179–1197

Grønlund M, Roussis A, Flemetakis E, Quaedvlieg NE, Schlaman HR, Umehara Y, Katinakis P, Stougaard J, Spaink HP (2005) Analysis of promoter activity of the early nodulin *Enod40* in *Lotus japonicus*. Mol Plant Microbe Interact 18:414–427

Grunewald W, van Noorden G, Van Isterdael G, Beeckman T, Gheysen G, Mathesius U (2009) Manipulation of auxin transport in plant roots during *Rhizobium* symbiosis and nematode parasitism. Plant Cell 21:2553–2562

Gultyaev AP, Roussis A (2007) Identification of conserved secondary structures and expansion segments in *enod40* RNAs reveals new enod40 homologues in plants. Nucleic Acids Res 35:3144–3152

Guo Y, Clark SE (2010) Membrane distributions of two ligand-binding receptor complexes in the CLAVATA pathway. Plant Signal Behav 5:1442–1445

Guo Y, Han L, Hymes M, Denver R, Clark SE (2010) CLAVATA2 forms a distinct CLE-binding receptor complex regulating Arabidopsis stem cell specification. Plant J 63:889–900

Guo Y, Ni Y, Denver R, Wang X, Clark SE (2011) Mechanisms of molecular mimicry of plant CLE peptide ligands by the parasitic nematode *Globodera rostochiensis*. Plant Physiol 157:476–484

Guzzo F, Portaluppi P, Grisi R, Barone S, Zampieri S, Franssen H, Levi M (2005) Reduction of cell size induced by *enod40* in *Arabidopsis thaliana*. J Exp Bot 56:507–513

Györgyey J, Vaubert D, Jimenez-Zurdo JI, Charon C, Troussard L, Kondorosi A, Kondorosi E (2000) Analysis of *Medicago truncatula* nodule expressed sequence tags. Mol Plant Microbe Interact 13:62–71

Hanai H, Matsuno T, Yamamoto M, Matsubayashi Y, Kobayashi T, Kamada H, Sakagami Y (2000) A secreted peptide growth factor, phytosulfokine, acting as a stimulatory factor of carrot somatic embryo formation. Plant Cell Physiol 41:27–32

Heidstra R, Yang WC, Yalcin Y, Peck S, Emons AM, van Kammen A, Bisseling T (1997) Ethylene provides positional information on cortical cell division but is not involved in Nod factor-induced root hair tip growth in *Rhizobium*-legume interaction. Development 124:1781–1787

Hirakawa Y, Shinohara H, Kondo Y, Inoue A, Nakanomyo I, Ogawa M, Sawa S, Ohashi-Ito K, Matsubayashi Y, Fukuda H (2008) Non-cell-autonomous control of vascular stem cell fate by a CLE peptide/receptor system. Proc Natl Acad Sci USA 105:15208–15213

Hirsch AM, Fang Y (1994) Plant hormones and nodulation: what's the connection? Plant Mol Biol 26:5–9

Hobe M, Muller R, Grunewald M, Brand U, Simon R (2003) Loss of CLE40, a protein functionally equivalent to the stem cell restricting signal CLV3, enhances root waving in Arabidopsis. Dev Genes Evol 213:371–381

Huang G, Allen R, Davis EL, Baum TJ, Hussey RS (2006a) Engineering broad root-knot nematode resistance in transgenic plants by RNAi silencing of a conserved and essential root-knot nematode parasitism gene. Proc Natl Acad Sci USA 103:14302–14306

Huang G, Dong R, Allen R, Davis EL, Baum TJ, Hussey RS (2006b) A root-knot nematode secretory peptide functions as a ligand for a plant transcription factor. Mol Plant Microbe Interact 19:463–470

Igasaki T, Akashi N, Ujino-Ihara T, Matsubayashi Y, Sakagami Y, Shinohara K (2003) Phytosulfokine stimulates somatic embryogenesis in *Cryptomeria japonica*. Plant Cell Physiol 44:1412–1416

Ito Y, Nakanomyo I, Motose H, Iwamoto K, Sawa S, Dohmae N, Fukuda H (2006) Dodeca-CLE peptides as suppressors of plant stem cell differentiation. Science 313:842–845

Jiang Q, Gresshoff PM (2002) Shoot control of hypernodulation and aberrant root formation in the *har1* mutant of *Lotus japonicus*. Funct Plant Biol 29:1371–1376

Kachroo A, Nasrallah ME, Nasrallah JB (2002) Self-incompatibility in the Brassicaceae: receptor-ligand signaling and cell-to-cell communication. Plant Cell 14(Suppl):S227–S238

Kaijalainen S, Schroda M, Lindstrom K (2002) Cloning of nodule-specific cDNAs of *Galega orientalis*. Physiol Plant 114:588–593

Kardailsky I, Yang WC, Zalensky A, van Kammen A, Bisseling T (1993) The pea late nodulin gene *PsNOD6* is homologous to the early nodulin genes *PsENOD3/14* and is expressed after the leghaemoglobin genes. Plant Mol Biol 23:1029–1037

Kato T, Kawashima K, Miwa M, Mimura Y, Tamaoki M, Kouchi H, Suganuma N (2002) Expression of genes encoding late nodulins characterized by a putative signal peptide and conserved cysteine residues is reduced in ineffective pea nodules. Mol Plant Microbe Interact 15:129–137

Kawaguchi M, ImaizumiAnraku H, Fukai S, Syono K (1996) Unusual branching in the seedlings of *Lotus japonicus* – Gibberellins reveal the nitrogen-sensitive cell divisions within the pericycle on roots. Plant Cell Physiol 37:1389–1396

Kinkema M, Gresshoff PM (2008) Investigation of downstream signals of the soybean autoregulation of nodulation receptor kinase GmNARK. Mol Plant Microbe Interact 21: 1337–1348

Kinoshita A, Betsuyaku S, Osakabe Y, Mizuno S, Nagawa S, Stahl Y, Simon R, Yamaguchi-Shinozaki K, Fukuda H, Sawa S (2010) RPK2 is an essential receptor-like kinase that transmits the CLV3 signal in *Arabidopsis*. Development 137:3911–3920

Kobayashi T, Eun C, Hanai H, Matsubayashi Y, Sakagami Y, Kamada H (1999) Phytosulfokine-a, a peptidyl plant growth factor, stimulates somatic embryogenesis in carrot. J Exp Bot 50: 1123–1128

Koltai H, Dhandaydham M, Opperman C, Thomas J, Bird D (2001) Overlapping plant signal transduction pathways induced by a parasitic nematode and a rhizobial endosymbiont. Mol Plant Microbe Interact 14:168–1177

Kondo T, Sawa S, Kinoshita A, Mizuno S, Kakimoto T, Fukuda H, Sakagami Y (2006) A plant peptide encoded by CLV3 identified by *in situ* MALDI-TOF MS analysis. Science 313:845–848

Kosslak RM, Bohlool BB (1984) Suppression of nodule development of one side of a split-root system of soybeans caused by prior inoculation of the other side. Plant Physiol 75:125–130

Kouchi H, Hata S (1993) Isolation and characterization of novel nodulin cDNAs representing genes expressed at early stages of soybean nodule development. Mol Gen Genet 238:106–119

Kouchi H, Takane K, So RB, Ladha JK, Reddy PM (1999) Rice *ENOD40*: isolation and expression analysis in rice and transgenic soybean root nodules. Plant J 18:121–129

Kouchi H, Imaizumi-Anraku H, Hayashi M, Hakoyama T, Nakagawa T, Umehara Y, Suganuma N, Kawaguchi M (2010) How many peas in a pod? Legume genes responsible for mutualistic symbioses underground. Plant Cell Physiol 51:1381–1397

Krusell L, Madsen LH, Sato S, Aubert G, Genua A, Szczyglowski K, Duc G, Kaneko T, Tabata S, de Bruijn F, Pajuelo E, Sandal N, Stougaard J (2002) Shoot control of root development and nodulation is mediated by a receptor-like kinase. Nature 420:422–426

Krusell L, Sato N, Fukuhara I, Koch BE, Grossmann C, Okamoto S, Oka-Kira E, Otsubo Y, Aubert G, Nakagawa T, Sato S, Tabata S, Duc G, Parniske M, Wang TL, Kawaguchi M, Stougaard J (2010) The *Clavata2* genes of pea and *Lotus japonicus* affect autoregulation of nodulation. Plant J 65:861–871

Kumagai H, Kinoshita E, Ridge RW, Kouchi H (2006) RNAi knock-down of ENOD40s leads to significant suppression of nodule formation in *Lotus japonicus*. Plant Cell Physiol 47:1102–1111

Larsen K (2003) Molecular cloning and characterization of a cDNA encoding a ryegrass (*Lolium perenne*) *ENOD40* homologue. J Plant Physiol 160:675–687

Lenhard M, Laux T (2003) Stem cell homeostasis in the *Arabidopsis* shoot meristem is regulated by intercellular movement of CLAVATA3 and its sequestration by CLAVATA1. Development 130:3163–3173

Lievens S, Goormachtig S, Den Herder J, Capoen W, Mathis R, Hedden P, Holsters M (2005) Gibberellins are involved in nodulation of *Sesbania rostrata*. Plant Physiol 139:1366–1379

Lim CW, Lee YW, Hwang CH (2011) Soybean nodule-enhanced CLE peptides in roots act as signals in GmNARK-mediated nodulation suppression. Plant Cell Physiol 52:1613–1627

Lohar DP, Schaff JE, Laskey JG, Kieber JJ, Bilyeu KD, Bird DM (2004) Cytokinins play opposite roles in lateral root formation, and nematode and Rhizobial symbioses. Plant J 38:203–214

Lu S-W, Chen S, Wang J, Yu H, Chronis D, Mitchum MG, Wang X (2009) Structural and functional diversity of *CLAVATA3/ESR* (*CLE*)-like genes from the potato cyst nematode *Globodera rostochiensis*. Mol Plant Microbe Interact 22:1128–1142

Magori S, Kawaguchi M (2009) Long-distance control of nodulation: molecules and models. Mol Cells 27:129–134

Mathesius U (2003) Conservation and divergence of signaling pathways between roots and soil microbes – the *Rhizobium*-legume symbiosis compared to the development of lateral roots, mycorrhizal interactions and nematode-induced galls. Plant Soil 255:105–119

Mathesius U, Schlaman HR, Spaink HP, Sautter C, Rolfe BG, Djordjevic MA (1998) Auxin transport inhibition precedes root nodule formation in white clover roots and is regulated by flavonoids and derivatives of chitin oligosaccharides. Plant J 14:23–34

Mathesius U, Charon C, Rolfe BG, Kondorosi A, Crespi M (2000) Temporal and spatial order of events during the induction of cortical cell divisions in white clover by *Rhizobium leguminosarum* bv. *trifolii* inoculation or localized cytokinin addition. Mol Plant Microbe Interact 13:617–628

Matsubayashi Y, Sakagami Y (1996) Phytosulfokine, sulfated peptides that induce the proliferation of single mesophyll cells of *Asparagus officinalis* L. Proc Natl Acad Sci USA 93: 7623–7627

Matsubayashi Y, Takagi L, Omura N, Morita A, Sakagami Y (1999) The endogenous sulfated pentapeptide phytosulfokine-alpha stimulates tracheary element differentiation of isolated mesophyll cells of Zinnia. Plant Physiol 120:1043–1048

Matvienko M, Van de Sande K, Pawlowski K, Van Kammen A, Franssen H, Bisseling T (1996) *Nicotiana tabacum* SR1 contains two *ENOD40* homologs. In: Stacey G, Mullin B, Gresshoff PM (eds) Biology of plant-microbe interactions. International Society for Molecular Plant-Microbe Interactions, St. Paul, MN, pp 387–391

Mayer KFX, Schoof H, Haecker A, Lenhard M, Jurgens G, Laux T (1998) Role of WUSCHEL in regulating stem cell fate in the Arabidopsis shoot meristem. Cell 95:805–815

McGurl B, Pearce G, Orozco-Cardenas M, Ryan CA (1992) Structure, expression, and antisense inhibition of the systemin precursor gene. Science 255:1570–1573

Meng L, Feldman LJ (2010) CLE14/CLE20 peptides may interact with CLAVATA2/CORYNE receptor-like kinases to irreversibly inhibit cell division in the root meristem of Arabidopsis. Planta 232:1061–1074

Meng L, Ruth KC, Fletcher JC, Feldman L (2010) The roles of different CLE domains in Arabidopsis CLE polypeptide activity and functional specificity. Mol Plant 3:760–772

Mergaert P, Nikovics K, Kelemen Z, Maunoury N, Vaubert D, Kondorosi A, Kondorosi E (2003) A novel family in *Medicago truncatula* consisting of more than 300 nodule-specific genes coding for small, secreted polypeptides with conserved cysteine motifs. Plant Physiol 132: 161–173

Minami E, Kouchi H, Cohn JR, Ogawa T, Stacey G (1996) Expression of the early nodulin, *ENOD40*, in soybean roots in response to various lipo-chitin signal molecules. Plant J 10: 23–32

Mitchum MG, Wang X, Davis EL (2008) Diverse and conserved roles of CLE peptides. Curr Opin Plant Biol 11:75–81

Miwa H, Sun J, Oldroyd GE, Downie JA (2006) Analysis of calcium spiking using a cameleon calcium sensor reveals that nodulation gene expression is regulated by calcium spike number and the developmental status of the cell. Plant J 48:883–894

Miwa H, Betsuyaku S, Iwamoto K, Kinoshita A, Fukuda H, Sawa S (2008) The receptor-like kinase SOL2 mediates CLE signaling in Arabidopsis. Plant Cell Physiol 49:1752–1757

Miwa H, Kinoshita A, Fukuda H, Sawa S (2009) Plant meristems: CLAVATA3/ESR-related signaling in the shoot apical meristem and the root apical meristem. J Plant Res 122:31–39

Miyazawa H, Oka-Kira E, Sato N, Takahashi H, Wu GJ, Sato S, Hayashi M, Betsuyaku S, Nakazono M, Tabata S, Harada K, Sawa S, Fukuda H, Kawaguchi M (2010) The receptor-like kinase KLAVIER mediates systemic regulation of nodulation and non-symbiotic shoot development in *Lotus japonicus*. Development 137:4317–4325

Mortier V, Den Herder G, Whitford R, Van de Velde W, Rombauts S, D'Haeseleer K, Holsters M, Goormachtig S (2010) CLE peptides control *Medicago truncatula* nodulation locally and systemically. Plant Physiol 153:222–237

Mortier V, Fenta BA, Martens C, Rombauts S, Holsters M, Kunert K, Goormachtig S (2011) Search for nodulation-related CLE genes in the genome of *Glycine max*. J Exp Bot 62 (8):2571–2583. doi:10.1093/jxb/erq426

Müller R, Borghi L, Kwiatkowska D, Laufs P, Simon R (2006) Dynamic and compensatory responses of *Arabidopsis* shoot and floral meristems to CLV3 signaling. Plant Cell 18: 1188–1198

Müller R, Bleckmann A, Simon R (2008) The receptor kinase CORYNE of Arabidopsis transmits the stem cell-limiting signal CLAVATA3 independently of CLAVATA1. Plant Cell 20: 934–946

Murray JD, Karas BJ, Sato S, Tabata S, Amyot L, Szczyglowski K (2007) A cytokinin perception mutant colonized by *Rhizobium* in the absence of nodule organogenesis. Science 315:101–104

Narita NN, Moore S, Horiguchi G, Kubo M, Demura T, Fukuda H, Goodrich J, Tsukaya H (2004) Overexpression of a novel small peptide ROTUNDIFOLIA4 decreases cell proliferation and alters leaf shape in *Arabidopsis thaliana*. Plant J 38:699–713

Ni J, Clark SE (2006) Evidence for functional conservation, sufficiency, and proteolytic processing of the CLAVATA3 CLE domain. Plant Physiol 140:726–733

Nishimura R, Hayashi M, Wu GJ, Kouchi H, Imaizumi-Anraku H, Murakami Y, Kawasaki S, Akao S, Ohmori M, Nagasawa M, Harada K, Kawaguchi M (2002) HAR1 mediates systemic regulation of symbiotic organ development. Nature 420:426–429

Nukui N, Ezura H, Yuhashi K, Yasuta T, Minamisawa K (2000) Effects of ethylene precursor and inhibitors for ethylene biosynthesis and perception on nodulation in *Lotus japonicus* and *Macroptilium atropurpureum*. Plant Cell Physiol 41:893–897

Nukui N, Ezura H, Minamisawa K (2004) Transgenic *Lotus japonicus* with an ethylene receptor gene *Cm-ERS1/H70A* enhances formation of infection threads and nodule primordia. Plant Cell Physiol 45:427–435

Nutman PS (1952) Studies on the physiology of nodule formation. III. Experiments on the excision of root-tips and nodules. Ann Bot 16:79–101

Oelkers K, Goffard N, Weiller GF, Gresshoff PM, Mathesius U, Frickey T (2008) Bioinformatic analysis of the CLE signaling peptide family. BMC Plant Biol 8:1

Ogawa M, Shinohara H, Sakagami Y, Matsubayashi Y (2008) Arabidopsis CLV3 peptide directly binds CLV1 ectodomain. Science 319:294

Ohyama K, Shinohara H, Ogawa-Ohnishi M, Matsubayashi Y (2009) A glycopeptide regulating stem cell fate in *Arabidopsis thaliana*. Nat Chem Biol 5:578–580

Oka-Kira E, Tateno K, Miura K, Haga T, Hayashi M, Harada K, Sato S, Tabata S, Shikazono N, Tanaka A, Watanabe Y, Fukuhara I, Nagata T, Kawaguchi M (2005) klavier (klv), a novel hypernodulation mutant of *Lotus japonicus* affected in vascular tissue organization and floral induction. Plant J 44:505–515

Okamoto S, Ohnishi E, Sato S, Takahashi H, Nakazono M, Tabata S, Kawaguchi M (2009) Nod factor/nitrate-induced CLE genes that drive HAR1-mediated systemic regulation of nodulation. Plant Cell Physiol 50:67–77

Oldroyd GE, Downie JA (2008) Coordinating nodule morphogenesis with rhizobial infection in legumes. Annu Rev Plant Biol 59:519–546

Oldroyd GE, Engstrom EM, Long SR (2001) Ethylene inhibits the Nod factor signal transduction pathway of *Medicago truncatula*. Plant Cell 13:1835–1849

Olsen AN, Mundy J, Skriver K (2002) Peptomics, identification of novel cationic Arabidopsis peptides with conserved sequence motifs. In Silico Biol 2:441–451

Pacios-Bras C, Schlaman HR, Boot K, Admiraal P, Langerak JM, Stougaard J, Spaink HP (2003) Auxin distribution in *Lotus japonicus* during root nodule development. Plant Mol Biol 52: 1169–1180

Papadopoulou K, Roussis A, Katinakis P (1996) *Phaseolus ENOD40* is involved in symbiotic and non-symbiotic organogenetic processes: expression during nodule and lateral root development. Plant Mol Biol 30:403–417

Patel N, Hamamouch N, Chunying L, Hussey R, Mitchum M, Baum T, Wang X, Davis EL (2008) Similarity and functional analyses of expressed parasitism genes in *Heterodera schachtii* and *Heterodera glycines*. J Nematol 40:299–310

Pearce G, Strydom D, Johnson S, Ryan CA (1991) A polypeptide from tomato leaves induces wound-inducible proteinase inhibitor proteins. Science 253:895–897

Pearce G, Moura DS, Stratmann J, Ryan CA Jr (2001) RALF, a 5-kDa ubiquitous polypeptide in plants, arrests root growth and development. Proc Natl Acad Sci USA 98:12843–12847

Penmetsa RV, Cook DR (1997) A legume ethylene-insensitive mutant hyperinfected by its rhizobial symbiont. Science 275:527–530

Penmetsa RV, Uribe P, Anderson J, Lichtenzveig J, Gish JC, Nam YW, Engstrom E, Xu K, Sckisel G, Pereira M, Baek JM, Lopez-Meyer M, Long SR, Harrison MJ, Singh KB, Kiss GB, Cook DR (2008) The *Medicago truncatula* ortholog of Arabidopsis *EIN2*, sickle, is a negative regulator of symbiotic and pathogenic microbial associations. Plant J 55:580–595

Pierce M, Bauer WD (1983) A rapid regulatory response governing nodulation in soybean. Plant Physiol 73:286–290

Plet J, Wasson A, Ariel F, Le Signor C, Baker D, Mathesius U, Crespi M, Frugier F (2011) MtCRE1-dependent cytokinin signaling integrates bacterial and plant cues to coordinate symbiotic nodule organogenesis in *Medicago truncatula*. Plant J 65:622–633

Reddy GV, Meyerowitz EM (2005) Stem-cell homeostasis and growth dynamics can be uncoupled in the Arabidopsis shoot apex. Science 310:663–667

Reid DE, Ferguson BF, Gresshoff PM (2011) Inoculation- and nitrate-induced CLE peptides of soybean control NARK-dependent nodule formation. Mol Plant Microbe Interact 24:606–618

Replogle A, Wang J, Bleckman A, Russey RS, Baum TJ, Sawa S, Davis EL, Wang X, Simon R, Mitchum MG (2011) Nematode CLE signaling in Arabidopsis requires CLAVATA2 and CORYNE. Plant J 65:430–440

Röhrig H, Schmidt J, Miklashevichs E, Schell J, John M (2002) Soybean ENOD40 encodes two peptides that bind to sucrose synthase. Proc Natl Acad Sci USA 99:1915–1920

Röhrig H, John M, Schmidt J (2004) Modification of soybean sucrose synthase by S-thiolation with ENOD40 peptide A. Biochem Biophys Res Commun 325:864–870

Rojo E, Sharma VK, Kovaleva V, Raikhel NV, Fletcher JC (2002) CLV3 is localized to the extracellular space, where it activates the Arabidopsis CLAVATA stem cell signaling pathway. Plant Cell 14:969–977

Ruttink T, Boot K, Kijne J, Bisseling T, Franssen H (2006) ENOD40 affects elongation growth in tobacco Bright Yellow-2 cells by alteration of ethylene biosynthesis kinetics. J Exp Bot 57:3271–3282

Saur IM, Oakes M, Djordjevic MA, Imin N (2011) Crosstalk between the nodulation signaling pathway and the autoregulation of nodulation in *Medicago truncatula*. New Phytol 190(4): 865–874. doi:10.1111/j.1469-8137.2011.03738.x

Sawa S, Tabata R (2011) RPK2 functions in diverged CLE signaling. Plant Signal Behav 6:86–88

Sawa S, Kinoshita A, Nakanomyo I, Fukuda H (2006) CLV3/ESR-related (CLE) peptides as intercellular signaling molecules in plants. Chem Rec 6:303–310

Scheer JM, Ryan CA (1999) A 160 kD systemin receptor on the surface of *Lycopersicon peruvianum* suspension-cultured cells. Plant Cell 221:667–674

Scheres B, van Engelen F, van der Knaap E, van de Wiel C, van Kammen A, Bisseling T (1990) Sequential induction of nodulin gene expression in the developing pea nodule. Plant Cell 2:687–700

Schnabel E, Journet EP, de Carvalho-Niebel F, Duc G, Frugoli J (2005) The *Medicago truncatula SUNN* gene encodes a CLV1-like leucine-rich repeat receptor kinase that regulates nodule number and root length. Plant Mol Biol 58:809–822

Schoof H, Lenhard M, Haecker A, Mayer KFX, Jurgens G, Laux T (2000) The stem cell population of Arabidopsis shoot meristems is maintained by a regulatory loop between the *CLAVATA* and *WUSCHEL* genes. Cell 100:635–644

Schopfer CR, Nasrallah ME, Nasrallah JB (1999) The male determinant of self-incompatibility in Brassica. Science 286:1697–1700

Schultze M, Kondorosi A (1998) Regulation of symbiotic root nodule development. Annu Rev Genet 32:33–57

Searle IR, Men AE, Laniya TS, Buzas DM, Iturbe-Ormaetxe I, Carroll BJ, Gresshoff PM (2003) Long-distance signaling in nodulation directed by a CLAVATA1-like receptor kinase. Science 299:109–112

Sharma VK, Ramirez J, Fletcher JC (2003) The Arabidopsis CLV3-like (CLE) genes are expressed in diverse tissues and encode secreted proteins. Plant Mol Biol 51:415–425

Smit G, de Koster CC, Schripsema J, Spaink HP, van Brussel AA, Kijne JW (1995) Uridine, a cell division factor in pea roots. Plant Mol Biol 29:869–873

Sousa C, Johansson C, Charon C, Manyani H, Sautter C, Kondorosi A, Crespi M (2001) Translational and structural requirements of the early nodulin gene enod40, a short-open reading frame-containing RNA, for elicitation of a cell-specific growth response in the alfalfa root cortex. Mol Cell Biol 21:354–366

Stahl A, Simon R (2009) Is the Arabidopsis root niche protected by sequestration of the CLE40 signal by its putative receptor ACR4? Plant Signal Behav 7:634–635

Stahl Y, Wink RH, Ingram GC, Simon R (2009) A signaling module controlling the stem cell niche in Arabidopsis root meristems. Curr Biol 19:909–914

Strabala TJ, O'Donnell PJ, Smit AM, Ampomah-Dwamena C, Martin EJ, Netzler N, Nieuwenhuizen NJ, Quinn BD, Foote HC, Hudson KR (2006) Gain-of-function phenotypes of many *CLAVATA3/ESR* genes, including four new family members, correlate with tandem variations in the conserved CLAVATA3/ESR domain. Plant Physiol 140:1331–1344

Sun J, Cardoza V, Mitchell DM, Bright L, Oldroyd G, Harris JM (2006) Crosstalk between jasmonic acid, ethylene and Nod factor signaling allows integration of diverse inputs for regulation of nodulation. Plant J 46:961–970

Sun J, Miwa H, Downie JA, Oldroyd GE (2007) Mastoparan activates calcium spiking analogous to Nod factor-induced responses in *Medicago truncatula* root hair cells. Plant Physiol 144:695–702

Takeda N, Okamoto S, Hayashi M, Murooka Y (2005) Expression of *LjENOD40* genes in response to symbiotic and non-symbiotic signals: *LjENOD40-1* and *LjENOD40-2* are differentially regulated in *Lotus japonicus*. Plant Cell Physiol 46:1291–1298

Talukdar T, Gorecka KM, de Carvalho-Niebel F, Downie JA, Cullimore J, Pikula S (2009) Annexins – calcium – and membrane-binding proteins in the plant kingdom: potential role in nodulation and mycorrhization in *Medicago truncatula*. Acta Biochim Pol 56:199–210

Timmers AC (2008) The role of the plant cytoskeleton in the interaction between legumes and rhizobia. J Microsc 231:247–256

Tirichine L, Sandal N, Madsen LH, Radutoiu S, Albrektsen AS, Sato S, Asamizu E, Tabata S, Stougaard J (2007) A gain-of-function mutation in a cytokinin receptor triggers spontaneous root nodule organogenesis. Science 315:104–107

Trotochaud AE, Hao T, Wu G, Yang Z, Clark SE (1999) The CLAVATA1 receptor-like kinase requires CLAVATA3 for its assembly into a signaling complex that includes KAPP and a Rho-related protein. Plant Cell 11:393–405

Trotochaud AE, Jeong S, Clark SE (2000) CLAVATA3, a multimeric ligand for the CLAVATA1 receptor-kinase. Science 289:613–617

van de Velde W, De Guerra JCP, Keyser A, De Rycke R, Rombauts S, Maunoury N, Mergaert P, Kondorosi E, Holsters M, Goormachtig S (2006) Aging in legume symbiosis. A molecular view on nodule senescence in *Medicago truncatula*. Plant Physiol 141:711–720

van de Velde W, Zehirov G, Szatmari A, Debreczeny M, Ishihara H, Kevei Z, Farkas A, Mikulass K, Nagy A, Tiricz H, Satiat-Jeunemaitre B, Alunni B, Bourge M, Kucho K, Abe M, Kereszt A, Maroti G, Uchiumi T, Kondorosi E, Mergaert P (2010) Plant peptides govern terminal differentiation of bacteria in symbiosis. Science 327:1122–1126 van Noorden GE, Ross JJ, Reid JB, Rolfe BG, Mathesius U (2006) Defective long-distance auxin transport regulation in the Medicago truncatula super numeric nodules mutant. Plant Physiol 140:1494–1506

van Noorden GE, Kerim T, Goffard N, Wiblin R, Pellerone FI, Rolfe BG, Mathesius U (2007) Overlap of proteome changes in *Medicago truncatula* in response to auxin and *Sinorhizobium meliloti*. Plant Physiol 144:1115–1131

van Spronsen PC, Gronlund M, Pacios Bras C, Spaink HP, Kijne JW (2001) Cell biological changes of outer cortical root cells in early determinate nodulation. Mol Plant Microbe Interact 14:839–847

Varkonyi-Gasic E, White DW (2002) The white clover *enod40* gene family. Expression patterns of two types of genes indicate a role in vascular function. Plant Physiol 129:1107–1118

Vleghels I, Hontelez J, Ribeiro A, Fransz P, Bisseling T, Franssen H (2003) Expression of *ENOD40* during tomato plant development. Planta 218:42–49

Wan X, Hontelez J, Lillo A, Guarnerio C, van de Peut D, Fedorova E, Bisseling T, Franssen H (2007) *Medicago truncatula ENOD40-1* and *ENOD40-2* are both involved in nodule initiation and bacteroid development. J Exp Bot 58:2033–2041

Wang X, Mitchum MG, Gao B, Li C, Diab H, Baum TJ, Hussey RS, Davis EL (2005) A parasitism gene from a plant-parasitic nematode with function similar to *CLAVATA3/ESR (CLE)* of *Arabidopsis thaliana*. Mol Plant Pathol 6:187–191

Wang J, Lee C, Replogle A, Joshi S, Korkin D, Hussey R, Baum TJ, Davis EL, Wang X, Mitchum MG (2010) Dual roles for the variable domain in protein trafficking and host-specific recognition of *Heterodera glycines* CLE effector proteins. New Phytol 187:1003–1017

Wang J, Replogle A, Hussey R, Baum T, Wang X, Davis EL, Mitchum MG (2011) Identification of potential host plant mimics of CLAVATA3/ESR (CLE)-like peptides from the plant-parasitic nematode *Heterodera schachtii*. Mol Plant Pathol 12:177–186

Wasson AP, Pellerone FI, Mathesius U (2006) Silencing the flavonoid pathway in *Medicago truncatula* inhibits root nodule formation and prevents auxin transport regulation by rhizobia. Plant Cell 18:1617–1629

Weerasinghe RR, Bird DM, Allen NS (2005) Root-knot nematodes and bacterial Nod factors elicit common signal transduction events in *Lotus japonicus*. Proc Natl Acad Sci USA 102:3147–3152

Wen J, Lease KA, Walker JC (2004) DVL, a novel class of small polypeptides: overexpression alters *Arabidopsis* development. Plant J 37:668–677

Whitford R, Fernandez A, De Groodt R, Ortega E, Hilson P (2008) Plant CLE peptides from two distinct functional classes synergistically induce division of vascular cells. Proc Natl Acad Sci USA 105:18625–18630

Wu J, Kurten EL, Monshausen G, Hummel GM, Gilroy S, Baldwin IT (2007) NaRALF, a peptide signal essential for the regulation of root hair tip apoplastic pH in *Nicotiana attenuata*, is required for root hair development and plant growth in native soils. Plant J 52:877–890

Yaginuma H, Hirakawa Y, Kondo Y, Ohashi-Ito K, Fukuda H (2011) A novel function of TDIF-related peptides: promotion of axillary bud formation. Plant Cell Physiol 52:1354–1364

Yamakawa S, Sakurai C, Matsubayashi Y, Sakagami Y, Kamada H, Satoh S (1998) The promotive effects of a peptidyl plant growth factor, phytosulfokine, on the formation of adventitious roots and expression of a gene for a root-specific cystatin in cucumber hypocotyls. J Plant Res 111:453–458

Yang WC, Katinakis P, Hendriks P, Smolders A, de Vries F, Spee J, van Kammen A, Bisseling T, Franssen H (1993) Characterization of *GmENOD40*, a gene showing novel patterns of cell-specific expression during soybean nodule development. Plant J 3:573–585

Yang WC, de Blank C, Meskiene I, Hirt H, Bakker J, van Kammen A, Franssen H, Bisseling T (1994) *Rhizobium* nod factors reactivate the cell cycle during infection and nodule primordium formation, but the cycle is only completed in primordium formation. Plant Cell 6:1415–1426

Yang H, Matsubayashi Y, Hanai H, Nakamura K, Sakagami Y (2000a) Molecular cloning and characterization of OsPSK, a gene encoding a precursor for phytosulfokine-alpha, required for rice cell proliferation. Plant Mol Biol 44:635–647

Yang H, Matsubayashi Y, Hanai H, Sakagami Y (2000b) Phytosulfokine-alpha, a peptide growth factor found in higher plants: its structure, functions, precursor and receptors. Plant Cell Physiol 41:825–830

Young RA (2000) Biomedical discovery with DNA arrays. Cell 102:9–15

Yu LP, Miller AK, Clark SE (2003) POLTERGEIST encodes a protein phosphatase 2 C that regulates CLAVATA pathways controlling stem cell identity at Arabidopsis shoot and flower meristems. Curr Biol 13:179–188

Zhu Y, Wan Y, Lin J (2010a) Multiple receptor complexes assembled for transmitting CLV3 signaling in Arabidopsis. Plant Signal Behav 5:300–302

Zhu Y, Wang Y, Li R, Song X, Wang Q, Huang S, Jin JB, Liu CM, Lin J (2010b) Analysis of interactions among the CLAVATA3 receptors reveals a direct interaction between CLAVATA2 and CORYNE in Arabidopsis. Plant J 61:223–233

Peptides as Danger Signals: MAMPs and DAMPs

Thomas Boller and Pascale Flury

Abstract In plants, active defense against pathogens and parasites relies substantially on the perception of danger. It has been well established that plants recognize invading microbes through the perception of "microbe-associated molecular patterns" (MAMPs), i.e., structures characteristic for whole classes of microbes. Among the MAMPs are highly conserved epitopes of important microbial proteins, such as bacterial flagellin or EF-Tu. These MAMPs are perceived by pattern recognition receptors (PRRs) on the plant cell surface. In *Arabidopsis*, the PRRs FLS2 and EFR, perceiving flagellin and EF-Tu, respectively, are leucine-rich repeat receptor-like kinases (LRR-RLKs). They bind their cognate ligands at the outside of the cell and relay the signal to the cytoplasm, inducing a potent defense response. In addition, alarm may also be caused by endogenous danger signals arising simply from cellular damage. In *Arabidopsis thaliana*, the small peptides *At*Pep1 to *At*Pep7 appear to act as such damage-associated molecular patterns (DAMPs), since they were found to induce various defense responses. *At*Peps are perceived by two receptors, PEPR1 and PEPR2, which are surprisingly similar to FLS2 and EFR. The *At*Pep/PEPR system is suggested to act as an amplifier of resistance, possibly during infections by pathogens as well as during responses to herbivore attack and other wounding events.

T. Boller (✉)
Zürich-Basel Plant Science Center, Botanisches Institut, Universität Basel, Hebelstrasse 1, CH-4056 Basel, Switzerland
e-mail: thomas.boller@unibas.ch

P. Flury
Zürich-Basel Plant Science Center, Institute of Integrative Biology, Plant Pathology, Swiss Federal Institute of Technology, CH-8092 Zurich, Switzerland

H.R. Irving and C. Gehring (eds.), *Plant Signaling Peptides*,
Signaling and Communication in Plants 16, DOI 10.1007/978-3-642-27603-3_9,

1 Introduction

In general terms, innate immunity of plants against potential pathogens relies on the perception of danger through specific "danger signals" [reviewed in Boller and Felix (2009)]. Peptides play an important role as such danger signals: on the one hand, highly conserved peptides derived from vital structures of microbes (such as flg22 representing bacterial flagellin or elf18/elf26 representing bacterial EF-Tu) are perceived by *Arabidopsis* plants with exquisite sensitivity. Such peptides can be considered as paradigms of microbe-associated molecular patterns (MAMPs), i.e., molecular patterns derived from whole classes of microbes, whether pathogenic or not (Mackey and McFall 2006; He et al. 2007; Boller and Felix 2009). On the other hand, peptides produced or released by the plant cell itself in response to wounding or to pathogen or parasite attack (such as systemin or the *At*Peps) may represent damage-associated molecular patterns (DAMPs), which are similarly recognized as danger signals [reviewed in Yamaguchi and Huffaker (2011)]. Plant cells perceive both MAMPs and DAMPs through pattern recognition receptors (PRRs) localized at the cell surface, just like animal cells (Ronald and Beutler 2010), and upon signal perception, various defense responses are initiated (Boller and Felix 2009).

This kind of defense, called basal or horizontal resistance in classic textbooks, is now termed "PTI" for "PAMP-triggered immunity" (Jones and Dangl 2006) or "pattern-triggered immunity" (Boller and Felix 2009). Typical responses to MAMP/DAMP perception are altered ion fluxes across the plasma membrane, production of reactive oxygen species (ROS), synthesis of the stress hormone ethylene, activation of a mitogen-activated protein kinase (MAPK) cascade, and induction of defense gene expression (Fig. 1).

PTI represents an important first line of defense. To overcome PTI, true pathogens evolved so-called effectors, which block MAMP perception or signal transduction (Jones and Dangl 2006). In the case of bacterial pathogens, a number of proteinaceous effectors are secreted into the plant cytoplasm via the type three secretion system (TTSS), and they suppress PTI through a number of different mechanisms (Boller and He 2009). However, in many cases, these effectors are recognized by the plant through the products of so-called *Resistance* (*R*) genes. Most of them belong to the NB-LRR, named after their characteristic nucleotide-binding (NB) and leucine-rich repeat (LRR) domains. Direct or indirect recognition of effectors leads to ETI, i.e., effector-triggered immunity (Jones and Dangl 2006). This type of resistance is often accompanied by the hypersensitive response (HR), characterized by rapid programmed cell death and local necrosis (Greenberg and Yao 2004). Considering the consequences of PTI and ETI, one can imagine that there is a constant arms race between plants and their pathogens.

The PTI/ETI model (Jones and Dangl 2006) gives the impression of two separate, evolutionarily consecutive types of innate immunity in plants. However, perception of MAMPs, DAMPs, and effectors may simply converge to a generalized perception of danger signals (Boller and Felix 2009), as aptly summarized for innate immunity in animals (Matzinger 2002). This may lead to a stereotypical defense program, as illustrated in Fig. 1. Current evidence indicates

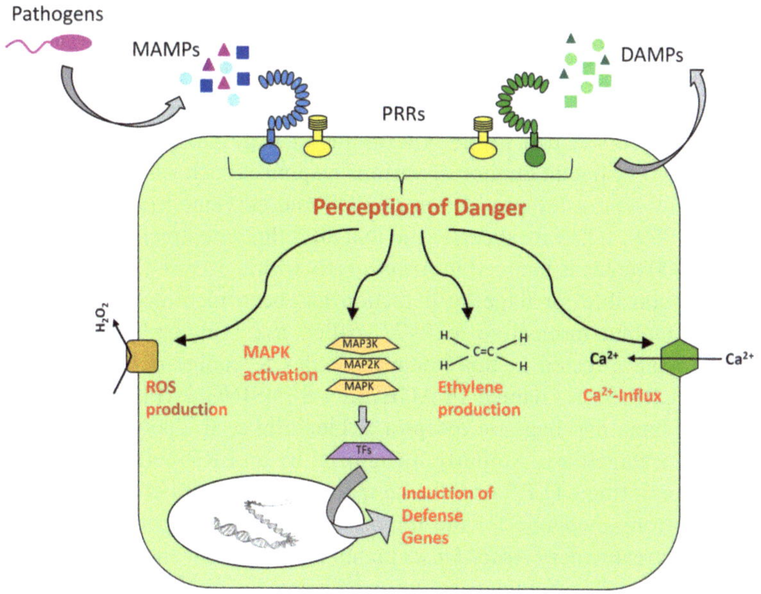

Fig. 1 Perception of danger signals induces various defense responses. Microbe-associated molecular patterns (MAMPs) derived from invading microbes and damage-associated molecular patterns (DAMPs) derived from the plant are perceived as danger signals through pattern recognition receptors (PRRs). Upon perception, various defense responses such as the production of reactive oxygen species (ROS), MAPK activation, Ca^{2+} influx, ethylene production, and defense gene expression are induced

that PTI and ETI have coexisted since the onset of seed plant evolution; this does not exclude fine tuning between the perception systems for MAMPs, DAMPs, and effectors and between PTI and ETI (Boller and Felix 2009).

2 Plant Danger Signals

2.1 Microbe-Associated Molecular Patterns

The most important group of signals for the initial perception of danger are probably the MAMPs, i.e., patterns highly conserved for whole classes of microbes. Since these patterns—carbohydrates, lipids, proteins, etc.—are essential for the microbes, they are difficult to alter in a way that they can escape the plant's recognition system. Typically, applied to the plant tissue, most MAMPs induce defense responses already at low nanomolar concentrations (Boller and Felix 2009). Among the known MAMPs are a number of peptides, as detailed below.

2.1.1 Flagellin and flg22

Currently, the best-known MAMP is flagellin, the main component of the bacterial flagellum. Most plants perceive flg22, a 22-amino-acid peptide corresponding to the most highly conserved part in the N terminus of flagellin. This peptide has been shown to be a very potent elicitor of various responses such as medium alkalinization in cell cultures or the production of ROS and ethylene in *Arabidopsis* plants (Felix et al. 1999). It has also been found that after flg22 treatment, the transcription of almost 1,000 genes is induced in *Arabidopsis* within 30 min (Zipfel et al. 2004). The PRR responsible for flagellin detection has been identified by analyzing an *Arabidopsis* mutant insensitive to flg22, called *fls2* (Gómez-Gómez and Boller 2000). The gene affected in this mutant encodes a leucine-rich repeat receptor-like kinase (LRR-RLK) named FLAGELLIN-SENSING 2 (FLS2). FLS2 turned out to be the bona fide flagellin receptor (Chinchilla et al. 2006). Interestingly, in mice and men, flagellin is similarly recognized by an LRR-containing transmembrane receptor, termed TLR5 or toll-like receptor 5 (Hayashi et al. 2001).

There is strong evidence for the importance of flagellin perception in disease resistance. Pretreatment of *Arabidopsis* plants with flg22 enhanced resistance to the bacterial pathogen *Pseudomonas syringae* pathovar *tomato* DC3000 (*Pst* DC3000) (Zipfel et al. 2004). Furthermore, Arabidopsis *fls2* mutants were more susceptible to *Pst* DC3000 after spray inoculation than wild-type plants (Zipfel et al. 2004).

2.1.2 Elongation Factor EF-Tu and elf18/elf26

Another well-studied MAMP is the bacterial elongation factor EF-Tu. EF-Tu itself and the peptides elf18 and elf26, originating from the N terminus of the bacterial elongation factor EF-Tu, induce characteristic defense responses in *Arabidopsis* (Kunze et al. 2004). In contrast to flg22, which is recognized in most plants, the elf18/26 peptides are recognized only in *Arabidopsis* and closely related species of the Brassicaceae. The receptor for EF-Tu, EFR, is a LRR-RLK resembling FLS2; mutation of EFR renders *Arabidopsis* more sensitive to *Agrobacterium* transformation (Zipfel et al. 2006). *Nicotiana benthamiana*, a plant completely blind to EF-Tu, has been shown to gain full responsiveness to elf18/26 when transiently transformed by the *Arabidopsis EFR* gene, indicating that signaling downstream of PRR activation is conserved between *Arabidopsis* and *Nicotiana* (Zipfel et al. 2006). Furthermore, tobacco and tomato expressing the *EFR* gene are less susceptible to bacterial diseases, pointing to the possibility that EFR and similar PRRs may be deployed in transgenic plants as a strategy to enhance their natural disease resistance (Lacombe et al. 2010; Segonzac and Zipfel 2011).

2.1.3 Peptide MAMPs with Unknown Receptors

One of the most well-known peptides identified as a MAMP is pep13, obtained from an elicitor preparation of the oomycete *Phytophthora megasperma*, a soybean pathogen. In parsley cells, pep13 has been found to elicit defense responses at the nanomolar level (Nürnberger et al. 1994). This peptide pep13 was subsequently found to represent a highly conserved, functionally important epitope of oomycete transglutaminase (Brunner et al. 2002).

The well-known elicitins represent another class of peptidic MAMPs of the oomycetes. These small proteins are essential for plant-pathogenic oomycetes as sterol-transfer proteins, and they act as potent elicitors in tobacco [reviewed in Blein et al. (2002)]. In addition, tobacco also perceives a cellulose-binding lectin from oomycetes (Gaulin et al. 2006). *Arabidopsis* is blind to both of these peptides. Tobacco also recognizes bacterial MAMPs that are not recognized by other species: Bacterial cold shock protein is a potent elicitor of defense responses in tobacco cell cultures (Felix and Boller 2003). The active epitope of cold shock protein is an N-terminal peptide represented by csp15, representing the highly conserved RNA-binding domain of cold shock protein (Felix and Boller 2003).

2.2 Damage-Associated Molecular Patterns

Plants appear to detect a situation of danger not only through MAMP perception but also by the recognition of DAMPs. These are plant-derived products, which are activated or produced in response to cellular damage [reviewed in Boller and Felix (2009)]. As a classic example, lytic enzymes of plant pathogens, such as polygalacturonases, may cause damage to a plant cell, releasing cell wall fragments such as oligogalacturonides (OGA). These fragments may be perceived by neighboring cells as a danger signal. Indeed, chemically pure OGAs have been described to elicit defense responses [reviewed by D'Ovidio et al. (2004)]. A number of plant peptides have also been proposed to act as DAMPs [reviewed in Yamaguchi and Huffaker (2011)]. These peptides are thought to be produced, released, or activated upon infection or wounding in order to initiate or enhance defense responses.

2.2.1 Systemin

In tomato, an 18-amino-acid peptide called systemin is known to induce defense responses against herbivores [reviewed in Ryan and Pearce (2003)]. Because its precursor, prosystemin, appears to be located in the cytoplasm, systemin is thought to be released only upon cell injury. Once outside the cell, systemin would act as DAMP and warn neighboring cells (Ryan and Pearce 2003). The receptor for systemin is still elusive, as discussed in more detail below (Sect. 5).

2.2.2 HypSys and RALF

Two hydroxyproline-containing glycopeptides (HypSys) (Pearce et al. 2001a) and the rapid alkalinization inducing factor (RALF) (Pearce et al. 2001b) are putative peptide DAMPs in tobacco. The HypSys precursor protein represents a secreted extracellular protein, and RALF is probably derived from an extracellular protein as well (see chapter "Processing of Peptides"). Thus, the precursors of these DAMPs are likely to be constitutively present in the cell wall. However, the DAMP peptides themselves are probably only released by microbial proteases or by intracellular proteases released upon cellular damage. The receptors for HypSys and RALF are unknown, but they are speculated to be PRRs (Ryan and Pearce 2003).

2.2.3 *At*Peps

An endogenous peptide of 23 amino acids, inducing defense responses in a similar way as systemin, has also been isolated and identified from wounded leaves of *Arabidopsis thaliana* (Huffaker et al. 2006). This peptide, named *At*Pep1, is derived from the C terminus of a larger precursor, and there are at least six additional *Arabidopsis* genes expected to encode for homologous precursors, as pointed out by Huffaker et al. (2006). *At*Pep1 is perceived by two homologous LRR-RLKs called PEPR1 (Yamaguchi et al. 2006) and PEPR2 (Yamaguchi et al. 2010; Krol et al. 2010). The AtPep/PEPR paradigm is discussed in more detail in Sect. 6.

2.2.4 Endogenous Peptides of Soybean

In soybean (*Glycine max*), a 12-amino-acid peptide was identified as a potent elicitor of defense responses and termed *Gm*SubPep (Pearce et al. 2010). It is processed from a unique region of an extracellular subtilase that is predicted to be secreted to the apoplast. The activity of *Gm*SubPep resembles that of other peptidic defense signaling peptides, for example, by inducing defense gene expression (Pearce et al. 2010). Recently, two other peptides of soybean, *Gm*Pep914 and *Gm*Pep890, have been identified and found to induce defense responses in soybean (Yamaguchi et al. 2011). The receptors for all of these peptides are unknown.

3 PRRs for MAMP and DAMP Perception

MAMPs and DAMPs are perceived by PRRs. Most of the PRRs identified up to now belong to the family of receptor-like kinases (RLK), which are membrane proteins with an extracellular receptor domain, a transmembrane domain, and an intracellular serine/threonine kinase domain (Morillo and Tax 2006). RLKs form the largest family of receptors in plants; they are not only involved in immunity but also in various developmental processes [reviewed by Tör et al. (2009); also see chapters "Plant Peptide Signaling: An Evolutionary Adaptation," "Peptides Regulating Apical Meristem Development," "Peptides Regulating Root Growth," "Peptides Regulating Plant Vascular Development," "The S-LOCUS CYSTEINE RICH PROTEIN (SCR): a small peptide with a high impact on the evolution of flowering plants," and "Peptides Modulating Development of Specialized Cells"). RLKs are classified into subfamilies according to their extracellular domains, including among others LRRs, lysine motifs (LysM), and wall-associated kinases (WAK) (see Tör et al. 2009). In *Arabidopsis* and rice, the LRR-RLKs are the largest subfamily by far (Shiu et al. 2004). It is this subfamily that comprises the currently most studied receptors for peptidic MAMPs and DAMPs, namely, FLS2, EFR, PEPR1, and PEPR2. Other types of MAMPs are recognized by different classes of RLKs: the putative chitin receptor CERK1 has an extracellular LysM domain (Miya et al. 2007), whereas the candidate receptors for OGA are members of the WAK family of RLKs (Brutus et al. 2010).

Within the LRR-RLKs, FLS2 and EFR are grouped together in subgroup LRR-XII (Shiu et al. 2004), while the PEPRs are structurally closely related and belong to the subgroup LRR-XI, together with developmentally associated receptors such as CLAVATA1 (Ryan et al. 2007). These two subgroups differ significantly in terms of their protein kinase domains: PEPR1 and 2, like most protein kinases, belong to the class of RD kinases; these are named after a conserved arginine (R) followed by an aspartate (D) which are required for catalytic activity (see Dardick and Ronald 2006). However, FLS2 and EFR are grouped in the small class referred to as non-RD kinases (Dardick and Ronald 2006). Non-RD kinases lack the conserved R in the kinase domain and appear to lack the capacity to autophosphorylate the activation loop. Most RLKs that function in MAMP perception appear to belong to this class, both in animals and plants (Dardick and Ronald 2006).

The LRR-RLKs FLS2, EFR, and PEPR1/PEPR2 do not act alone in activating downstream signaling. Instead, upon stimulation by their ligand, they all appear to associate very rapidly with a common interaction partner, BRI1-associated kinase 1 (BAK1), as initially shown for FLS2 (Chinchilla et al. 2007; Heese et al. 2007), and later for the other receptors (Schulze et al. 2010). Both the PRRs and BAK1 are rapidly phosphorylated upon ligand-stimulated interaction (Schulze et al. 2010). BAK1 was first described as a coreceptor for the brassinosteroid receptor BRI1, an LRR-RLK (Li et al. 2002a; Nam and Li 2002). This raises intriguing questions about possible interferences between hormonal signaling and innate immune signaling due to coreceptor sharing (Chinchilla et al. 2009).

4 Responses Induced by MAMP and DAMP Perception

4.1 Ion Fluxes

In response to MAMPs and DAMPs, ion fluxes across the plasma membrane are altered within minutes. An increased influx of H^+ and Ca^{2+} and efflux of K^+ and anions can be observed. In suspension-cultured cells, these ion fluxes result in a readily measurable alkalinization of the culture medium (Boller 1995; Boller and Felix 2009). The rapid increase of cytosolic Ca^{2+} is regarded as particularly significant since Ca^{2+} acts as a second messenger in diverse cellular processes (Lecourieux et al. 2006). Recently, four calcium-dependent protein kinases (CDPKs) were identified as Ca^{2+} sensors involved in regulating innate immunity in *Arabidopsis* (Boudsocq et al. 2010). Activation of these CDPKs results in transcriptional reprogramming of the cell. CDPKs also seem to regulate the production of ROS, possibly by direct phosphorylation of the ROS-producing enzyme NADPH oxidase (Kobayashi et al. 2007).

4.2 Production of ROS

The production of ROS plays a central role in plant pathogen defense. ROS has been suggested to directly harm or kill attacking pathogens, as it has been reported for ROS produced by animal phagocytes and B lymphocytes [reviewed by Apel and Hirt (2004) and Mittler et al. (2011)]. However, ROS might also harm the host cells and induce programmed cell death, which may be in favor of the plant as in the case of the hypersensitive response (HR). Moreover, ROS can cause the establishment of a physical barrier by cell wall cross-linking. In addition, ROS may act as secondary stress signals in inducing various defense responses such as MAPK activation and defense gene expression (see Apel and Hirt 2004; Mittler et al. 2011).

During pathogen defense, ROS are predominantly produced by plasma-membrane-bound NADPH oxidases (Rboh), but also by cell wall-bound peroxidases and amine oxidases in the apoplast (Torres 2010). Simultaneously, ROS scavenging capacities are decreased (Apel and Hirt 2004). The oxidative burst can be measured easily by making use of the H_2O_2-dependent oxidation of luminol and detection of the resulting luminescence (see, e.g., Chinchilla et al. 2007).

4.3 MAPK Activation

In plants, like in other eukaryotes, mitogen-activated protein kinase (MAPK) cascades are key elements in transducing various environmental and developmental signals into intracellular responses, such as altered gene expression (Rodriguez et al. 2010). In general, the initiation of a MAP kinase cascade starts with the phosphorylation and activation of a MAP kinase kinase kinase (MAPKKK). Subsequently, MAP kinase kinases (MAPKK) are phosphorylated which in turn phosphorylate one or more downstream MAP kinases (MAPK). Activated MAPKs then target various downstream proteins in the cytoplasm or nucleus, such as other kinases, enzymes, or transcription factors (Rodriguez et al. 2010). MAPK cascades are important signaling components in innate immunity and are activated upon MAMP and DAMP treatment (Nühse et al. 2000; Pearce et al. 2001a; Chinchilla et al. 2007). The activation of the *Arabidopsis* MAPKs MPK3 and MPK6 by flg22 through FLS2 has been shown to culminate in the activation of WRKY-type transcription factors (Asai et al. 2002). Analogous to *Arabidopsis*, two MAPKs, termed as wound-induced protein kinase (WIPK) and salicylic acid (SA)-induced protein kinase (SIPK), are involved in defense responses in tobacco (Zhang and Klessig 2001).

4.4 Ethylene Production

The gaseous hormone ethylene plays a role in many aspects of plant growth and development, but also in responses to abiotic and biotic stresses. Upon MAMP or DAMP treatment, it is produced within the first hours (Boller and Felix 2009; Krol et al. 2010). Ethylene is formed from methionine via the Yang cycle with a rate-limiting step at the conversion of *S*-adenosyl-L-methionine to 1-aminocyclopropane-1-carboxylic acid (ACC) by the enzyme ACC synthase (ACS); this key enzyme has been shown to be activated by phosphorylation within minutes upon MAMP treatment (Spanu et al. 1994). During the last few years, major advances have been made in understanding the complex ethylene signaling pathway [reviewed by Stepanova and Alonso (2009)].

By mutant analysis, ethylene signaling has been shown to be a critical component for the oxidative burst in response to flg22 and to be required for normal FLS2 accumulation in *Arabidopsis* (Mersmann et al. 2010; Boutrot et al. 2010). Moreover, an intact ethylene signaling pathway seems to be required for callose deposition (Clay et al. 2009), a typical defense response establishing within hours to days upon MAMP treatment [reviewed in Boller and He (2009)].

4.5 Transcriptional Changes

Perception of MAMPs and DAMPs alters gene expression dramatically. Upon flg22 and elf26 treatment, about 1,000 genes are induced within 30–60 min (Zipfel et al. 2004, 2006). Remarkably, almost the same set of genes is induced by both MAMPs. This supports the current view that signaling pathways of different MAMP/PRR pairs converge upstream of a MAP kinase cascade (Zipfel et al. 2006). Notably, marker genes for these MAMPs are similarly induced by DAMPs. This indicates a stereotypical gene activation response to MAMPs and DAMPs (Boller and Felix 2009; Yamaguchi et al. 2010). It has been reported (Zipfel et al. 2004, 2006) that among the genes induced by MAMPs are over 100 of the 610 RLKs present in the *Arabidopsis* genome (Shiu et al. 2004), including the PRRs FLS2, EFR, PEPR1, and PEPR2. This implies a positive feedback loop to amplify the perception system for danger signals (Zipfel et al. 2004; Boller and Felix 2009). Other main groups of MAMP/DAMP-induced genes include the pathogenesis-related (PR) genes, transcription regulatory factors, kinases, and phosphatases (Boller and Felix 2009; Yamaguchi et al. 2010).

4.6 Seedling Growth Inhibition

The addition of an elicitor to the growth medium can result in reduced seedling growth. This may be due to a switch from a growth to a defense program. However, the exact mechanisms still remain unclear (Boller and Felix 2009). Interestingly, a recent study in *Arabidopsis* found a connection between development and immunity by means of CLAVATA3 peptide (CLV3p). This key regulator of stem cell homeostasis in the shoot apical meristem via CLV1 and CLV2 receptors was shown to induce defense responses via FLS2 (Lee et al. 2011). Consequently, CLV3p-FLS2 signaling was proposed to provide the constitutive immune protection observed in the shoot apical meristem (Lee et al. 2011).

5 Systemin and Its Elusive Receptor

It has been known since a long time that pathogen and parasite attacks as well as wounding (damage) can activate responses systemically, for example, in leaves distant to the site of attack (Shah 2009). In this respect, the peptide "systemin" has received much attention early on. In a landmark publication, systemin, an 18-amino-acid peptide, was identified from wounded tomato leaves and shown to induce the accumulation of proteinase inhibitors (a hallmark of the wound-induced systemic response) at nanomolar concentrations (Pearce et al. 1991). Initially, systemin was believed to be the mobile signal formed in a wounded leaf and transported systemically throughout the plant. However, a study with reciprocal grafts showed that jasmonate (or a jasmonate metabolite) was the systemic signal rather than systemin itself (Li et al. 2002b).

Interestingly, in suspension-cultured *Lycopersicon peruvianum* cells, systemin was found to induce medium alkalinization, ethylene production, and phenylalanine ammonia lyase activity in a similar way as microbial elicitors (Felix and Boller 1995). These cell cultures were found to possess high-affinity binding sites for systemin, indicative of a systemin receptor (Meindl et al. 1998). This experimental system was subsequently used by the group of the late C.A. Ryan in an attempt to identify the systemin receptor in an approach based on cross-linking of the radioactive ligand to its receptor. As reviewed before (Boller 2005), the putative receptor identified in this way appeared to be Cu-3/Curl-3, the tomato ortholog of BRI1, the brassinosteroid receptor of *Arabidopsis*. This caused quite a stir because of the notion that there might be bifunctional receptors in plants, perceiving peptides and steroid-type hormone signals at the same time (Wang and He 2004). However, subsequent work demonstrated that the *cu-3/curl-3* mutants of tomato, which lacked the brassinosteroid receptor and were thus dwarf in phenotype, were still fully responsive to systemin (Holton et al. 2007; Lanfermeijer et al. 2008). Thus, the search for the systemin receptor must go on (Boller and Felix 2009).

6 The *At*Pep/PEPR System

6.1 At*Peps*

In 2006, a new potential DAMP was isolated from an extract of entire intact *A. thaliana* plants. *At*Pep1, a small peptide of 23 amino acids, was found to elicit various defense responses when exogenously applied to *Arabidopsis* plants. It caused alkalinization of suspension-cultured *Arabidopsis* cells at subnanomolar concentrations (Huffaker et al. 2006), and it induced the activation of Ca^{2+} channels—leading to a depolarization of the membrane—the production of ROS, and the biosynthesis of the stress hormone ethylene (Krol et al. 2010). Seedlings grown in liquid medium supplied with *At*Pep1 showed reduced growth (Krol et al. 2010), as it was shown before for MAMPs such as flg22 and elf18 (Boller and Felix 2009). Furthermore, *At*Pep1 activated the expression of different defense-related genes such as *PDF1.2*, *PR1*, *MPK3*, and several *WRKY* genes encoding transcription factors (Huffaker et al. 2006; Yamaguchi et al. 2010).

There is evidence that *At*Pep1-induced defense responses may be biologically relevant in case of an infection. Plants overexpressing *PROPEP1*, the *At*Pep1 precursor peptide, showed enhanced resistance to the fungal pathogen *Pythium irregulare* (Huffaker et al. 2006), and wild-type plants pretreated with *At*Pep1 became more resistant to the bacterial pathogen *Pseudomonas syringae* pathovar *tomato* DC3000 (Yamaguchi et al. 2010).

As far as currently known, *At*Peps are only active in *A. thaliana* and very closely related species within the Brassicaceae, but potential orthologs have been identified in various species (Huffaker et al. 2006). Structure activity studies (Pearce et al. 2008) have shown that the C terminus is decisive for activity. By alanine substitution, two particularly important residues have been identified, serine (15) and glycine (17), as shown in Fig. 2. It seems that the glycine residue is conserved in putative *At*Pep homologs in other species (Pearce et al. 2008).

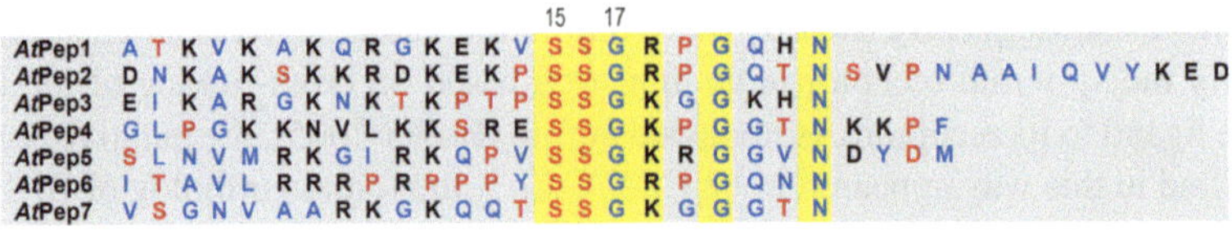

Fig. 2 Structures of *At*Peps. Proline (P), threonine (T), and serine (S) residues are in *red*, other charged amino acids are in *black*, and neutral amino acids are in *blue*. Conserved residues of the SSGR/KxGxxN are highlighted in *yellow*

6.2 PROPEPs

Six paralogs of *At*Pep1 have been identified and termed *At*Pep2-7 (Huffaker et al. 2006; Huffaker and Ryan 2007; Pearce et al. 2008). Despite generally low sequence similarity, they all possess the conserved residues SSGR/KxGxxN (Fig. 2).

They are all encoded as larger proteins, named PROPEP1–7. The gene *PROPEP1*, for example, encodes a 92-amino-acid peptide, which is highly positively charged and lacks a signal sequence. This indicates that it is produced on cytoplasmic ribosomes and not via the secretory pathway. The *At*Pep1 peptide is derived from the C terminus of its precursor and is not posttranslationally modified (Huffaker et al. 2006). The cleavage mechanism has not been unraveled yet. In fact, it is currently not known whether the propeptides are cleaved in their natural environment. However, an extracellular aspartic protease, CDR1, involved in *Arabidopsis* disease resistance has been proposed to be important for the release an endogenous peptide elicitors (Xia et al. 2004). The involvement of this enzyme in PROPEP cleavage remains to be investigated. Apart from that, it is an open question whether the PROPEPs are also acting as full-length peptides. So far, it remains unclear how the peptide exits the cell to be perceived by the extracellular ligand-binding domain of its receptors. One could think of a yet unknown active export mechanism or of release of *At*Peps upon cell damage by pathogens or herbivores.

The *PROPEP1* gene is induced by wounding, methyl jasmonate (MeJA), and ethephon but not by drought, cold stress, UV-B irradiation, methyl salicylic acid (MeSA), and abscisic acid (Huffaker et al. 2006). In contrast, *PROPEP2* and *3* transcription is induced by MeSA (Huffaker and Ryan 2007). These genes are also the most strongly expressed PROPEP paralogs in response to a fungal, a bacterial, and an oomycete pathogen (Huffaker et al. 2006; Toufighi et al. 2005). *PROPEP* expression is also induced by *At*Peps themselves (Huffaker and Ryan 2007, also see chapter "Computational-Based Analysis to Associate the Function of Plant Signaling Peptides with Distinct Biological Processes"). Conversely, the accumulation of the *PROPEP1* gene product is only strongly upregulated by *At*Pep1, whereas accumulation of *PROPEP2* and *3* gene products are induced by all peptides. *PROPEP4* and *5* are only weakly induced, and *PROPEP6* is not induced upon *At*Pep treatment (Huffaker and Ryan 2007). No expression data are available yet for *PROPEP7*.

6.3 PEPRs

Yamaguchi and collaborators were able to identify a receptor candidate for the *At*Peps by photoaffinity labeling; they called it PEPR1 (Yamaguchi et al. 2006). PEPR1 was shown to be a membrane-associated protein of about 170 kD belonging to the class XI of the LRR-RLK family. PEPR1 has been proven to be a receptor for *At*Pep1 by gain-of-function experiments in tobacco. After transformation with *PEPR1*, *Nicotiana tabaccum*, which is naturally insensitive to *At*Pep1, was able to respond to *At*Pep1 in the alkalinization assay (Yamaguchi et al. 2006). In these assays, synthetic *At*Pep2, 3, 5, and 6 have also been shown to be fully active. They are good competitors for *At*Pep1 for binding the receptor PEPR1, as demonstrated by photoaffinity labeling. The *pepr1* mutant plants still responded *At*Pep1 by the induction of various defense responses, although at a somewhat reduced level. This led to the hypothesis that a second receptor might be involved in *At*Pep perception. Recently, a plasma membrane receptor kinase with 76% similarity to *PEPR1* was identified and termed as *PEPR2* (Krol et al. 2010; Yamaguchi et al. 2010). Using transgenic tobacco cells expressing either *PEPR1* or *PEPR2*, it was shown that PEPR2 acts as a receptor for *At*Pep1 and *At*Pep2 only, whereas PEPR1 is a receptor for *At*Pep1–6 (Yamaguchi et al. 2010). PEPR2 is about 150 kD, slightly smaller than PEPR1. SALK lines for *pepr1* and *pepr2* single mutants were investigated and shown to be still responsive to *At*Pep1, whereas the double mutant *pepr1 pepr2* is totally insensitive to *At*Peps (Krol et al. 2010; Yamaguchi et al. 2010).

PEPR1 and *PEPR2* are transcriptionally induced by wounding, MeJA, and MAMPs (Yamaguchi et al. 2010; Zipfel et al. 2004, 2006), but also by *At*Peps (Krol et al. 2010; Yamaguchi et al. 2010). They are not induced by MeSA. Thus, their expression pattern resembles that of *At*Pep1.

6.4 Downstream Signals of a Microbial Attack

*At*Pep1 may act upstream of the jasmonic acid and ethephon and the salicylic acid pathway as mutants impaired in these pathways are incapable of upregulating the defense genes *PDF1.2* and *PR1* after *At*Pep1 application. Moreover, the production of H_2O_2 appears to be essential for the induction of *PDF1.2* and *PR1* by *At*Pep1, as demonstrated using the NADPH oxidase inhibitor diphenyleneiodonium chloride (DPI) (Huffaker et al. 2006). It has been shown that for full activation of defense gene expression upon *At*Pep treatment, a "calcium burst," i.e., a transient elevation of cytosolic Ca^{2+}, is required. When Ca^{2+} inwardly conducting channels were blocked, *At*Pep-dependent expression of the defense genes *WRKY33* and *MPK3* was reduced. It has recently been suggested that *At*PEPR1 has guanylyl cyclase activity, generating cGMP from GTP (Qi et al. 2010). This cGMP could activate cyclic nucleotide-gated channel (CNGC2) that allows Ca^{2+} influx. However,

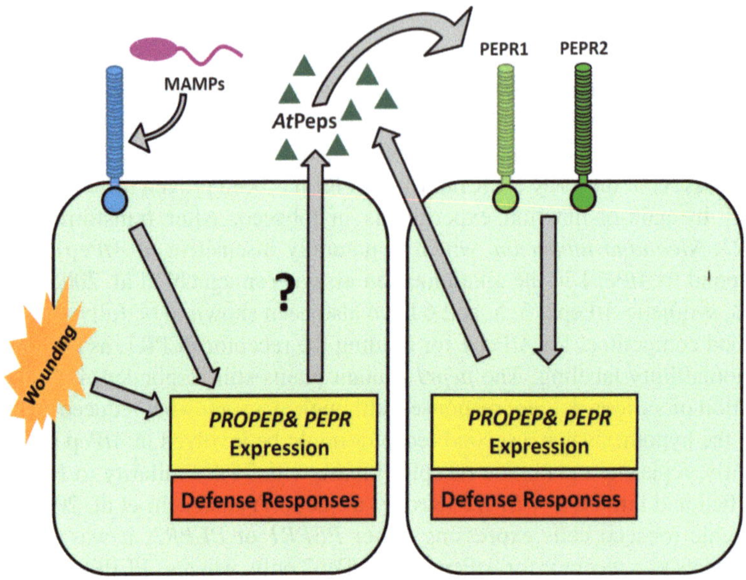

Fig. 3 A model for the action of *At*Peps. Wounding or pathogen recognition leads to defense responses and activates *At*Peps. The released *At*Peps themselves also induce defense responses as well as upregulated expression of their precursors and receptors. In this way, *At*Peps may contribute to the amplification and maintenance of defense responses

enhanced levels of cGMP upon *At*Pep application were not experimentally confirmed so far (Qi et al. 2010).

*6.5 At*Peps as Amplifiers**

One current view of the *At*Pep/PEPR system is to see it as an amplification mechanism within the innate immune system. MAMP recognition or wounding leads to the induction of defense responses but also induces the expression of *PROPEPs* (Yamaguchi et al. 2010). This may provide additional peptide signals that behave like MAMPs. Hence, the released *At*Peps would induce further defense responses, but also activate the transcription of their precursor and receptor genes. Acting like this in a feedback signaling system to amplify defense responses, *At*Peps could contribute to rapid, strong, and maintained defense responses against invaders (Fig. 3). Signaling amplification might occur in addition through both the jasmonic acid/ethephon and the salicylic acid pathways (Huffaker and Ryan 2007).

6.6 AtPeps in Systemic Resistance

It has been suggested that *At*Peps may not only amplify local defense mechanisms but may also be involved in systemic defense responses (Huffaker and Ryan 2007). However, as in the case of systemin, there is currently no evidence that *At*Peps are transported systemically. Nevertheless, *At*Peps show many similarities to systemin. First, *At*Peps and systemin are small peptides of about 20 amino acids that can induce defense responses (Huffaker et al. 2006). Second, they derive from larger precursor peptides and lack a signal sequence, because of which they are thought to be released upon cell injury (Ryan and Pearce 2003). Third, expression of *PROSYSTEMIN* (McGurl and Ryan 1992) as well as of *PROPEP* (Huffaker et al. 2006) is induced by wounding.

6.7 ZmPep

A *PROPEP*-like gene has been identified in maize has been investigated and shown to have properties similar to the *PROPEP* genes of *Arabidopsis* (Huffaker et al. 2011). The derived C terminus of the gene product encodes *Zm*Pep1, a 23-amino-acid-long peptide with only low identity with *At*Pep1 but shared conserved residues, including the glycine 17 which is known to be crucial for *At*Pep1 functionality. Like *AtPROPEP1*, the precursor *ZmPROPEP1* is induced by fungal infection, jasmonic acid, and *Zm*Pep1 application. It is able to activate de novo synthesis of jasmonic acid and ethephon and to induce the expression of defense genes. Pretreatment with *Zm*Pep1 enhances resistance to the fungal infections, southern leaf blight and anthracnose stalk rot, and induces SerPIN, a serine proteinase inhibitor (Huffaker et al. 2011). This could have a direct defense effect as it could inhibit digestive proteases from insects and microbial invaders. The induction of proteinase inhibitors is the classic hallmark of systemin action in tomato plants (Pearce et al. 1991).

7 Conclusion

There is a striking similarity between the perception, signal transduction, and response of exogenous and endogenous peptides acting as "danger signals," i.e., microbial peptides such as flg22 and elf18/elf25 which represent MAMPs, and endogenous peptides such as the *At*Peps which are believed to act as DAMPs (see Boller and Felix 2009). Importantly, both are perceived by similar PRRs, namely, by LRR-RLKs acting as receptors in concert with BAK1 (Schulze et al. 2010). However, we still do not know exactly how these PRRs work in the context of plant defense. For example, although it has been shown that FLS2 and EFR contribute to disease resistance against bacterial pathogens (Zipfel et al. 2004, 2006), it remains a riddle how the active epitopes, flg22 (which is buried inside the flagellum in its natural environment) and elf18/elf25 (which belongs to an intracellular bacterial protein), can be exposed to the corresponding PRRs of the plant (see Boller and Felix 2009). Moreover, there is currently no evidence that *Arabidopsis* plants lacking *PEPR1* and *PEPR2* are more resistant to pathogens or parasites. It is a challenge for the future to find out how these highly specific and sensitive perception systems function in vivo.

References

Apel K, Hirt H (2004) Reactive oxygen species: metabolism, oxidative stress, and signal transduction. Annu Rev Plant Biol 55:373–399

Asai T, Tena G, Plotnikova J, Willmann MR, Chiu WL, Gómez-Gómez L, Boller T, Ausubel FM, Sheen J (2002) MAP kinase signalling cascade in *Arabidopsis* innate immunity. Nature 415:977–983

Blein JP, Coutos-Thevenot P, Marion D, Ponchet M (2002) From elicitins to lipid-transfer proteins: a new insight in cell signalling involved in plant defence mechanisms. Trends Plant Sci 7:293–296

Boller T (1995) Chemoperception of microbial signals in plant cells. Annu Rev Plant Physiol Mol Biol 46:189–214

Boller T (2005) Peptide signalling in plant development and self/non-self perception. Curr Opin Cell Biol 17:116–122

Boller T, Felix G (2009) A renaissance of elicitors: perception of microbe-associated molecular patterns and danger signals by pattern-recognition receptors. Annu Rev Plant Biol 60:379–406

Boller T, He SY (2009) Innate immunity in plants: an arms race between pattern recognition receptors in plants and effectors in microbial pathogens. Science 324:742–744

Boudsocq M, Willmann MR, McCormack M, Lee H, Shan LB, He P, Bush J, Cheng SH, Sheen J (2010) Differential innate immune signalling via Ca^{2+} sensor protein kinases. Nature 464:418–422

Boutrot F, Segonzac C, Chang KN, Qiao H, Ecker JR, Zipfel C, Rathjen JP (2010) Direct transcriptional control of the Arabidopsis immune receptor FLS2 by the ethylene-dependent transcription factors EIN3 and EIL1. Proc Natl Acad Sci USA 107:14502–14507

Brunner F, Rosahl S, Lee J, Rudd JJ, Geiler C, Kauppinen S, Rasmussen G, Scheel D, Nürnberger T (2002) Pep-13, a plant defense-inducing pathogen-associated pattern from *Phytophthora* transglutaminases. EMBO J 21:6681–6688

Brutus A, Sicilia F, Macone A, Cervone F, De Lorenzo G (2010) A domain swap approach reveals a role of the plant wall-associated kinase 1 (WAK1) as a receptor of oligogalacturonides. Proc Natl Acad Sci USA 107:9452–9457

Chinchilla D, Bauer Z, Regenass M, Boller T, Felix G (2006) The *Arabidopsis* receptor kinase FLS2 binds flg22 and determines the specificity of flagellin perception. Plant Cell 18:465–476

Chinchilla D, Zipfel C, Robatzek S, Kemmerling B, Nürnberger T, Jones JDG, Felix G, Boller T (2007) A flagellin-induced complex of the receptor FLS2 and BAK1 initiates plant defence. Nature 448:497–500

Chinchilla D, Shan L, He P, de Vries S, Kemmerling B (2009) One for all: the receptor-associated kinase BAK1. Trends Plant Sci 14:535–541

Clay NK, Adio AM, Denoux C, Jander G, Ausubel FM (2009) Glucosinolate metabolites required for an Arabidopsis innate immune response. Science 323:95–101

D'Ovidio R, Mattei B, Roberti S, Bellincampi D (2004) Polygalacturonases, polygalacturonase-inhibiting proteins and pectic oligomers in plant-pathogen interactions. Biochim Biophys Acta Protein Proteomics 1696:237–244

Dardick C, Ronald P (2006) Plant and animal pathogen recognition receptors signal through non-RD kinases. PLos Pathog 2:14–28

Felix G, Boller T (1995) Systemin induces rapid ion fluxes and ethylene biosynthesis in *Lycopersicon peruvianum* cells. Plant J 7:381–389

Felix G, Boller T (2003) Molecular sensing of bacteria in plants – the highly conserved RNA-binding motif RNP-1 of bacterial cold shock proteins is recognized as an elicitor signal in tobacco. J Biol Chem 278:6201–6208

Felix G, Duran JD, Volko S, Boller T (1999) Plants have a sensitive perception system for the most conserved domain of bacterial flagellin. Plant J 18:265–276

Gaulin E, Dramé N, Lafitte C, Torto-Alalibo T, Martinez Y, Ameline-Torregrosa C, Khatib M, Mazarguil H, Villalba-Mateos F, Kamoun S, Mazars C, Dumas B, Bottin A, Esquerre-Tugaye MT, Rickauer M (2006) Cellulose binding domains of a *Phytophthora* cell wall protein are novel pathogen-associated molecular patterns. Plant Cell 18:1766–1777

Gómez-Gómez L, Boller T (2000) FLS2: an LRR receptor-like kinase involved in the perception of the bacterial elicitor flagellin in *Arabidopsis*. Mol Cell 5:1003–1011

Greenberg JT, Yao N (2004) The role and regulation of programmed cell death in plant-pathogen interactions. Cell Microbiol 6:201–211

Hayashi F, Smith KD, Ozinsky A, Hawn TR, Yi EC, Goodlett DR, Eng JK, Akira S, Underhill DM, Aderem A (2001) The innate immune response to bacterial flagellin is mediated by Toll-like receptor 5. Nature 410:1099–1103

He P, Shan L, Sheen J (2007) Elicitation and suppression of microbe-associated molecular pattern-triggered immunity in plant-microbe interactions. Cell Microbiol 9:1385–1396

Heese A, Hann DR, Gimenez-Ibanez S, Jones AME, He K, Li J, Schroeder JI, Peck SC, Rathjen JP (2007) The receptor-like kinase SERK3/BAK1 is a central regulator of innate immunity in plants. Proc Natl Acad Sci USA 104:12217–12222

Holton N, Caño-Delgado A, Harrison K, Montoya T, Chory J, Bishop GJ (2007) Tomato *BRASSINOSTEROID INSENSITIVE1* is required for systemin-induced root elongation in *Solanum pimpinellifolium* but is not essential for wound signaling. Plant Cell 19:1709–1717

Huffaker A, Ryan CA (2007) Endogenous peptide defense signals in *Arabidopsis* differentially amplify signaling for the innate immune response. Proc Natl Acad Sci USA 104:10732–10736

Huffaker A, Pearce G, Ryan CA (2006) An endogenous peptide signal in *Arabidopsis* activates components of the innate immune response. Proc Natl Acad Sci USA 103:10098–10103

Huffaker A, Dafoe NJ, Schmelz EA (2011) ZmPep1, an ortholog of Arabidopsis elicitor peptide 1, regulates maize innate immunity and enhances disease resistance. Plant Physiol 155:1325–1338

Jones JDG, Dangl JL (2006) The plant immune system. Nature 444:323–329

Kobayashi M, Ohura I, Kawakita K, Yokota N, Fujiwara M, Shimamoto K, Doke N, Yoshioka H (2007) Calcium-dependent protein kinases regulate the production of reactive oxygen species by potato NADPH oxidase. Plant Cell 19:1065–1080

Krol E, Mentzel T, Chinchilla D, Boller T, Felix G, Kemmerling B, Postel S, Arents M, Jeworutzki E, Al-Rasheid KAS, Becker D, Hedrich R (2010) Perception of the Arabidopsis danger signal peptide 1 involves the pattern recognition receptor AtPEPR1 and its close homologue AtPEPR2. J Biol Chem 285:13471–13479

Kunze G, Zipfel C, Robatzek S, Niehaus K, Boller T, Felix G (2004) The N terminus of bacterial elongation factor Tu elicits innate immunity in *Arabidopsis* plants. Plant Cell 16:3496–3507

Lacombe S, Rougon-Cardoso A, Sherwood E, Peeters N, Dahlbeck D, van Esse HP, Smoker M, Rallapalli G, Thomma B, Staskawicz B, Jones JDG, Zipfel C (2010) Interfamily transfer of a plant pattern-recognition receptor confers broad-spectrum bacterial resistance. Nat Biotechnol 28:365–369

Lanfermeijer FC, Staal M, Malinowski R, Stratmann JW, Elzenga JTM (2008) Micro-electrode flux estimation confirms that the *Solanum pimpinellifolium* cu3 mutant still responds to systemin. Plant Physiol 146:129–139

Lecourieux D, Raneva R, Pugin A (2006) Calcium in plant defence-signalling pathways. New Phytol 171:249–269

Lee H, Chah O-K, Sheen J (2011) Stem-cell-triggered immunity through CLV3p-FLS2 signalling. Nature 473:376–379

Li J, Wen JQ, Lease KA, Doke JT, Tax FE, Walker JC (2002a) BAK1, an *Arabidopsis* LRR receptor-like protein kinase, interacts with BRI1 and modulates brassinosteroid signaling. Cell 110:213–222

Li L, Li CY, Lee GI, Howe GA (2002b) Distinct roles for jasmonate synthesis and action in the systemic wound response of tomato. Proc Natl Acad Sci USA 99:6416–6421

Mackey D, McFall AJ (2006) MAMPs and MIMPs: proposed classifications for inducers of innate immunity. Mol Microbiol 61:1365–1371

Matzinger P (2002) The danger model: a renewed sense of self. Science 296:301–305

McGurl B, Ryan CA (1992) The organization of the prosystemin gene. Plant Mol Biol 20:405–409

Meindl T, Boller T, Felix G (1998) The plant wound hormone systemin binds with the N-terminal part to its receptor but needs the C-terminal part to activate it. Plant Cell 10:1561–1570

Mersmann S, Bourdais G, Rietz S, Robatzek S (2010) Ethylene signaling regulates accumulation of the FLS2 receptor and is required for the oxidative burst contributing to plant immunity. Plant Physiol 154:391–400

Mittler R, Vanderauwera S, Suzuki N, Miller G, Tognetti VB, Vandepoele K, Gollery M, Shulaev V, Van Breusegem F (2011) ROS signaling: the new wave? Trends Plant Sci 16:300–309

Miya A, Albert P, Shinya T, Desaki Y, Ichimura K, Shirasu K, Narusaka Y, Kawakami N, Kaku H, Shibuya N (2007) CERK1, a LysM receptor kinase, is essential for chitin elicitor signaling in *Arabidopsis*. Proc Natl Acad Sci USA 104:19613–19618

Morillo SA, Tax FE (2006) Functional analysis of receptor-like kinases in monocots and dicots. Curr Opin Plant Biol 9:460–469

Nam KH, Li JM (2002) BRI1/BAK1, a receptor kinase pair mediating brassinosteroid signaling. Cell 110:203–212

Nühse TS, Peck SC, Hirt H, Boller T (2000) Microbial elicitors induce activation and dual phosphorylation of the *Arabidopsis thaliana* MAPK 6. J Biol Chem 275:7521–7526

Nürnberger T, Nennstiel D, Jabs T, Sacks WR, Hahlbrock K, Scheel D (1994) High affinity binding of a binding of a fungal oligopeptide elicitor to parsley plasma membranes triggers multiple defense responses. Cell 78:449–460

Pearce G, Strydom D, Johnson S, Ryan CA (1991) A polypeptide from tomato leaves induces woun-inducible proteinase-inhibitor proteins. Science 253:895–898

Pearce G, Moura DS, Stratmann J, Ryan CA (2001a) Production of multiple plant hormones from a single polyprotein precursor. Nature 411:817–820

Pearce G, Moura DS, Stratmann J, Ryan CA (2001b) RALF, a 5-kDa ubiquitous polypeptide in plants, arrests root growth and development. Proc Natl Acad Sci USA 98:12843–12847

Pearce G, Yamaguchi Y, Munske G, Ryan CA (2008) Structure-activity studies of AtPep1, a plant peptide signal involved in the innate immune response. Peptides 29:2083–2089

Pearce G, Yamaguchi Y, Barona G, Ryan CA (2010) A subtilisin-like protein from soybean contains an embedded, cryptic signal that activates defense-related genes. Proc Natl Acad Sci USA 107:14921–14925

Qi Z, Verma R, Gehring C, Yamaguchi Y, Zhao YC, Ryan CA, Berkowitz GA (2010) Ca(2+) signaling by plant *Arabidopsis thaliana* Pep peptides depends on AtPepR1, a receptor with guanylyl cyclase activity, and cGMP-activated Ca(2+) channels. Proc Natl Acad Sci USA 107:21193–21198

Rodriguez MCS, Petersen M, Mundy J (2010) Mitogen-activated protein kinase signaling in plants. Annu Rev Plant Biol 61:621–649

Ronald PC, Beutler B (2010) Plant and animal sensors of conserved microbial signatures. Science 330:1061–1064

Ryan CA, Pearce G (2003) Systemins: a functionally defined family of peptide signal that regulate defensive genes in Solanaceae species. Proc Natl Acad Sci USA 100:14577–14580

Ryan CA, Huffaker A, Yamaguchi Y (2007) New insights into innate immunity in Arabidopsis. Cell Microbiol 9:1902–1908
Schulze B, Mentzel T, Jehle AK, Mueller K, Beeler S, Boller T, Felix G, Chinchilla D (2010) Rapid heteromerization and phosphorylation of ligand-activated plant transmembrane receptors and their associated kinase BAK1. J Biol Chem 285:9444–9451
Segonzac C, Zipfel C (2011) Activation of plant pattern-recognition receptors by bacteria. Curr Opin Microbiol 14:54–61
Shah J (2009) Plants under attack: systemic signals in defence. Curr Opin Plant Biol 12:459–464
Shiu SH, Karlowski WM, Pan RS, Tzeng YH, Mayer KFX, Li WH (2004) Comparative analysis of the receptor-like kinase family in Arabidopsis and rice. Plant Cell 16:1220–1234
Spanu P, Grosskopf DG, Felix G, Boller T (1994) The apparent turnover of 1-aminocyclopropane-1-carboxylate synthase in tomato cells is regulated by protein phosphorylation and dephosphorylation. Plant Physiol 106:529–535
Stepanova AN, Alonso JM (2009) Ethylene signaling and response: where different regulatory modules meet. Curr Opin Plant Biol 12:548–555
Tör M, Lotze MT, Holton N (2009) Receptor-mediated signalling in plants: molecular patterns and programmes. J Exp Bot 60:3645–3654
Torres MA (2010) ROS in biotic interactions. Physiol Plant 138:414–429
Toufighi K, Brady SM, Austin R, Ly E, Provart NJ (2005) The botany array resource: e-Northerns, expression angling, and promoter analyses. Plant J 43:153–163
Wang ZY, He JX (2004) Brassinosteroid signal transduction – choices of signals and receptors. Trends Plant Sci 9:91–96
Xia YJ, Suzuki H, Borevitz J, Blount J, Guo ZJ, Patel K, Dixon RA, Lamb C (2004) An extracellular aspartic protease functions in Arabidopsis disease resistance signaling. EMBO J 23:980–988
Yamaguchi Y, Huffaker A (2011) Endogenous peptide elicitors in higher plants. Curr Opin Plant Biol 14:351–357
Yamaguchi Y, Pearce G, Ryan CA (2006) The cell surface leucine-rich repeat receptor for AtPep1, an endogenous peptide elicitor in *Arabidopsis*, is functional in transgenic tobacco cells. Proc Natl Acad Sci USA 103:10104–10109
Yamaguchi Y, Huffaker A, Bryan AC, Tax FE, Ryan CA (2010) PEPR2 Is a second receptor for the Pep1 and Pep2 peptides and contributes to defense responses in Arabidopsis. Plant Cell 22:508–522
Yamaguchi Y, Barona G, Ryan CA, Pearce G (2011) GmPep914, an eight-amino acid peptide isolated from soybean leaves, activates defense-related genes. Plant Physiol 156:932–942
Zhang SQ, Klessig DF (2001) MAPK cascades in plant defense signaling. Trends Plant Sci 6:520–527
Zipfel C, Robatzek S, Navarro L, Oakeley EJ, Jones JDG, Felix G, Boller T (2004) Bacterial disease resistance in *Arabidopsis* through flagellin perception. Nature 428:764–767
Zipfel C, Kunze G, Chinchilla D, Caniard A, Jones JDG, Boller T, Felix G (2006) Perception of the bacterial PAMP EF-Tu by the receptor EFR restricts *Agrobacterium*-mediated transformation. Cell 125:749–760

Peptides and the Regulation of Plant Homeostasis

Chris Gehring and Helen R. Irving

Abstract Homeostasis defines the property of living organisms to keep a stable internal environment despite considerable external fluctuations, and this process is of considerable importance to plants as they are trapped unmoving in their changing surroundings. In addition to plant hormones, signaling peptides also have role(s) in maintaining plant homeostasis. Plant natriuretic peptides (PNPs) modulate ion channels and water uptake and have been implicated as compounds important in maintaining homeostasis. PNP and molecular mimics produced by pathogens modulate photosynthesis and the chloroplast proteome, thus indicating that PNP has widespread effects on homeostasis and could be considered a prototype homeostatic peptide. We propose that other peptides such as phytosulfokines (PSKs) and rapid alkalinization factors (RALFs) that have recognized roles in development and defense also have functions in plant homeostasis.

1 The Role of Cellular Homeostasis

Homeostasis from the Greek similar (ὅμοιος) and stasis (στάσις) defines an unmoving state. It defines a property of an enclosed system capable of keeping its internal environment in a stable equilibrium often despite considerable external fluctuations. Given the nature of living systems with their membrane-delineated compartments, creating "insides" and "outsides," it is only natural that the concept of homeostasis is central. One of the most obvious and best-studied examples is the

C. Gehring (✉)
Division of Chemistry, Life Science and Engineering, 4700 King Abdullah University of Science and Technology, Thuwal 23955-6900, Kingdom of Saudi Arabia
e-mail: christoph.gehring@kaust.edu.sa

H.R. Irving
Monash Institute of Pharmaceutical Sciences, Monash University, 381 Royal Parade, Parkville, VIC 3052, Australia

H.R. Irving and C. Gehring (eds.), *Plant Signaling Peptides*, Signaling and Communication in Plants 16, DOI 10.1007/978-3-642-27603-3_10, homeostasis of the body temperature in warm-blooded animals where the maintenance of near constant temperatures is critical for survival. However, homeostasis is a principle with far wider scope, having implications not just for whole organisms, but also for tissues, cells, and subcellular compartments. All of these need to maintain biological and biochemical equilibria, a controlled ionic environment and metabolic conditions conducive to the generation and maintenance of ion gradients as well as the maintenance of the structural components required for the formation of the different cellular components and compartments. In short, the functioning of biological systems is critically dependent on homeostasis and hence the regulation of the latter, directly or indirectly by hormones.

There is a vast body of literature dealing with animal homeostasis, covering aspects ranging from pH homeostasis and ion homeostasis to thermoregulation and not least energy homeostasis. The role of glucose and fat metabolism and its regulation by insulin in particular predominates in the literature. One of the key cellular energy sensors in animals is AMP-activated protein kinase (AMPK) which has a homologous role to sucrose non-fermenting 1 (SNF1) protein kinase occurring in yeast. In plants, the literature is considerably less extensive; however, the increasing emphasis on systems analyses, and in particular metabolomics, is resulting in a better and more integrated understanding of regulatory processes. It is now evident that plants also employ a key energy-sensing protein kinase, SNF1-related kinase 1 (SnRK1), at the cellular level to direct extensive transcriptional responses to maintain homeostasis (Baena-Gonzalez and Sheen 2008).

Given that plants cannot usually actively move to places of more favorable environmental conditions, we would expect that homeostatic regulation is particularly critical for the survival of the individual as well as the species. Land plants in particular are subject to large long-term and short-term fluctuations in environmental conditions—light, temperature, nutrient availability, and soil water content—that require rapid and coordinated responses at both the cellular and whole-plant level (Fig. 1). Plant hormones have a major role in the regulation of coordinated whole-plant responses to the environment, and mutant plants with defects in hormone synthesis or hormone perception can display phenotypes that are indicative of severe disturbances of homeostasis. A point in case is the *Arabidopsis thaliana* abscisic acid (ABA)-deficient (*aba*) and ABA-insensitive (*abi*) mutants that are wilty due to severely disturbed water homeostasis. Plant hormones also are involved in coordinating development and plant growth through intricate interactions where the hormones act to promote or antagonize growth and development (Wang and Irving 2011). Both development and plant growth do in fact involve modulations in homeostasis events, but in this chapter, we wish to focus on the daily needs of plants to maintain homeostasis at a whole plant and cellular level.

The five "classical" hormones (Kende and Zeevaart 1997)—auxin, ethylene, ABA, cytokinins, and gibberellic acid (GA)—are small molecules. They are well studied both in terms of their effects and the signal transduction pathways used to coordinately regulate plant responses to environmental fluctuations and development. This is particularly so in *Arabidopsis* as it offers many advantages of a model

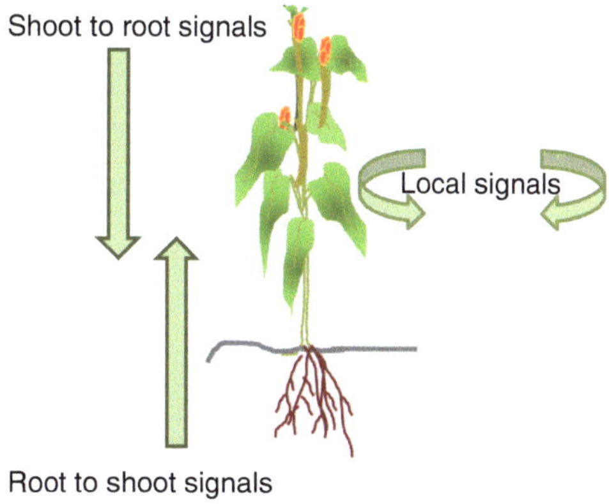

Fig. 1 Overview of whole-plant homeostasis. Whole-plant homeostasis is continually modulated in response to environmental changes (e.g., day to night, soil water levels). Cells sensing the changes generate signals that act locally and move long distances to maintain homeostatic control

organism such as the availability of the complete genome, a large amount of transcriptional data in the public domain, and a vast number of mutants that enable a systems analysis of hormone-dependent responses (e.g., Bastian et al. 2010; Pandey et al. 2010). This has led to increased understanding of the essential cellular components that are involved in maintaining homeostasis (Fig. 2). Hormones or signaling molecules or light are perceived by specific receptors that can be located at the membrane or intracellularly. These receptors initiate a signal cascade involving second messengers that can directly or indirectly affect ion channels which in turn have a key role in maintaining cellular homeostasis. Examples of second messengers affecting ion channels are calcium and cyclic nucleotides (e.g., cGMP) (Leng et al. 1999; Kaplan et al. 2007; Qi et al. 2010). Second messengers can also, through specifically phosphorylated intermediates and/or transcription factors, modulate the transcriptional program of the cell. Diurnal changes in light for instance are accompanied by extensive changes in photosynthesis and production of metabolites which result in massive changes in transcription of over 1,000 genes in leaf mesophyll cells (Baena-Gonzalez and Sheen 2008). In addition, hormones or signaling molecules may regulate their own transcription and processing and so permit signal amplification or dampening.

Signaling peptides also have role(s) in maintaining plant homeostasis and form the focus of the remainder of this chapter. First, we discuss the evidence implicating plant natriuretic peptides (PNPs) in maintaining homeostasis, and then we speculate that other peptides with recognized roles in development and defense also have functions in plant homeostasis.

2 PNPs and the Systemic Regulation of Plant Homeostasis

Natriuretic peptides (NPs) are a family of peptide hormones in vertebrates that have a role in osmoregulation (Toop and Donald 2004; Potter et al. 2006). Atrial NPs (ANPs) are synthesized in the atrium and released into the bloodstream at elevated

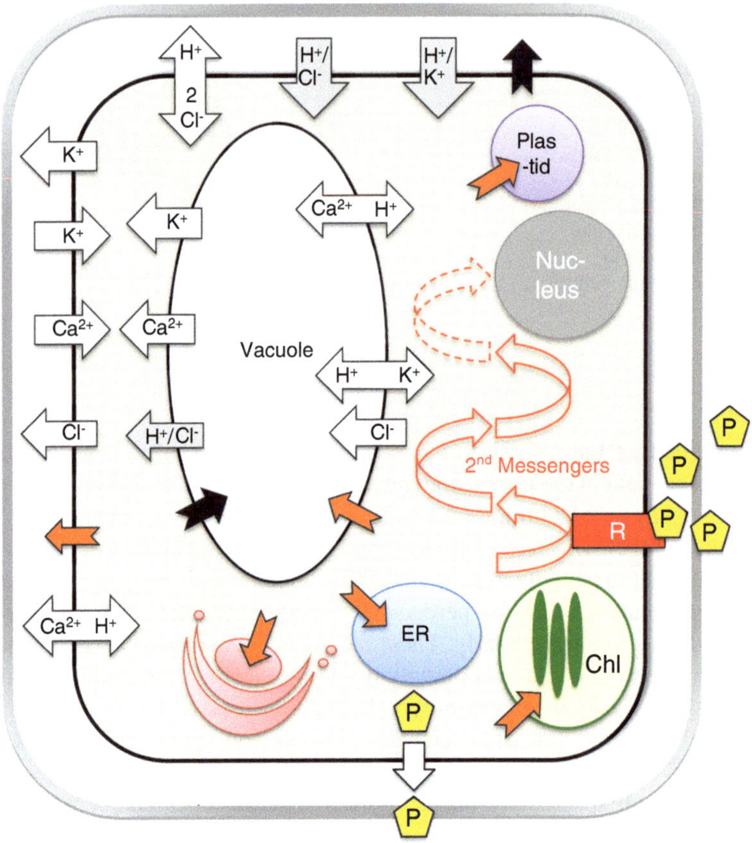

Fig. 2 Overview of major channels and processes with a role in the regulation of cellular homeostasis. Peptide hormones or signaling molecules (P) are likely to be perceived by specific receptors (R) and initiate a signal cascade involving second messengers. Some of them (e.g., Ca^{2+} or cGMP) will directly or indirectly affect ion channels which in turn have a key role in cellular homeostasis. Second messengers can also, through specifically phosphorylated intermediates and/or transcription factors, modulate the transcriptional program of the cell. In addition, peptide hormones or signaling molecules may regulate their own transcription and processing and so permit signal amplification or dampening. *Double arrows* are antiporters, *gray arrows* are symporters, *black solid arrows* are proton ATPases, and *orange arrows* signify calcium ATPases; all of which contribute to cellular homeostasis and can directly or indirectly be affected by peptide hormones or signaling molecules

blood pressure. ANP acts as ligand for ANP receptors in the kidney where they trigger guanosine 3′,5′-cyclic monophosphate (cGMP)-dependent natriuresis to counteract the renin-angiotensin-aldosterone system. Mouse knockout mutants of proANP (the ANP precursor) and NPR-A (the ANP receptor) both show hypertension and increased cardiac hypertrophy providing genetic evidence that ANP plays important roles in the homeostasis of body fluid and control of blood pressure (Kuhn 2005; Potter et al. 2006).

The first evidence for the presence of NP in plants was that radioimmunoassays recognized molecules from extracts of Florida Beauty (Vesely and Giordano 1991) and that exogenous application of human ANP could increase plant transpiration and solute flow (Vesely et al. 1993). Subsequently, rat ANP was demonstrated to bind to isolated leaf membranes (Gehring et al. 1996; Suwastika et al. 2000) and induce stomatal opening in a concentration, conformation, and cGMP-dependent manner (Pharmawati et al. 1998b). Plant NPs (PNPs) have been immunoaffinity purified from several different species (Billington et al. 1997; Maryani et al. 2001; Rafudeen et al. 2003). Partial protein sequencing of these immunoreactant PNPs led to the identification of two *A. thaliana* sequences that have weak similarity to the cell wall loosening expansins (Ludidi et al. 2002; Kende et al. 2004). However, while PNPs share some sequence homology with expansins, they have lost the domain thought to be involved in cell wall binding, thus affording them increased mobility (Ludidi et al. 2002). Consistent with the absence of a wall-binding domain is also the fact that there is no evidence of expansin activity in PNPs (Gehring and Irving 2003). The fact that PNPs also act on protoplasts provides additional evidence that PNPs have different roles to expansins. Interestingly, the 33-amino-acid active region of PNP-A is the only region with similarity to ANP, and this region forms part of a loop that juts out from the protein (Wang et al. 2007). Recombinant PNP can mimic the ability of synthetic ANP to induce apoptosis in cardiomyoblasts (Wang et al. 2010) as ANP can mimic some of the actions of PNP in plant cells. Phylogenetic analysis suggests that the similarities between the animal and plant NPs are the result of convergent evolution (Gehring and Irving 2003).

PNPs have an N-terminal signal peptide directing them to the extracellular space—a location that is consistent with the identification of AtPNP-A in the apoplast proteome (Boudart et al. 2004). More recently, it has been demonstrated that *A. thaliana* PNP-A is mainly expressed in leaf mesophyll cells and mesophyll protoplast assays using AtPNP-A: GFP reporter constructs and flow cytometry have proved that it is secreted (Wang et al. 2011a). PNP-A expression is transiently induced by osmoticum-mimicking drought and also saline conditions at both the RNA and protein level (Rafudeen et al. 2003; Wang et al. 2011a). Interestingly, PNP-A also regulates its own expression, suggesting that it acts at adjacent cells to where it is made (paracrine) or possibly on the cell where it is produced (autocrine) and so modulates the amount of PNP present in the leaf and potentially amplifies levels when the plant is under abiotic stress (Rafudeen et al. 2003; Wang et al. 2011a, b). Co-expression analysis of publicly available microarray data of wild type and mutants has revealed that the transcription of PNP-A is strongly induced in response to pathogen challenges and correlates with genes in the systemic-acquired resistance pathway (Meier et al. 2008; also see chapter "Computational-Based Analysis to Associate the Function of Plant Signaling Peptides with Distinct Biological Processes").

Exogenous application of recombinant and native PNPs promotes a number of physiological responses including radial water movements out of the xylem (Suwastika and Gehring 1998), osmoticum-dependent protoplast swelling (Maryani et al. 2001; Morse et al. 2004), and transient increases in the second messenger cGMP (Pharmawati et al. 1998a, 2001; Maryani et al. 2001; Wang et al. 2007). PNPs are mainly expressed in leaf mesophyll cells where they are secreted and directly act in a paracrine (or autocrine) fashion on nearby cells to induce cell water uptake (swelling) and changes in cellular cGMP levels (Wang et al. 2007, 2011a). Importantly, studies with noninvasive ion-selective vibrating microelectrodes have shown PNP-dependent H^+, K^+, and Na^+ fluxes (Pharmawati et al. 1999) in *Zea mays* root conductive tissue. PNP can cause immediate net H^+ influx and delayed net K^+ and Na^+ uptake, and such delayed net K^+ influx was also observed in response to the cell permeable cGMP analogue, 8-Br-cGMP; however, this was not accompanied by significant changes in net H^+ fluxes. In addition, 8-Br-cGMP does not appear to stimulate the plasma membrane H^+-ATPase, thus implying that cGMP directly affects cation channels. PNP was also shown to cause rapid H^+ influx in the elongation zone of *Arabidopsis* roots but not in the mature zone and significant K^+ and Na^+ efflux in the mature root zone only (Ludidi et al. 2004). Taken together, these observations suggest that PNP can elicit rapid and tissue-specific ion responses that point to a complex role in plant growth and homeostasis.

Perhaps surprisingly, the presence of a *PNP-like* gene was reported in the bacterial plant pathogen *Xanthomonas axonopodis* (XacPNP) (Nembaware et al. 2004). Given its paradoxical phylogeny—the closest relative of XacPNP are PNPs from many different plant species—it was hypothesized that it was acquired as a result of an ancient horizontal gene transfer and that the bacterial pathogen was using the plant-like PNP to modify host homeostasis to its advantage. This form of molecular mimicry was later demonstrated in a series of experiments (Gottig et al. 2008).

Firstly, expressed and purified XacPNP was shown to alter physiological responses including stomatal opening in plants. Secondly, XacPNP is not expressed under standard nutrient-rich culture conditions but is strongly induced under conditions that mimic the nutrient poor intercellular apoplastic environment in the host and also in infected plant tissue, and this suggests that *XacPNP* transcription is inducible by the host environment (Gottig et al. 2008). In addition, XacPNP deletion mutants induce more necrotic lesions than those observed with the wild type, and bacterial cell death occurred earlier. These observations suggest that the plant-like bacterial PNP enables the biotrophic plant pathogen to modify host responses to create conditions favorable to its own survival. It also establishes changes in host homeostasis as a potentially important mechanism in the plant-pathogen arms race (Gottig et al. 2009).

Given the link between photosynthetic activity and homeostasis, we have recently investigated the effects of XacPNP on chlorophyll fluorescence parameters and the water potential of citrus leaves (Garavaglia et al. 2010a). We showed that XacPNP enhances the performance of photosystem II and that this effect is mirrored in a proteomic analysis of the PNP response that shows an upregulation notably of Rubisco activase, ATP synthase CF1 alpha subunit, maturase K, and alpha- and beta-tubulin (Garavaglia et al. 2010a). Furthermore, changes in the proteome were also observed in citrus leaves infected with the wild-type

X. axonopodis and XacPNP deletion mutants. Both the wild-type and the mutant pathogen cause downregulation of enzymes related to photosynthesis and chloroplast ribosomal proteins, while proteins related to defense responses are upregulated. However, leaves infiltrated with the XacPNP deletion mutant show a pronounced decrease in photosynthetic proteins with no reduction in defense-related proteins when compared to the wild-type pathogen. This might indicate that XacPNP serves the pathogen to maintain—at least in part—host photosynthetic efficiency during pathogenesis and prevents the host from starving the pathogen (Garavaglia et al. 2010b). It also suggests that PNP normally expressed in the leaves can have a role in modulating photosynthesis levels to maintain plant homeostasis.

Recently, we demonstrated that PNP, much like the peptide hormone systemin (McGurl et al. 1994), also signals long distances at a systemic level. Recombinant AtPNP-A rapidly increases the rate of dark respiration in treated leaves of a perennial forest sage (*Plectranthus ecklonii*) after 5 min. In addition, increases in dark respiration were also registered in lower leaves, and with a lag time of about 10 min, the effect spreads to the upper leaves and subsequently (after 15 min) to the opposite leaves. This response signature is indicative for phloem mobility of the signal, a hypothesis that is further strengthened by the fact that cold girdling, which affects phloem but not xylem or apoplastic processes (Peuke et al. 2006), delays the long-distance AtPNP-A effect (Ruzvidzo et al. 2011). It was therefore concluded that locally applied AtPNP-A can induce a phloem-mobile signal that in turn is capable of rapidly modifying plant homeostasis in distal parts (Ruzvidzo et al. 2011).

The exact nature of the long-distance signal is still under investigation; however, it appears unlikely to be PNP itself due to the velocity of the response. Systemin is a smaller peptide, and it took 1–2 h for it to move into the phloem when applied to tomato leaves (Pearce et al. 1991). Hence, we hypothesize that the signal mediating the long-distance PNP response could be electrical. Long-distance electrical signaling may take the form of action potentials or variation potentials (Fromm and Lautner 2007). Action potentials are phloem-transmitted signals driven by voltage-gated channels generating rapid self-propagating waves of membrane depolarization with constant velocity and amplitude. The membrane depolarization is probably due to an initial Ca^{2+} influx and intensified by Cl^{-} efflux followed by K^{+} efflux necessary for repolarization. The amplitude and speed of variation potentials varies, and they result from local changes in membrane potential possibly caused by loss of hydraulic pressure from the xylem- or ligand-activated channels but do not involve voltage-gated channels. The "electrical signal" hypothesis is supported by previous findings that PNPs stimulate rapid membrane depolarization (Maryani et al. 2000), induce rapid ion fluxes (Pharmawati et al. 1998a; Ludidi et al. 2004), as well as promote the movement of water out of the xylem (Suwastika and Gehring 1998). We have therefore proposed that local PNP treatment induces an electrical signal that leads to the observed distal response such as changes in the rate of dark respiration in systemic leaves (Ruzvidzo et al. 2011).

The various ways that PNP modulates plant homeostasis responses are summarized in Fig. 3. At the cellular level, PNP expression is induced by abiotic

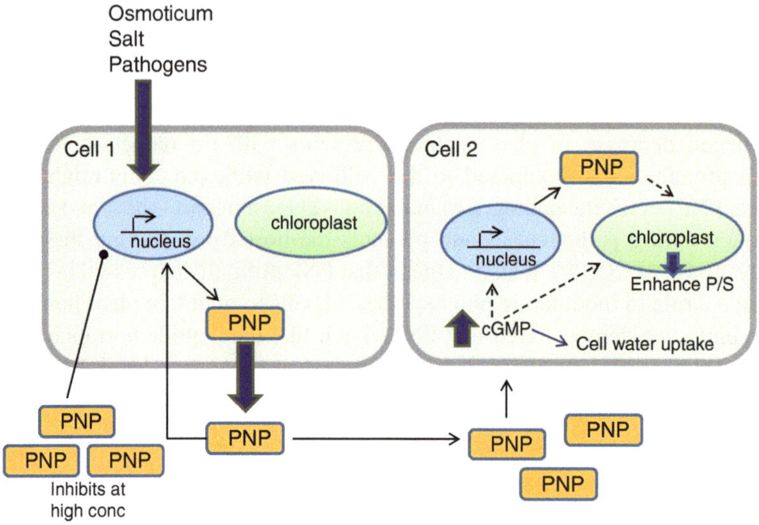

Fig. 3 The local action of PNP in leaf mesophyll cells. Abiotic and biotic stimuli enhance PNP-A expression resulting in PNP being exported from the cell where it will act on itself (autocrine) or adjacent cells (paracrine) as it moves through the apoplast. PNP-A causes increases in cGMP and directly enhances water uptake and also affects photosynthesis. PNP-A also directly enhances its own expression; however, higher concentrations inhibit transcription and protein production. *Solid lines* indicate that the effects are established and *dotted lines* indicate that circumstantial evidence is supporting the direct activation by either cGMP or PNP

and biotic stimuli resulting in an increase in PNP in the apoplast which can act either autocrinely or paracrinely. PNP raises intracellular levels of cGMP possibly via a receptor containing guanylate cyclase activity or via a receptor-stimulated cascade activating a separate guanylate cyclase. PNP also induces cellular water uptake, and if the responding cells are guard cells, this results in stomatal opening which can occur even in the presence of ABA (Wang et al. 2007). PNP also induces increases in photosynthetic capacity both at a local level and distantly, possibly by electrical signals (Garavaglia et al. 2010a, b; Ruzvidzo et al. 2011). At the local level in the leaves, PNP also modulates its own expression to maintain PNP levels within defined ranges in part to maintain whole-plant homeostasis (Wang et al. 2011a, b). How PNP modulates changes in photosynthesis is not understood at present, but it may act directly or through second messengers such as cGMP and/or calcium. There is limited evidence available to support either possibility. We have obtained some evidence of association of PNP-A with isolated crude chloroplast preparations (Wang and Irving, unpublished observations). Light treatments result in changes in cGMP, and this regulates phytochrome-dependent chloroplast and anthocyanin development (Bowler et al. 1994). PNP also indirectly interacts with Ca^{2+}, and this has been attributed to cross talk between cGMP and Ca^{2+} due to increases in cGMP mediated by PNP (Pharmawati et al. 2001).

3 A Role for Phytosulfokines in Plant Homeostasis

Phytosulfokines (PSKs) are members of the growing family of sulfated peptides that have roles in localized paracrine signaling (see chapter "Plant Peptide Signaling: An Evolutionary Adaptation"). PSK-α is a pentapeptide sulfated on the two tyrosine residues [Y(SO_3H)IY(SO_3H)TQ], and this is a requirement for cellular activity as non-sulfated PSKs have activity reduced by over 100-fold (Matsubayashi and Sakagami 1996). PSKs were first discovered as a cell proliferation agent necessary to maintain low-density cell cultures (Matsubayashi and Sakagami 1996). Several precursor proteins for PSK (proPSK) have been identified in different species that contain an N-terminal secretory signal sequence and a common PSK sequence near the C terminus (Yang et al. 1999, 2001; Lorbiecke and Sauter 2002). Processing of PSKs is likely to involve sulfating of proPSK by tyrosylprotein sulfotransferase as the protein is processed through the Golgi network (Hanai et al. 2000) before secretion when specific subtilisin serine proteases in the apoplast will cleave proPSK at dibasic residues upstream from the PSK sequence (Srivastava et al. 2008) before further processing to the active pentapeptide (see chapter "Processing of Peptides"). PSK acts in a paracrine or autocrine fashion on nearby cells expressing the receptor, and this has been most clearly shown in roots where PSK enhances root elongation by controlling cell size (Matsubayashi et al. 2006; Kutschmar et al. 2009; also see chapter "Peptides Regulating Root Growth"). PSKs influence gene expression and attenuate expression of stress response genes during differentiation of tracheary elements (Motose et al. 2009). PSK-α promotes tracheary element differentiation in *Zinnia elegans* mesophyll cells, and this process is inhibited by CLE41-44 to regulate organogenesis (Matsubayashi et al. 1999; Ito et al. 2006; Motose et al. 2009; also see chapter "Peptides Regulating Plant Vascular Development").

PSK-α is the natural ligand of the PSK receptor (PSKR1) (Matsubayashi et al. 2002) belonging to the large family of leucine-rich repeat receptor-like kinases (LRR-RLK). Recently, a guanylate cyclase (GC, enzymes that catalyze the conversion of GTP into cGMP) catalytic center was discovered in PSKR1 embedded within the cytosolic kinase domain, and it was shown that this dual kinase-GC domain has both serine/threonine kinase activity as well as Mg^{2+}- or Mn^{2+}-dependent GC activity (Kwezi et al. 2011). Overexpression of the full-length AtPSKR1 receptor in *Arabidopsis* leaf protoplasts raised endogenous basal cGMP levels over 20-fold indicating that the receptor has GC activity in vivo. In addition, PSK-α itself, but not the non-sulfated peptide, rapidly raises cGMP levels in protoplasts. It was argued that this elevation of the second messenger cGMP is critical for PSK downstream signaling possibly via activation of cyclic nucleotide-gated channels (Ma et al. 2010), cyclic nucleotide-dependent phosphorylation, and direct or indirect activation of cGMP-dependent transcription (Maathuis 2006). It is becoming increasingly clear that cyclic nucleotides and in particular cGMP play a key role in ion and water homeostasis during biotic (Meier et al. 2009, 2010) and abiotic stresses (Donaldson et al. 2004). Hence, it is reasonable to expect that PSK (and other plant-signaling peptides) will also affect cellular homeostasis particularly during changes in growth rate when there is a need for rapid adaptation to changes in ion and water balance.

4 Expanding Roles of the Rapid Alkalinization Factor

Rapid alkalinization factor (RALF) is a secreted 5-kDa polypeptide that was first isolated from tobacco leaves where it induced rapid alkalinization of the culture medium of tobacco suspension-cultured cells and activation of an intracellular mitogen-activated protein kinase (Pearce et al. 2001). Alkalinization is presumably caused by a blockage of a plasmalemma H^+ pump (Pearce et al. 2010). As RALF was identified through the alkalinization screen to identify plant peptides involved in activating defense responses, an early focus was on its role in defense. It turns out that RALFs and RALF-like (RALFL) proteins constitute a family of different related peptides found throughout the plant kingdom, and it has been suggested that they may have a role in plant development given that, in poplar and potato, different RALF-encoding genes were found to be expressed in a tissue-specific manner (Haruta and Constabel 2003; Germain et al. 2005). Application of exogenous RALF to *Arabidopsis* and tomato seedlings inhibited root growth (Pearce et al. 2001) and silencing of RALF in tobacco-disrupted root hair development as RALF is necessary for regulating root hair extracellular pH and the transition from root hair initiation to tip growth (Wu et al. 2007; also see chapter "Peptides Regulating Root Growth"). In sugarcane, RALF transcripts are found in the elongation zones of roots and leaves (Mingossi et al. 2010) where they presumably modulate extracellular pH and growth. RALFs also interact with hormonal signaling as overexpression of AtRALF23 impairs brassinosteroid effects on growth, and brassinosteroids themselves downregulate AtRALF23 expression in wild type (Srivastava et al. 2009).

Cytoplasmic Ca^{2+} transients that initiate and or modulate many cellular-signaling pathways in response to developmental and environmental cues in plants have also been noted in response to AtRALF1. They occur within 40 s of treatment with nanomolar concentrations of the natural or synthetic version of the peptide, and a model has been proposed that foresees that the peptide mediates Ca^{2+}-dependent signaling through a cell surface receptor. By doing so, it may play a role in eliciting stress responses or the modulation of growth (Haruta et al. 2008).

Given that RALFs can affect growth and developmental processes, which by their nature involve rapid homeostatic adjustments, and particularly in the light of the effect of RALF on plasmalemma H^+ pumps (Pearce et al. 2010), it is likely that RALFs affect homeostasis. How exactly this occurs, and what the effect of the blockage of the proton pump does to other ions and to the cytosolic pH remains to be investigated. It is however likely that changes in the transmembrane pH gradient will affect proton-gated and/or pH-regulated ion channels. Indeed, both inward- and outward-rectifying K channels ($I_{K,in}$ and $I_{K,out}$, respectively) are acutely sensitive to extracellular (Blatt 1992; Ilan et al. 1994; Yu et al. 2001) as well as intracellular pH variations (Blatt and Armstrong 1993). This is even more relevant as intracellular alkalinization was shown to be an important step in the signaling mechanism by which ABA gates open $I_{K,out}$ (Blatt and Armstrong 1993). In animals, typical examples of proton-gated channels (members of the amiloride-sensitive ENaC/degenerin family) involved in pain sensing triggered by damaged or inflamed tissue are expressed in sensory neurons: the acid-sensing ionic channel (ASIC) and the dorsal root ASIC (DRASIC), sensory neurons (Chen et al. 1998). Intracellular pH change is also the main signal involved in spermatozoa-"capacitating" process. Indeed, once in the female reproductive tract, the spermatozoan cytoplasm alkalinizes and activates a weakly rectifying outwardly potassium current (I_{KSper}) that rapidly hyperpolarizes the cell. This in turn sets the membrane to negative potentials where Ca^{2+} entry via another channel ($I_{CatSper}$) is maximized resulting in hyperactivated motility (Navarro et al. 2007). Incidentally, K_{Sper} channel is thought to be composed of subunits encoded by the *Slo3* gene as genetic deletion of *Slo3* results in the abolition of alkalization-activated K^+ current in the corpus epididymal sperm (Zeng et al. 2011).

In addition, if RALFs do cause cytosolic pH changes, these changes in turn will affect cellular phosphorylation events. There are well-documented examples of pH dependence (in the physiological range) of kinases in plants, e.g., the histidine kinase activity of the *Arabidopsis* ethylene receptor ETR1 (Gamble et al. 1998). Finally, it is conceivable that RALF-dependent cytosolic pH changes can reprogram cells via activity changes of a number of diverse kinases with critical regulatory functions. Such a hypothesis is amenable to experimental verification using phospho-proteomics that affords a systems view of the phosphorylation states of proteome.

5 Conclusions

As we discover more about peptide-signaling molecules, it becomes ever more apparent that their roles are intricately integrated with classical hormones to regulate plant growth and development as well as responses to the environment at a local (paracrine) and long-distance level (Wang and Irving 2011). PNP may be a prototype homeostatic peptide-signaling molecule as it has clear effects on photosynthesis and photorespiration that are mediated both locally and afar. However, at this stage, the receptor for PNP has not been identified, and it remains to be elucidated how the intracellular signals induced by PNP such as raised cGMP modulate chloroplast function. Moreover, although PNP has long-distance activity which we speculate is electrical in nature, it is unclear how this is perceived at the distal locations. We have raised the possibility that other peptides such as PSK and RALF also have roles in homeostasis in addition to the roles already attributed to these peptides in development and plant growth.

Acknowledgments This work was supported by the Australian Research Council's Discovery project funding scheme (DP0557561, DP0878194).

References

Baena-Gonzalez E, Sheen J (2008) Convergent energy and stress signaling. Trends Plant Sci 13:474–482

Bastian R, Dawe A, Meier S, Ludidi N, Bajic VB, Gehring C (2010) Gibberelic acid and cGMP-dependent transcriptional regulation in *Arabidopsis thaliana*. Plant Signal Behav 5:224–232

Billington T, Pharmawati M, Gehring CA (1997) Isolation and immunoaffinity purification of biologically active plant natriuretic peptide. Biochem Biophys Res Commun 235:722–725

Blatt MR (1992) K^+ channels of stomatal guard cells – characteristics of the inward rectifier and its control by pH. J Gen Physiol 99:615–644

Blatt MR, Armstrong F (1993) K^+ channels of stomatal guard cells – abscisic-acid-evoked control of the outward rectifier mediated by cytoplasmic pH. Planta 191:330–341

Boudart G, Jamet E, Rossignol M, Lafitte C, Borderies G, Jauneau A, Esquerre-Tugaye M, Pont-Lezica R (2004) Cell wall proteins in apoplastic fluids of *Arabidopsis thaliana* rosettes: identification by mass spectrometry and bioinformatics. Proteomics 5:212–221

Bowler C, Neuhaus G, Yamagata H, Chua N-H (1994) Cyclic GMP and calcium mediate phytochrome phototransduction. Cell 77:73–81

Chen CC, England S, Akopian AN, Wood JN (1998) A sensory neuron-specific, proton-gated ion channel. Proc Natl Acad Sci USA 95:10240–10245

Donaldson L, Ludidi N, Knight MR, Gehring C, Denby K (2004) Salt and osmotic stress cause rapid increases in *Arabidopsis thaliana* cGMP levels. FEBS Lett 569:317–320

Fromm J, Lautner S (2007) Electrical signals and their physiological significance in plants. Plant Cell Environ 30:249–257

Gamble RL, Coonfield ML, Schaller GE (1998) Histidine kinase activity of the ETR1 ethylene receptor from Arabidopsis. Proc Natl Acad Sci USA 95:7825–7829

Garavaglia BS, Thomas L, Zimaro T, Gottig N, Daurelio LD, Ndimba B, Orellano EG, Ottado J, Gehring C (2010a) A plant natriuretic peptide-like molecule of the pathogen *Xanthomonas axonopodis* pv. *citri* causes rapid changes in the proteome of its citrus host. BMC Plant Biol 10:51

Garavaglia BS, Thomas L, Gottig N, Dunger G, Garofalo CG, Daurelio LD, Ndimba B, Orellano EG, Gehring C, Ottado J (2010b) A eukaryotic-acquired gene by a biotrophic phytopathogen allows prolonged survival on the host by counteracting the shut-down of plant photosynthesis. PLoS One 5:e8950

Gehring CA, Irving HR (2003) Natriuretic peptides – a class of heterologous molecules in plants. Int J Biochem Cell Biol 35:1318–1322

Gehring CA, Khalid KM, Toop T, Donald JA (1996) Rat natriuretic peptide binds specifically to plant membranes and induces stomatal opening. Biochem Biophys Res Commun 228:739–744

Germain H, Chevalier E, Caron S, Matton DP (2005) Characterization of five RALF-like genes from *Solanum chacoense* provides support for a developmental role in plants. Planta 220:447–454

Gottig N, Garavaglia BS, Daurelio LD, Valentine A, Gehring C, Orellano EG, Ottado J (2008) *Xanthomonas axonopodis* pv. *citri* uses a plant natriuretic peptide-like protein to modify host homeostasis. Proc Natl Acad Sci USA 105:18631–18636

Gottig N, Garavaglia BS, Daurelio LD, Valentine A, Gehring C, Orellano EG, Ottado J (2009) Modulating host homeostasis as a strategy in the plant-pathogen arms race. Commun Integr Biol 2:89–90

Hanai H, Nakayama D, Yang H, Matsubayashi Y, Hirota Y, Sakagami Y (2000) Existence of a plant tyrosylprotein sulfotransferase: novel plant enzyme catalyzing tyrosine O-sulfation of preprophytosulfokine variants in vitro. FEBS Lett 470:97–101

Haruta M, Constabel CP (2003) Rapid alkalinization factors in poplar cell cultures. Peptide isolation, cDNA cloning, and differential expression in leaves and methyl jasmonate-treated cells. Plant Physiol 131:814–823

Haruta M, Monshausen G, Gilroy S, Sussman MR (2008) A cytoplasmic Ca^{2+} functional assay for identifying and purifying endogenous cell signaling peptides in Arabidopsis seedlings: identification of AtRALF1 peptide. Biochemistry 47:6311–6321

Ilan N, Schwartz A, Moran N (1994) External pH effects on the depolarization-activated K-channels in guard cell protoplasts of *Vicia faba*. J Gen Physiol 103:807–831

Ito Y, Nakanoyo I, Motose H, Iwamoto K, Sawa S, Dohmae N, Fukuda H (2006) Dodeca-CLE peptides as suppressors of plant stem cell differentiation. Science 313:842–845

Kaplan B, Sherman T, Fromm H (2007) Cyclic nucleotide-gated channels in plants. FEBS Lett 581:2237–2246

Kende H, Zeevaart J (1997) The five “classical” plant hormones. Plant Cell 9:1197–1210

Kende H, Bradford K, Brummell D, Cho HT, Cosgrove D, Fleming A, Gehring C, Lee Y, McQueen-Mason S, Rose J, Voesenek LA (2004) Nomenclature for members of the expansin superfamily of genes and proteins. Plant Mol Biol 55:311–314

Kuhn M (2005) Cardiac and intestinal natriuretic peptides: insights from genetically modified mice. Peptides 26:1078–1085

Kutschmar A, Rzewuski G, Stuhrwohldt N, Beemster GT, Inze D, Sauter M (2009) PSK-alpha promotes root growth in Arabidopsis. New Phytol 181:820–831

Kwezi L, Ruzvidzo O, Wheeler JI, Govender K, Iacuone S, Thompson PE, Gehring C, Irving HR (2011) The phytosulfokine (PSK) receptor is capable of guanylate cyclase activity and enabling cyclic GMP-dependent signalling in plants. J Biol Chem 286:22580–22588

Leng Q, Mercier RW, Yao W, Berkowitz GA (1999) Cloning and first functional characterization of a plant cyclic nucleotide-gated cation channel. Plant Physiol 121:753–761

Lorbiecke R, Sauter M (2002) Comparative analysis of PSK peptide growth factor precursor homologs. Plant Sci 163:321–332

Ludidi NN, Heazlewood JL, Seoighe C, Irving HR, Gehring CA (2002) Expansin-like molecules: novel functions derived from common domains. J Mol Evol 54:587–594

Ludidi N, Morse M, Sayed M, Wherrett T, Shabala S, Gehring C (2004) A recombinant plant natriuretic peptide causes rapid and spatially differentiated K^+, Na^+ and H^+ flux changes in *Arabidopsis thaliana* roots. Plant Cell Physiol 45:1093–1098

Ma W, Yoshioka K, Gehring C, Berkowitz G (2010) The function of cyclic nucleotide-gated channels in biotic stress. In: Demidchik V, Maathuis F (eds) Ion channels and plant stress responses, signaling and communication in plants. Springer, Berlin, pp 159–174

Maathuis FJ (2006) cGMP modulates gene transcription and cation transport in Arabidopsis roots. Plant J 45:700–711

Maryani MM, Shabala SN, Gehring CA (2000) Plant natriuretic peptide immunoreactants modulate plasma-membrane H(+) gradients in *Solanum tuberosum* L. leaf tissue vesicles. Arch Biochem Biophys 376:456–458

Maryani MM, Bradley G, Cahill DM, Gehring CA (2001) Natriuretic peptides and immunoreactants modify osmoticum-dependent volume changes in *Solanum tuberosum* L. mesophyll cell protoplasts. Plant Sci 161:443–452

Matsubayashi Y, Sakagami Y (1996) Phytosulfokine, sulfated peptides that induce the proliferation of single mesophyll cells of Asparagus officinalis L. Proc Natl Acad Sci USA 93:7623–7627

Matsubayashi Y, Takagi L, Omura N, Morita A, Sakagami Y (1999) The endogenous sulfated pentapeptide phytosulfokine-alpha stimulates tracheary element differentiation of isolated mesophyll cells of zinnia. Plant Physiol 120:1043–1048

Matsubayashi Y, Ogawa M, Morita A, Sakagami Y (2002) An LRR receptor kinase involved in perception of a peptide plant hormone, phytosulfokine. Science 296:1470–1472
Matsubayashi Y, Ogawa M, Kihara H, Niwa M, Sakagami Y (2006) Disruption and overexpression of Arabidopsis phytosulfokine receptor gene affects cellular longevity and potential for growth. Plant Physiol 142:45–53
McGurl B, Pearce G, Ryan CA (1994) Polypeptide signalling for plant defence genes. Biochem Soc Symp 60:149–154
Meier S, Bastian R, Donaldson L, Murray S, Bajic V, Gehring C (2008) Co-expression and promoter content analyses assign a role in biotic and abiotic stress responses to plant natriuretic peptides. BMC Plant Biol 8:24
Meier S, Madeo L, Ederli L, Donaldson L, Pasqualini S, Gehring C (2009) Deciphering cGMP signatures and cGMP-dependent pathways in plant defence. Plant Signal Behav 4:307–309
Meier S, Ruzvidzo O, Morse M, Donaldson L, Kwezi L, Gehring C (2010) The Arabidopsis wall associated kinase-like 10 gene encodes a functional guanylyl cyclase and is co-expressed with pathogen defense related genes. PLoS One 5:e8904
Mingossi FB, Matos JL, Rizzato AP, Medeiros AH, Falco MC, Silva-Filho MC, Moura DS (2010) SacRALF1, a peptide signal from the grass sugarcane (Saccharum spp.), is potentially involved in the regulation of tissue expansion. Plant Mol Biol 73:271–281
Morse M, Pironcheva G, Gehring C (2004) AtPNP-A is a systemically mobile natriuretic peptide immunoanalogue with a role in *Arabidopsis thaliana* cell volume regulation. FEBS Lett 556:99–103
Motose H, Iwamoto K, Endo S, Demura T, Sakagami Y, Matsubayashi Y, Moore KL, Fukuda H (2009) Involvement of phytosulfokine in the attenuation of stress response during the transdifferentiation of Zinnia mesophyll cells into tracheary elements. Plant Physiol 150:437–447
Navarro B, Kirichok Y, Clapham DE (2007) KSper, a pH-sensitive K^+ current that controls sperm membrane potential. Proc Natl Acad Sci USA 104:7688–7692
Nembaware V, Seoighe C, Sayed M, Gehring C (2004) A plant natriuretic peptide-like gene in the bacterial pathogen *Xanthomonas axonopodis* may induce hyper-hydration in the plant host: a hypothesis of molecular mimicry. BMC Evol Biol 4:10
Pandey S, Wang R-S, Wilson L, Li S, Zhao Z, Gookin TE, Assmann SM, Albert R (2010) Boolean modeling of transcriptome data reveals novel modes of heterotrimeric G-protein action. Mol Syst Biol 6:372
Pearce G, Strydom D, Johnson S, Ryan CA (1991) A polypeptide from tomato leaves induces wound-inducible proteinase inhibitor proteins. Science 253:895–898
Pearce G, Moura DS, Stratmann J, Ryan CA Jr (2001) RALF, a 5-kDa ubiquitous polypeptide in plants, arrests root growth and development. Proc Natl Acad Sci USA 98:12843–12847
Pearce G, Munske G, Yamaguchi Y, Ryan CA (2010) Structure-activity studies of GmSubPep, a soybean peptide defense signal derived from an extracellular protease. Peptides 31:2159–2164
Peuke AD, Windt C, Van As H (2006) Effects of cold-girdling on flows in the transport phloem in *Ricinus communis*: is mass flow inhibited? Plant Cell Environ 29:15–25
Pharmawati M, Gehring CA, Irving HR (1998a) An immunoaffinity purified plant natriuretic peptide analogue modulates cGMP levels in the *Zea mays* root stele. Plant Sci 137:107–115
Pharmawati M, Billington T, Gehring CA (1998b) Stomatal guard cell responses to kinetin and natriuretic peptides are cGMP-dependent. Cell Mol Life Sci 54:272–276
Pharmawati M, Shabala SN, Newman IA, Gehring CA (1999) Natriuretic peptides and cGMP modulate K^+, Na^+, and H^+ fluxes in *Zea mays* roots. Mol Cell Biol Res Commun 2:53–57
Pharmawati M, Maryani MM, Nikolakopoulos T, Gehring CA, Irving HR (2001) Cyclic GMP modulates stomatal opening induced by natriuretic peptides and immunoreactive analogues. Plant Physiol Biochem 39:385–394
Potter LR, Abbey-Hosch S, Dickey DM (2006) Natriuretic peptides, their receptors, and cyclic guanosine monophosphate-dependent signaling functions. Endocr Rev 27:47–72
Qi Z, Verma R, Gehring C, Yamaguchi Y, Zhao Y, Ryan CA, Berkowitz GA (2010) Ca^{2+} signaling by plant *Arabidopsis thaliana* Pep peptides depends on AtPepR1, a receptor with guanylyl

cyclase activity, and cGMP-activated Ca^{2+} channels. Proc Natl Acad Sci USA 107:21193–21198

Rafudeen S, Gxaba G, Makgoke G, Bradley G, Pironcheva G, Raitt L, Irving H, Gehring C (2003) A role for plant natriuretic peptide immuno-analogues in $NaCl^-$ and drought-stress responses. Physiol Planta 119:554–562

Ruzvidzo O, Donaldson L, Valentine A, Gehring C (2011) The *Arabidopsis thaliana* natriuretic peptide AtPNP-A is a systemic regulator of leaf dark respiration and signals via the phloem. J Plant Physiol 168:1710–1714

Srivastava R, Liu JX, Howell SH (2008) Proteolytic processing of a precursor protein for a growth-promoting peptide by a subtilisin serine protease in Arabidopsis. Plant J 56:219–227

Srivastava R, Liu JX, Guo H, Yin Y, Howell SH (2009) Regulation and processing of a plant peptide hormone, AtRALF23, in Arabidopsis. Plant J 59:930–939

Suwastika I, Gehring C (1998) Natriuretic peptide hormones promote radial water movements from the xylem of Tradescantia shoots. Cell Mol Life Sci 54:1161–1167

Suwastika IN, Toop T, Irving HR, Gehring CA (2000) In situ and in vitro binding of natriuretic peptide hormones in *Tradescantia multiflora*. Plant Biol 2:1–3

Toop T, Donald JA (2004) Comparative aspects of natriuretic peptide physiology in non-mammalian vertebrates: a review. J Comp Physiol B 174:189–204

Vesely DL, Giordano AT (1991) Atrial natriuretic peptide hormonal system in plants. Biochem Biophys Res Commun 179:695–700

Vesely DL, Gower WR Jr, Giordano AT (1993) Atrial natriuretic peptides are present throughout the plant kingdom and enhance solute flow in plants. Am J Physiol 265:E465–E477

Wang YH, Irving HR (2011) Developing a model of plant hormone interactions. Plant Signal Behav 6:494–500

Wang YH, Gehring C, Cahill DM, Irving HR (2007) Plant natriuretic peptide active site determination and effects on cGMP and cell volume regulation. Funct Plant Biol 34:645–653

Wang YH, Ahmar H, Irving HR (2010) Induction of apoptosis by plant natriuretic peptides in rat cardiomyoblasts. Peptides 31:1213–1218

Wang YH, Gehring C, Irving HR (2011a) Plant natriuretic peptides are apoplastic and paracrine stress response molecules. Plant Cell Physiol 52:837–850

Wang YH, Donaldson L, Gehring C, Irving HR (2011b) Plant natriuretic peptides: control of synthesis and systemic effects. Plant Signal Behav 6:1606–1608

Wu J, Kurten EL, Monshausen G, Hummel GM, Gilroy S, Baldwin IT (2007) NaRALF, a peptide signal essential for the regulation of root hair tip apoplastic pH in *Nicotiana attenuata*, is required for root hair development and plant growth in native soils. Plant J 52:877–890

Yang H, Matsubayashi Y, Nakamura K, Sakagami Y (1999) *Oryza sativa* PSK gene encodes a precursor of phytosulfokine-alpha, a sulfated peptide growth factor found in plants. Proc Natl Acad Sci USA 96:13560–13565

Yang H, Matsubayashi Y, Nakamura K, Sakagami Y (2001) Diversity of Arabidopsis genes encoding precursors for phytosulfokine, a peptide growth factor. Plant Physiol 127:842–851

Yu L, Moshelion M, Moran N (2001) Extracellular protons inhibit the activity of inward-rectifying potassium channels in the motor cells of *Samanea saman* pulvini. Plant Physiol 127:1310–1322

Zeng X-H, Yang C, Kim ST, Lingle CJ, Xia X-M (2011) Deletion of the *Slo3* gene abolishes alkalization-activated K(+) current in mouse spermatozoa. Proc Natl Acad Sci USA 108:5879–5884

Processing of Peptides

Renu Srivastava and Stephen H. Howell

Abstract Plant peptide signaling molecules are subject to proteolytic processing and/or to various forms of posttranslational modifications. Most signaling peptides are secreted and made as prepropeptides from which signal peptides are removed as they are translocated across the endoplasmic reticulum membrane. Proteolytic processing of propeptides may be required to regulate their activity, for peptides to fold properly, to navigate the secretory pathway, and to diffuse freely in the apoplast. Posttranslational modifications of peptides in plants include tyrosine sulfation, proline hydroxylation, and hydroxyproline arabinosylation. The proteases responsible for processing some plant peptides have been determined and shown considerable specificity for their propeptide substrates. New high-throughput screening technologies can be used to match up peptides to their processing machinery or identify propeptide substrates of various proteases.

1 Introduction

Peptides are well-known growth factors in animal systems, but they have only recently been recognized in plants for their widespread role in cell-to-cell signaling (Ryan et al. 2002). In animal cells, most peptide hormones and neuropeptides are synthesized as larger proteins, propeptides, which are then subject to posttranslational processing, including proteolytic processing. Extensive knowledge has been generated on animal protein precursor processing through the contribution of various studies. On the contrary, knowledge about processing of plant signaling peptides has been experimentally established in only a few cases. Inference from peptide sequence data clearly indicates that they are proteolytically processed from precursors. Wheeler and Irving (2010) recently reviewed the various classes of *Arabidopsis* peptide hormones in

R. Srivastava • S.H. Howell (✉)
Plant Sciences Institute, Iowa State University, Ames, IA 50011, USA
e-mail: shh@iastate.edu

H.R. Irving and C. Gehring (eds.), *Plant Signaling Peptides*,
Signaling and Communication in Plants 16, DOI 10.1007/978-3-642-27603-3_11,

terms of their function, proteolytic processing, and nitrogen scores. They also showed through phylogenetic analysis that different peptide classes are not related to each other (also see chapter "Plant Peptide Signaling: An Evolutionary Adaptation").

There are a number of reasons why signaling peptides in both animal and plant systems are proteolytically processed from larger protein propeptides. First, some peptides can be stored in a more stable inactive state as propeptides. Their activation by proteolytic cleavage or by other modifications may control their activity allowing them to regulate some aspect of growth and development or respond to certain environmental conditions. Second, the production of a peptide in the context of a propeptide may assist in peptide folding. As such, the propeptide sequence may act as a chaperone for the peptide. Third, most biologically active peptides are secreted and, as such, have to navigate the secretory pathway. To make their way into and through the secretory pathway, peptides may require other properties conferred by the propeptide, such as targeting signals or other features to pass the strict endoplasmic reticulum (ER) quality control system. Fourth, most plant signaling peptides must be small enough to operate in the apoplast. Cell walls are porous but their porosity is limited. For example, *Chenopodium* cells growing in culture have cell walls with a mean size exclusion limit of 3.3–3.7 nm (Fleischer et al. 1999). With pores that size, small peptides would be expected to diffuse freely, but larger proteins would be excluded.

Similar arguments can be made for why some plant peptides are posttranslationally modified. In certain cases, it has been shown that modified peptides are more active biologically. For example, CLAVATA3 (CLV3), a 13 amino acid peptide regulating meristem development, is posttranslationally modified by arabinosylation. The modification is critical for its biological activity and high-affinity binding to its receptor, CLV1 (Ohyama et al. 2009; see chapters "Peptides Regulating Apical Meristem Development" and "Peptides Regulating Root Growth"). Also, a recently described root meristem growth factor 1 (RGF1) is a tyrosine-sulfated peptide. The desulfated peptide is less active than the sulfated peptide in stimulating root growth in *Arabidopsis* (Matsuzaki et al. 2010).

Posttranslational modifications may also influence peptide structure and/or folding through steric interactions with the peptide backbone (Matsubayashi 2011). With respect to the possible influence of protein modification on protein structure, it is interesting that plant signaling peptides fall into two broad categories—the small posttranslationally modified peptides and the ones that have multiple intramolecular disulfide bonds (Shinohara and Matsubayashi 2010). It is as if peptide structure is conferred by modifications or by intramolecular bridges, but not by both.

2 Peptide Processing Proteases

Since most plant signaling peptides are secreted, it is logical that factors which process propeptides or peptide would be found in the secretory pathway. In animal cells, proteases that process secreted signaling peptides are called proprotein or prohormone convertases (Steiner 1998). They are related to the yeast processing protease, Kex2p, required for proteolytically processing the killer substance in killer strains of yeast (Wickner 1974). Kex2p is a subtilisin-like serine protease (Mizuno et al. 1988), and the prohormone convertases in animal cells are also subtilases. Seven members of the subtilisin-/Kex2p-like family are recognized in animal systems, furin, PC1/PC3, PC2, PC4, PACE4, PC5, and PC7.

The subtilases in the secretory pathway are multidomain proteins synthesized as prepropeptides. They have a presequence targeting them to the secretory pathway; a prodomain which may act as a chaperone for proper folding; an N-terminal catalytic domain including a substrate-binding pocket, a mid protein "P" domain that contributes to Ca^{2+} dependence, specificity, stability, and pH dependence of the enzyme; and a C-terminal variable domain that differs among members of the prohormone convertase family (Zhou et al. 1998; Brix et al. 2001; Henrich et al. 2005). PC1, PC2, and PC4 are soluble proteases, while furin, PC5, and PC7 are transmembrane proteases with a transmembrane domain in the C-terminal region of the protein. The crystal structure of furin in association with an inhibitor peptide reveals a deep substrate-binding pocket with the characteristic serine protease catalytic triad (S368, H194, and D153) at the active site (Henrich et al. 2003). Furin has a strict specificity for RXR/KR motifs, and contoured surface loops in the active site cleft accommodate the RXR/KR motif in highly charged complementary pockets. Specificities of the various PCs for different substrates are attributed to differences in global charge compensation and surface distribution of charges surrounding the core region of the binding pocket (Tian and Jianhua 2010).

Recent studies have implicated the role of subtilases in the processing of plant signaling peptides. Several propeptide precursors have conserved dibasic residues, characteristic of subtilase recognition sites, a few amino acids upstream from their mature peptide sequences. However, matching up the processing of specific propeptides to particular subtilases is challenging in plants because in *Arabidopsis*, for example, subtilases are encoded by a large family of genes (56 members in 6 different subfamilies) (Rautengarten et al. 2005). The functionality of the different subtilases has been difficult to assess, because in *Arabidopsis*, T-DNA insertions in 55 of the subtilase genes have no obvious phenotype except for AtSBT1.2 which has an altered pattern of stomates (Berger and Altmann 2000; also see chapter "Peptides Modulating Development of Specialized Cells"). In addition, 46 members of the *Arabidopsis* subtilase gene family are predicted to be targeted to the secretory pathway (Rautengarten et al. 2005).

Other than subtilases, Casamitjana-Martínez et al. (2003) have also shown the role of carboxypeptidases in the processing of CLE peptides.

3 Signaling Peptides

3.1 Systemins

Systemin, the first signaling peptide found in plants, was discovered by Pearce et al. (1991). Tomato systemin (TomSys) was isolated from tomato leaves as an 18 amino acid peptide derived from the C terminus of a 200 amino acid propeptide precursor called prosystemin (McGurl et al. 1992; McGurl and Ryan 1992). TomSys was characterized by its ability to elicit the synthesis of proteinase inhibitors in solanaceous plants. Two other 18 amino acid peptides, unrelated in sequence to TomSys, were isolated from tobacco and called tobacco systemins I and II (TobHypSys I and II shown in Table 1). TobHypSys I and II were derived from the same 165 amino acid precursor and shown to induce proteinase inhibitor production in tobacco and to cause rapid alkalization in tobacco suspension culture cells (Pearce et al. 2001a). Both TobHypSys I and II were modified by the addition of 8–17 pentose units, and the carbohydrate residues were essential for activity in a suspension culture alkalinization assay (Pearce and Ryan 2003).

3.2 CLE Peptides

Cell-to-cell signaling mediated by peptide ligands is key to many plant growth and development processes (Mitchum et al. 2008). For example, CLE peptides are known to control the fate of stem cells (see chapters "Peptides Regulating Apical Meristem Development," "Peptides Regulating Root Growth," and "Peptides Regulating Plant Vascular Development"). *CLAVATA 3* (*CLV3*) was the founding member of the Arabidopsis 31-member CLE gene family, which are characterized by a conserved stretch of 14 amino acids known as the CLE domain located close to or at the carboxyl terminus of the precursor (Table 1). It has been shown experimentally that the CLE domain confers biological activity (Kondo et al. 2006). The CLE peptides are posttranslationally hydroxylated at two proline residues, but the modifications do not appear necessary for activity (Ito et al. 2006).

Biochemical evidence for the proteolytic release of the CLE domain from the CLV3 protein was provided by Ni and Clark (2006) who showed that a non-membrane-associated protease activity can process both CLV3 and CLE1. When a GST-tagged form of CLV3 (GST-mCLV3) was incubated with protein extracts of cauliflower (*Brassica oleracea*), two smaller forms of GST-mCLV3 were detected with anti-GST antibodies, indicating that C-terminal proteolytic processing had occurred. Mass spectrometry analysis of the two processed forms of CLV3, called p1 and p2, indicated that p1 processing occurred at a conserved arginine at the beginning of CLE1 motif and p2 processing occurred at a string of methionines in the variable domain.

Similarly, GST-tagged forms of CLE1 (GST-CLE1) were also subject to in vitro processing by a cauliflower extract in a similar manner as CLV3, and two processed forms of GST-mCLE1 appeared as products in the assay. Ni et al. (2011) characterized the processing activity of the CLV3 CLE domain and showed that substituting A for R70 blocks the processing of the CLAVATA peptide. Using chimeric constructs and site-directed mutagenesis, they demonstrated that R70 is critical for processing and may have a role in receptor binding. Evidence provided by Ni et al. (2011) showed that CLV3 is processed by a secretory serine protease, though the specific protease has not yet been identified. The protease activity was

Table 1 Proteases involved in processing various plant propeptides

Name of peptide	Propeptide (size/length)	Active peptide (length)	Sequence of active peptide	Protease involved
Systemin	Prosystemin (200AA)	TomSys I and II (18AAaa)	RGANLPPOSOASSOOSKE NRKPLSOOSOKPADGQRP	Secretory protease?
CLE and CLV3	7.4–14.5 kDa	CLE/mCLV3 (12–14 AA)	RTVPSGPDPLHH	Serine protease/carboxypeptidase
PSK	80 AA	PSKα (5 AA)	YIYTQ	AtSBT1.1 and other
RALF/RALFL	7–14 kDa	(25–30 AA)	ATTRRY. . . . . .RRS	AtSBT6.1
IDA and IDL	8.4–13 kDa	EPIP-C (28 AA)	FGYLPKGVPIPPSAPSKRHNSFNSLPH	Soluble protease
GmSubPEP	789 AA	GmSubPEP (12 AA)	NTPPRRAKSRPH	Subtilase?
CEP	82–126 AA	CEP1 (12 AA)	DFRPTNPGNSPGVGHS	not established
ProPEP	9.3–12.3 kDa	PEP1 (23 AA)	ATKVKAKQRGKEKVSSGRPGQHN	Apoplastic aspartic protease
PSY	79 kDa	PSY1 (18 AA)	DYGDPSANPKHDPGVPS	not established

detected in the culture media of *Arabidopsis* and tobacco BY-2 cells. Inhibitor profiling studies showed that the p1 cleavage at R70 was mediated by a serine protease, while the p2 processing in the variable arginine domain may be mediated by another enzyme, possibly also a serine protease.

Djordjevic et al. (2011) showed in *Medicago truncatula* that extracellular subtilisins play a role in processing the border sequences of CLE36 propeptide, MtCLE36. This peptide differs from other CLE peptides by being embedded in a larger propeptide and requiring proteolysis at both ends to produce the mature peptide. When a synthetic peptide mirroring the sequence of a segment of the propeptide was incubated with extracellular extracts from *Medicago* and *Glycine*, a 15 amino acid peptide that is highly conserved across plant species was generated. The proteolytic activity giving rise to the peptide was inhibited by PMSF, suggesting that subtilisins might be involved in proteolysis. Since the active peptide requires processing at both ends, it was believed that two subtilisins may play a role in releasing the CLE36 peptide from its precursor. The conserved border sequences of the peptide appear to be functional as indicated by the fact that CLE36 peptides of different lengths affected primary root growth differentially.

In searching for activities that proteolytically process CLE peptides, genetic studies were carried out to find suppressors of the CLE19 overexpression phenotype (Casamitjana-Martínez et al. 2003). One of the suppressors, SOL1, encoded a Zn^{2+}-carboxypeptidase with a region homologous to the conserved domains of animal carboxypeptidase D and carboxypeptidase E (Aloy et al. 2001). SOL1 homologues in animals CPD and CPE have been shown to play a key role in cleaving the terminal R and K residues of prohormones and neuropeptides. Therefore, it is possible that SOL1 processes inactive CLE peptides. However, in vitro assays with purified SOL1 have not yet been carried to demonstrate a direct proteolytic reaction between SOL1 carboxypeptidase and CLE19.

3.3 Phytosulfokines

Phytosulfokines (PSKs) are a group of plant peptides that function as growth factors mediating density-dependent growth in cell cultures (also see chapters "Peptides Regulating Root Growth," "The Role of Plant Peptides in Symbiotic Interactions," and "Peptides and the Regulation of Plant Homeostasis"). Matsubayashi and Sakagami (1996) isolated and identified a sulfated peptide, called a phytosulfokine (or PSK), from conditioned culture medium of *Asparagus* mesophyll cells. PSK is a small (five amino acid) sulfated peptide. It was later found that PSKs in *Arabidopsis* are encoded by a small family of genes and that the mature PSK peptides are derived from the conserved C-termini of the propeptides (Table 1). A PSK receptor was first identified in carrot as a leucine-rich receptor kinase (LRR-RK) by Matsubayashi et al. (2002).

Subtilase-mediated processing of an *Arabidopsis* phytosulfokine propeptide (PSK4) was described by Srivastava et al. (2008). PSK4 is one of the more abundantly expressed members of the five-membered PSK gene family in *Arabidopsis*, and the mature peptide promotes callus production in tissue culture medium. The predicted propeptide sequence of PSK4 had a canonical subtilase site (RRXL), a few amino acids upstream of the mature peptide sequence (YIYTQ). PSK4 was expressed (as judged by the activity of a PSK promoter: GUS fusion) at the cut ends of root segments when callus was induced on tissue culture medium. A subtilase gene, AtSBT1.1 (monitored through a promoter: GUS fusion), was also expressed at the cut ends of roots under similar culture conditions. These correlations prompted further studies to look for the possible involvement of AtSBT1.1 in the processing of PSK4.

When epitope-tagged forms of the PSK4 propeptide were introduced into transgenic plants, the construct was processed when root segments were explanted on callus induction medium (CIM). However, the propeptide was not processed under similar culture conditions when callus was not induced. Also, the PSK4 propeptide was not processed in the T-DNA mutant with defects in AtSBT1.1 production, indicating that, indeed, AtSBT1.1 was required for PSK4 processing in vivo.

In vitro assays were conducted to determine the specificity of AtSBT1.1 for the PSK4 propeptide. To do so, a myc-tagged version of AtSBT1.1 (AtSBT1.1-myc) was produced in and isolated from transgenic plants and incubated with a fluorogenic peptide representing the putative subtilase recognition sequence on PSK4. The peptide, Abz-RRSLVLLHTDY(NO2)D-OH, where Abz is a 2-aminobenzoyl fluorescent group and Y(NO2)D-OH is a 3-nitrotyrosine quencher, was cleaved in the in vitro assay, and one of the products was shown by mass spectrometry analysis to be the N-terminal Abz-RRSLVL sequence, indicating cleavage at RRSLVL/LHTDY. An alanine scan through the peptide established that the P2–P4 positions were most important for subtilase recognition. When AtSBT1.1 was challenged with other fluorogenic peptides representing the putative cleavage sites in other PSK propeptides, it was found that AtSBT1.1 had preference for the PSK4 propeptide. The cleavage site deduced from the in vitro assay described above did not correspond to the junction between the propeptide and the mature PSK peptide. Therefore, it was concluded that other proteases may be involved subsequently in trimming the PSK peptide to its mature form. It is assumed that since both the protease and the peptide are present in the apoplast, the processing takes place in the apoplastic region of the plant cell.

3.4 Rapid Alkalinization Factors

Rapid alkalinization factor (RALF) is a 5-kD ubiquitous plant peptide hormone first isolated from tobacco leaves (also see chapters “The Role of Plant Peptides in Symbiotic Interactions” and “Peptides and the Regulation of Plant Homeostasis”). RALF induces a rapid alkalinization of cell suspension cultures and activates a MAP kinase (Pearce et al. 2001b). Exogenous application of the peptide to *Arabidopsis* seedlings inhibited root growth and development. Olsen et al. (2002)

identified a family of 34 RALF-like genes in *Arabidopsis* and found that the different genes were expressed in different tissues. Attempts are being made by several groups to identify RALF receptors.

RALFs are also thought to be processed from propeptide precursors (Table 1). Alignment of the *Arabidopsis* RALF genes showed the presence of dibasic residues, possible subtilase recognition sites, upstream of many of the putative mature peptides (Matos et al. 2008). Evidence showing that the dibasic sites are essential for in vivo processing of RALF1 in *Arabidopsis* was provided by Matos et al. (2008). Plants overexpressing the RALF1 gene showed a semidwarf phenotype. However, plants overexpressing a mutated RALF1 propeptide with an R to A substitution at the conserved dibasic site (R69A) did not show the dwarf phenotype. This was taken to mean that RALF1 is likely processed by a subtilase in vivo. RALF1 was also shown to be processed in vitro. Radiolabeled wild-type preproRALF1 and mutated preproRALF1 (R69A) were produced by in vitro transcription and translation and incubated with microsomal fractions of *Arabidopsis* cell suspension cultures. The results showed that the normal form of RALF1 was processed by the plant extracts but not the R69A form.

Srivastava et al. (2009) demonstrated that a subtilisin-like protease was likely involved in the processing of RALF23 from *Arabidopsis*. RALF23 encodes a 138 amino acid preproprotein which was thought to be processed to a bioactive peptide (Matsubayashi and Sakagami 2006). RALF23 has a canonical subtilase recognition site, RRIL, immediately upstream the mature RALF peptide. Overexpression of RALF23 produced plants with shorter roots and a bushier and stunted appearance in a wild-type background. RALF23 did not produce the bushy phenotype in the background of a T-DNA mutant which fails to produce SITE-1 PROTEASE (S1P), a Golgi-localized subtilase (Liu et al. 2007). Epitope-tagged forms of the RALF23 prepropeptide (AtRALF23-myc) were processed in a wild-type background but not the *s1p* mutant background—again demonstrating the role of AtS1P in the processing of RALF23.

An in vitro assay using fluorogenic peptides was also developed to demonstrate directly that AtS1P can cleave RALF23 and to determine the site of cleavage. The analysis demonstrated that the canonical subtilase recognition site (RRXL) was required for cleavage and that the propeptide is cleaved just upstream from the mature peptide. Unlike the processing of PSK, AtS1P cleavage cleanly cuts out RALF23 from the propeptide with no extra amino acid residues on the N terminus of the mature peptide.

Since AtS1P is localized in the Golgi apparatus (Liu et al. 2007), it is likely that the AtRALF23 propeptide is processed in the Golgi on its way through the secretory pathway. Escobar et al. (2003) tracked RALF after transfecting *Nicotiana benthamiana* leaves with NtRALF-GFP via a viral vector. They first found NtRALF-GFP in the ER and later in the apoplast. It would seem possible that some AtS1P is secreted and that AtRALF23 propeptide is proteolytically processed in the apoplast. However, AtS1P has a C-terminal transmembrane domain that is thought to anchor the enzyme to the membranes of the secretory pathway. In mammalian cells, it was found that a small quantity of S1P undergoes autocatalytic cleavage to release the truncated enzyme, which is then secreted into the media (Cheng et al. 1999; Espenshade et al. 1999). Nonetheless, it is unlikely that proAtRALF23 is cleaved by AtS1P in the apoplast, because the *Arabidopsis* enzyme has a pH optimum of 7.5 (R Srivastava, unpublished data), and little or no activity at the acidic pH of the apoplast. However, if a small quantity of proAtRALF23 is processed and begins to promote the alkalization of the apoplast, then that might generate an apoplastic environment conducive to the processing of AtRALF23 by a released form of AtS1P.

3.5 EPIP Peptide

Stenvik et al. (2008) showed that plants overexpressing INFLORESCENCE DEFICIENT IN ABSCISSION (IDA) display an earlier abscission phenotype. IDA/IDA-like (IDL) proteins have N-terminal signal peptides, which direct them to the secretory pathway, and a C-terminal conserved motif [extended PIP (EPIP) and at the C terminus (EPIP-C)]. The EPIP-C of IDA and IDL can replace IDA function in vivo. GFP fusions showed that the protein is localized to the apoplastic space, and synthetic EPIP peptides can induce abscission demonstrating their functionality. Stenvik et al. (2008) demonstrated that IDA can be processed to release the EPIP-C peptide. An assay similar to the one developed for CLV3 by Ni and Clark (2006) was used to study the processing of IDA proteins. A GST version of IDA was expressed in *Escherichia coli* and incubated with cauliflower extracts in in vitro reactions. Mobility shifts were detected on protein blots in a manner showing C-terminal processing of IDA, which probably released the EPIP peptide. Since IDA proteins are localized in the apoplast, it is likely that the processing may be catalyzed by a protease localized in the apoplast. Determination of the pH optimum for the cleavage reaction and mass spectrometry analysis of the digested fragments should throw light on the nature of the protease.

3.6 Other Peptides with Evidence of Processing

3.6.1 GmSubPep

Pearce et al. (2010) described an interesting example in which a soybean peptide that plays a unique role in defense signaling is derived from a subtilase itself. *Glycine max* subtilase peptide (GmSubPep) is a 12 amino acid peptide that promotes alkalinization of soybean suspension cultures, a response typical of pathogen-derived elicitors (Table 1). The peptide is derived from a unique region of an extracellular subtilisin-like protease present only in legume plants. GmSubPep is encoded by Glyma18g48580 and not upregulated by methyl jasmonate or ethephon. It is possible that Glyma18g48580 is activated upon pathogen attack or wounding and posttranscriptional events play a role in regulation. It is predicted that the PA domain of GmSubPep may be made accessible for processing by proteases that are activated in response to defense signaling. It would be interesting to know if the subtilase is assisted by other proteases to release the peptide or has the capacity to process itself.

3.6.2 AtPep1

Huffaker et al. (2006) isolated and characterized a 23 amino acid peptide from *Arabidopsis* called AtPep1. AtPep1 is derived from the C terminus of a 92 amino acid protein called AtPROPEP1 which is a highly charged protein precursor protein molecule. This peptide is not posttranslationally modified and activates the synthesis of H_2O_2 and transcription of PDF1.2 (a defense-related gene), which are components of the innate immune response. Six paralogs of PROPEP1 are present in *Arabidopsis*, and other plant species have several orthologs of the same gene. Since AtPep1 is derived from its precursor, a proteolytic event must be requisite for release of the mature peptide. It has been speculated that AtPROPEP1 is processed by an apoplastic aspartic protease (Vlot et al. 2008).

3.6.3 CEP1: C-Terminally Encoded Peptide 1

Ohyama et al. (2008) identified a 12 amino acid mature CLE44 peptide by liquid chromatography–mass spectrometry analysis of the culture medium of transgenic plants over expressing CLE44. The 12 amino acid peptide has two proline residues that are posttranslationally modified to hydroxyproline (Hyp) residues. It is specifically expressed in the lateral root primordia, and its overexpression or external application significantly arrests root growth. C-terminally encoded peptide 1 (CEP1) is derived from a family of five genes encoding 82–126 amino acid-secreted peptides that exhibit significant sequence diversity with the exception of the C-terminal conserved 17 amino acid domain. It is believed that the CEP1 peptide is derived by proteolytic processing of its posttranslationally modified precursor.

3.6.4 PSY1: Plant Peptide Containing Sulfated Tyrosine

Another plant peptide that appears to function as a growth factor is an 18 amino acid tyrosine-sulfated glycopeptide isolated from *Arabidopsis* cell suspension culture medium (Amano et al. 2007). This peptide was named plant peptide containing sulfated tyrosine (PSY1), and it significantly promotes cellular proliferation and expansion at nanomolar concentrations. It was shown that posttranslational modifications are important to confer functionality on the peptide. The peptide is derived from the C terminus of a polypeptide precursor called PSY (encoded by an

Arabidopsis gene with the AGI designation, AT5G58650). It has an atypical secretion signal sequence at its N terminus, and proline residues P16 and P17 are hydroxylated. They also showed that the plant peptide is glycosylated with three pentose units. Overexpression of PSY1 resulted in plants with longer roots and larger cotyledons compared to wild-type plants.

4 Posttranslational Modification Machinery

For the plant peptides described above, they are cleaved and trimmed by various protease-processing activities, some of which are known (Table 1). However, for a number of plant peptides, the processing proteases required for their maturation are not yet known. In mammalian systems, both soluble and membrane-bound proteases have been shown to be involved in processing peptide hormones. In plants, mainly soluble proteases, except for AtS1P, have been documented or implicated in peptide precursor processing. Future studies may be able to link other soluble or membrane-bound proteases to peptide precursor processing.

In mammalian systems, cleaved peptides may undergo additional biochemical modifications such as acetylation, amidation, sulfation, etc., to generate biologically active hormones. These modifications occur sequentially in the distal secretory pathway, including the trans-Golgi network (TGN) and immature and mature secretory granules (Fisher and Scheller 1988; Orci et al. 1987; Schnabel et al. 1989; Tooze et al. 1987; Xu and Shields 1993). Consequently, the biosynthesis of peptide hormones represents a good model to understand protein processing and membrane trafficking.

Matsubayashi (2011) categorized biologically active plant peptides partly on the basis of their posttranslational modifications:

1. Peptides characterized by the presence of posttranslational modifications mediated by specific transferases and by proteolytic processing into smaller sizes (<20 amino acids), for example, PSK (Matsubayashi and Sakagami 1996), tobacco hydroxyproline-rich systemin (TobHypSys) (Pearce et al. 2001a), PSY1 (Amano et al. 2007), TDIF (Ito et al. 2006), CLV3 (Fletcher et al. 1999; Ohyama et al. 2009), CLAVATA3/ESR-Related 2 (CLE2) (Ohyama et al. 2009), CEP1 (Ohyama et al. 2008), and RGF1 (Matsuzaki et al. 2010).
2. Peptides characterized by the presence of an even number of cysteine residues referred to as "cysteine-rich peptides." They typically have six or eight cysteine residues that participate in the formation of intramolecular disulfide bonds, for example, *S*-LOCUS CYSTEINE-RICH PROTEIN/*S*-LOCUS PROTEIN 11 (SCR/SP11), a male determinant of self-incompatibility (Takayama et al. 2000). SCR/SP11 is a secreted peptide active after disulfide bond formation, but is not proteolytically processed (Takayama et al. 2001; also see chapters "The S-LOCUS CYSTEINE RICH PROTEIN (SCR): A Small Peptide with a High Impact on the Evolution of Flowering Plants" and "Plant Antimicrobial Peptides").

3. Another group of cysteine-rich peptides such as RALF (four cysteine residues) (Covey et al. 2010; Pearce et al. 2001b; Wu et al. 2007) and STOMAGEN (six cysteine residues) (Kondo et al. 2010; Sugano et al. 2010; also see chapter "Peptides Modulating Development of Specialized Cells") involve proteolytic processing even though they have intramolecular disulfide bonds.

The following enzymatic activities have been shown to be involved in the posttranslational processing of plant peptides:

Tyrosine sulfation. This modification is mediated by enzymes such as tyrosylprotein sulfotransferases (TPSTs), which catalyze the transfer of sulfate from 3′-phosphoadenosine 5′-phosphosulfate (PAPS) to the phenolic groups of tyrosines (Moore 2003). The three known tyrosine-sulfated peptide hormones from plants are PSK, PSY1, and RGF1.

Proline hydroxylation. This modification is mediated by prolyl 4-hydroxylase (P4H) which belongs to a family of 2-oxoglutarate-dependent dioxygenases (Myllyharju 2003). P4H catalyzes the oxidation of proline residues exclusively at the fourth position carbon. Some hydroxyproline residue containing peptides with evidence of subsequent proteolytic processing are PSY1, CLV3, CLE2, and CEP1.

Hydroxyproline arabinosylation. Biochemical analysis of PSY1, CLV3, and CLE2 has shown that the hydroxyproline residues on these peptides are further modified with an *O*-linked L-arabinose chain (Amano et al. 2007; Ohyama et al. 2009). *O*-glycosylation is believed to occur by the successive addition of nucleotide-activated sugars catalyzed by glycosyltransferases localized in the Golgi apparatus.

5 Tools for Future Plant Peptide Processing Studies

Our present knowledge about the involvement of proteases in processing plant peptide precursor molecules has been acquired through the use of various bioassays and biochemical genetic tools. Several high-throughput strategies have been used in mammalian systems to study peptide hormone processing, which have not been employed in plant systems. If some of these approaches were applied to plant systems, they may be better able to link proteases to their peptide precursor substrates.

5.1 Peptide Libraries

Turk and Cantley (2003) described the use of peptide libraries to identify optimal cleavage motifs for proteolytic enzymes (Fig. 1). The method they described first involves determining the specificity of the proteolytic enzyme at positions carboxy-terminal to the scissile bond (known as primed or P position) and the subsequent

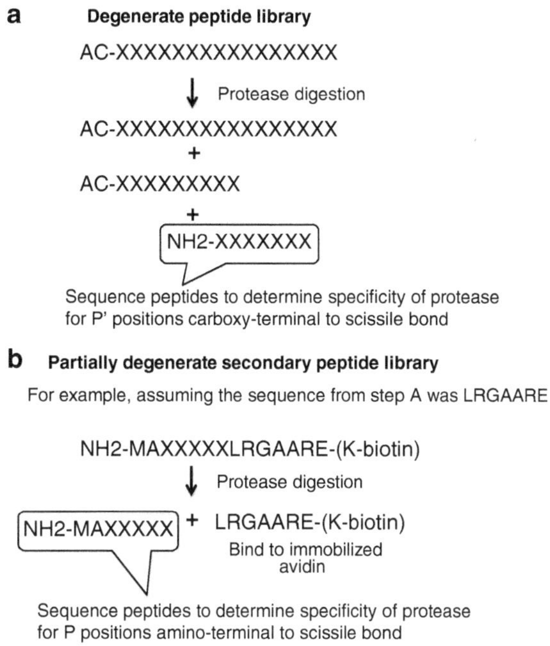

Fig. 1 Schematic diagram depicting a peptide library method for determining the substrate specificity of proteases (after Turk and Cantley 2003)

determination of specificity amino-terminal to the cleavage site (at the unprimed, or P, positions). Then a degenerate peptide mixture, amino-terminally acetylated, is digested with the protease of interest followed by amino-terminal sequencing of the mixture. Since the sequencer is unable to detect unreacted peptides with blocked amino-termini, the carboxy-terminal fragments are sequenced and read. The enzyme's preference for each amino acid at a given position relative to the cleavage site is ascertained based on the molar proportion of each residue present in a given cycle. The data derived from this initial library screen is then used to design additional peptide libraries to determine the cleavage motif for the unprimed positions. The secondary peptide libraries are prepared with a number of degenerate positions followed by positions in which the optimal primed residues are fixed. Secondary libraries are made with a biotin group at the carboxy-terminus and with a free (unblocked) amino-terminus. The library is digested with the enzyme of interest, and then the unreacted peptides and carboxy-terminal fragments are removed with immobilized avidin. The amino-terminal fragments that remain in solution are sequenced by Edman degradation as above, allowing for the determination of the unprimed side preferences. A source of highly pure enzyme is essential, as even trace levels of contaminating peptidases can give spurious results. This method does not require costly instrumentation since conventional peptide synthesis and sequencing methodology are applied.

5.2 Peptide Microarrays

Ghosalia et al. (2005) described a novel microarray-based proteolytic profiling assay for the rapid determination of protease substrate specificities using fluorogenic substrate microarrays. Microarray-based substrate specificity determination allows for rapid profiling of different proteases in a high-density/high-throughput format. This strategy involves a miniaturized assay and requires minimal sample and enzyme usage. Large libraries containing thousands of fluorogenic substrates with the general format of Ac-A-X-X-(R/K)-coumarin were arrayed on a microscope slide. Serine proteases such as thrombin and Factor XA and plasmin were chosen as enzymes for the assay and profiling. The arrays provided complete maps of protease specificity and cooperativity of the substrates tested.

5.3 Phage Display Substrate

Deperthes (2002) reviewed several phage display strategies to search for protease substrates. Smith (1985) pioneered phage display technology when he expressed short peptides on the surface of the filamentous bacteriophage M13 by inserting synthetic DNA sequences, coding for peptides, into the phage coat protein gene, gIIIp. This was accomplished without affecting phage infectivity and viability. To do so, large numbers of random peptide sequences were fused to gIIIp and screened in vitro for target protein binding. Matthews and Wells (1993) modified this method to identify proteolysis recognition sites on peptides substrates, a technique now known as "phage display substrate." A library of fusion proteins was constructed containing an amino-terminal domain used to bind to an affinity support, followed by a randomized protease substrate sequence and the carboxyl-terminal domain of M13 gene III. Each fusion protein was displayed as a single copy on filamentous phagemid particles (substrate phage). Phage were then bound to an affinity support and treated with the protease of interest. Phage with sites susceptible to proteolysis were released, whereas phage with substrates that resisted proteolysis remained bound. After several rounds of binding, proteolysis, and phagemid propagation, sensitive and resistant substrate sequences were identified. Substrates for several serine proteases, metalloproteases, and aspartic proteases have been already identified by this method.

The above mentioned are some of the well-established approaches that may allow the screening of plant peptide substrates for secretory and other proteases which may serve as processing enzymes for plant peptide precursors. Linking proteases and posttranslational modification enzymes to their cognate propeptide and peptides would provide new insights into the regulatory role of plant peptides. This would help to further our understanding of the role of plant peptides in development, pathogen resistance, and in different physiological responses in plants.

References

Aloy P, Companys V, Vendrell J, Aviles FX, Fricker LD, Coll M, Gomis-Ruth FX (2001) The crystal structure of the inhibitor-complexed carboxypeptidase D domain II and the modelling of regulatory carboxypeptidases. J Biol Chem 276:16177–16184

Amano Y, Tsubouchi H, Shinohara H, Ogawa M, Matsubayashi Y (2007) Tyrosine-sulfated glycopeptide involved in cellular proliferation and expansion in Arabidopsis. Proc Natl Acad Sci USA 104:18333–18338

Berger D, Altmann T (2000) A subtilisin-like serine protease involved in the regulation of stomatal density and distribution in *Arabidopsis thaliana*. Genes Dev 14:1119–1131

Brix K, Linke M, Tepel C, Herzog V (2001) Cysteine proteinases mediate extracellular prohormone. Biol Chem 382:717–725

Casamitjana-Martínez E, Hofhuis HF, Xu J, Liu CM, Heidstra R, Scheres B (2003) Root-specific CLE19 overexpression and the sol1/2 suppressors implicate a CLV-like pathway in the control of Arabidopsis root meristem maintenance. Curr Biol 13:1435–1441

Cheng D, Espenshade PJ, Slaughter C, Jaen JC, Brown MS, Goldstein JL (1999) Secreted site-1 protease cleaves peptides corresponding to luminal loop of sterol regulatory element-binding proteins. J Biol Chem 274:22805–22812

Covey PA, Subbaiah CC, Parsons RL, Pearce G, Lay FT, Anderson MA, Ryan CA, Bedinger PA (2010) A pollen-specific RALF from tomato that regulates pollen tube elongation. Plant Physiol 153:703–715

Deperthes D (2002) Phage display substrate: a blind method for determining protease specificity. Biol Chem 383:1107–1112

Djordjevic MA, Oakes M, Wong CE, Singh M, Bhalla P, Kusumawati L, Imin N (2011) Border sequences of *Medicago truncatula* CLE36 are specifically cleaved by endoproteases common to extracellular fluids of Medicago and soybean. J Exp Bot 62:4649–4659

Escobar NM, Haupt S, Thow G, Boevink P, Chapman S, Oparka K (2003) High-throughput viral expression of cDNA-green fluorescent protein fusions reveals novel subcellular addresses and identifies unique proteins that interact with plasmodesmata. Plant Cell 15:1507–1523

Espenshade PJ, Cheng D, Goldstein JL, Brown MS (1999) Autocatalytic processing of site-1 protease removes propeptides and permits cleavage of sterol regulatory element-binding proteins. J Biol Chem 274:22795–22804

Fisher JM, Scheller RH (1988) Prohormone processing and the secretory pathway. J Biol Chem 263:16515–16518

Fleischer A, O'Neill MA, Ehwald R (1999) The pore size of non-graminaceous plant cell walls is rapidly decreased by borate ester cross-linking of the pectic polysaccharide rhamnogalacturonan II. Plant Physiol 121:829–838

Fletcher JC, Brand U, Running MP, Simon R, Meyerowitz EM (1999) Signaling of cell fate decisions by CLAVATA3 in Arabidopsis shoot meristems. Science 283:1911–1914

Ghosalia DN, Salisbury CM, Maly DJ, Ellman JA, Diamond SL (2005) Profiling serine protease substrate specificity with solution phase fluorogenic peptide microarrays. Proteomics 5:1292–1298

Henrich S, Cameron A, Bourenkov GP, Kiefersauer R, Huber R, Lindberg I, Bode W, Than ME (2003) The crystal structure of the proprotein processing proteinase furin explains its stringent specificity. Nat Struct Biol 10:520–526

Henrich S, Lindberg I, Bode W, Than ME (2005) Proprotein convertase models based on the crystal structures of furin and kexin: explanation of their specificity. J Mol Biol 345:211–227

Huffaker A, Pearce G, Ryan CA (2006) An endogenous peptide signal in Arabidopsis activates components of the innate immune response. Proc Natl Acad Sci USA 103:2610098–2610103

Ito Y, Nakanomyo I, Motose H, Iwamoto K, Sawa S, Dohmae N, Fukuda H (2006) Dodeca-CLE peptides as suppressors of plant stem cell differentiation. Science 313:842–845

Kondo T, Sawa S, Kinoshita A, Mizuno S, Kakimoto T, Fukuda H, Sakagami Y (2006) A plant peptide encoded by CLV3 identified by in situ MALDI-TOF MS analysis. Science 13:845–848

Kondo T, Kajita R, Miyazaki A, Hokoyama M, Nakamura-Miura T, Mizuno S, Masuda Y, Irie K, Tanaka Y, Takada S, Kakimoto T, Sakagami Y (2010) Stomatal density is controlled by a mesophyll-derived signaling molecule. Plant Cell Physiol 51:1–8

Liu JX, Srivastava R, Che P, Howell SH (2007) Salt stress responses in Arabidopsis utilize a signal transduction pathway related to endoplasmic stress signaling. Plant J 51:897–909

Matos JL, Fiori CS, Silva-Filho MC, Moura DS (2008) A conserved dibasic site is essential for correct processing of the peptide hormone AtRALF1 in *Arabidopsis thaliana*. FEBS Lett 582:3343–3347

Matsubayashi Y (2011) Posttranslational modifications in peptide hormone. Plant Cell Physiol 52:5–13

Matsubayashi Y, Sakagami Y (1996) Phytosulfokine, sulfated peptides that induce the proliferation of single mesophyll cells of Asparagus officinalis L. Proc Natl Acad Sci USA 93:7623–7627

Matsubayashi Y, Sakagami Y (2006) Peptide hormones in plants. Annu Rev Plant Biol 57:649–674

Matsubayashi Y, Ogawa M, Morita A, Sakagami Y (2002) An LRR receptor kinase involved in perception of a peptide plant hormone, phytosulfokine. Science 24:1470–1472

Matsuzaki Y, Ogawa-Ohnishi M, Mori A, Matsubayashi Y (2010) Secreted peptide signals required for maintenance of root stem cell niche in Arabidopsis. Science 329:1065–1067

Matthews DJ, Wells JA (1993) Substrate phage: selection of protease substrates by monovalent phage display. Science 260:1113–1117

McGurl B, Ryan CA (1992) The organization of the prosystemin gene. Plant Mol Biol 20:405–409

McGurl B, Pearce G, Orozco Cardenas M, Ryan CA (1992) Structure expression and antisense inhibition of the system in precursor gene. Science 255:1570–1573

Mitchum MG, Wang X, Davis EL (2008) Diverse and conserved roles of CLE peptides. Curr Opin Plant Biol 11:75–81

Mizuno K, Nakamura T, Ohshima T, Tanaka S, Matsuo H (1988) Yeast KEX2 gene encodes an endopeptidase homologous to subtilisin-like serine proteases. Biochem Biophys Res Commun 156:246–254

Moore KL (2003) The biology and enzymology of protein tyrosine O-sulfation. J Biol Chem 278:24243–24246

Myllyharju J (2003) Prolyl 4-hydroxylases, the key enzymes of collagen biosynthesis. Matrix Biol 22:15–24

Ni J, Clark SE (2006) Evidence for functional conservation, sufficiency, and proteolytic processing of the CLAVATA3 CLE domain. Plant Physiol 140:726–733

Ni J, Guo Y, Jin H, Hartsell J, Clark SE (2011) Characterization of a CLE processing activity. Plant Mol Biol 75:67–75

Ohyama K, Ogawa M, Matsubayashi Y (2008) Identification of a biologically active, small, secreted peptide in Arabidopsis by in silico gene screening, followed by LC-MS-based structure analysis. Plant J 55:152–160

Ohyama K, Shinohara H, Ogawa-Ohnishi M, Matsubayashi Y (2009) A glycopeptide regulating stem cell fate in *Arabidopsis thaliana*. Nat Chem Biol 5:578–580

Olsen AN, Mundy J, Skriver K (2002) Peptomics, identification of novel cationic Arabidopsis peptides with conserved sequence motifs. In Silico Biol 2:441–451

Orci L, Ravazzola M, Storch MJ, Anderson RGW, Vassalli JD, Perrelet A (1987) Proteolytic maturation of insulin is a post-Golgi event which occurs in acidifying clathrin-coated secretory vesicles. Cell 49:865–868

Pearce G, Ryan CA (2003) Systemic signaling in tomato plants for defense against herbivores. J Biol Chem 278:30044–30050

Pearce G, Strydom D, Johnson S, Ryan CA (1991) A polypeptide from tomato leaves induce wound-inducible inhibitor proteins. Science 253:895–898

Pearce G, Moura DS, Stratmann J, Ryan CA (2001a) Production of multiple plant hormones from a single polyprotein precursor. Nature 411:817–820

Pearce G, Moura DS, Stratmann J, Ryan CA Jr (2001b) RALF, a 5-kDa ubiquitous polypeptide in plants, arrests root growth and development. Proc Natl Acad Sci USA 98:12843–12847

Pearce G, Yamaguchi Y, Barona G, Ryan CA (2010) A subtilisin-like protein from soybean contains an embedded, cryptic signal that activates defense related genes. Proc Natl Acad Sci USA 107:14921–14925
Rautengarten C, Steinhauser D, Büssis D, Stintzi A, Schaller A, Kopka J, Altmann T (2005) Inferring hypotheses on functional relationships of genes: analysis of the *Arabidopsis thaliana* subtilase gene family. PLoS Comput Biol 1:e40
Ryan CA, Pearce G, Scheer J, Moura DS (2002) Polypeptide hormones. Plant Cell 14:S251–S264
Schnabel E, Mains RE, Farquhar MG (1989) Proteolytic processing of pro-ACTH/endorphin begins in the Golgi complex of pituitary corticotropes and AtT-20 cells. Mol Endocrinol 3:1223–1235
Shinohara H, Matsubayashi Y (2010) Arabinosylated glycopeptide hormones: new insights into CLAVATA3 structure. Curr Opin Plant Biol 13:515–551
Smith GP (1985) Filamentous fusion phage: novel expression vectors that display cloned antigens on the virion surface. Science 228:1315–1317
Srivastava R, Liu JX, Howell SH (2008) Proteolytic processing of a precursor protein for a growth-promoting peptide by a subtilisin serine protease in Arabidopsis. Plant J 56:219–227
Srivastava R, Liu JX, Guo H, Yin Y, Howell SH (2009) Regulation and processing of a plant peptide hormone, AtRALF23, in Arabidopsis. Plant J 59:930–939
Steiner DF (1998) The proprotein convertases. Curr Opin Chem Biol 2:31–39
Stenvik GE, Tandstad NM, Guo Y, Shi CL, Kristiansen W, Holmgren A, Clark SE, Aalen RB, Butenko MA (2008) The EPIP peptide of INFLORESCENCE DEFICIENT IN ABSCISSION is sufficient to induce abscission in Arabidopsis through the receptor-like kinases HAESA and HAESA-LIKE2. Plant Cell 20:1805–1817
Sugano SS, Shimada T, Imai Y, Okawa K, Tamai A, Mori M, Hara-Nishimura I (2010) Stomagen positively regulates stomatal density in Arabidopsis. Nature 463:241–244
Takayama S, Shiba H, Iwano M, Shimosato H, Che FS, Kai N, Watanabe M, Suzuki G, Hinata K, Isogai A (2000) The pollen determinant of self-incompatibility in Brassica campestris. Proc Natl Acad Sci USA 97:1920–1925
Takayama S, Shimosato H, Shiba H, Funato M, Che FS, Watanabe M, Iwano M, Isogai A (2001) Direct ligand-receptor complex interaction controls Brassica self-incompatibility. Nature 413:534–538
Tian S, Jianhua W (2010) Comparative study of the binding pockets of mammalian proprotein convertases and its implications for the design of specific small molecule inhibitors. Int J Biol Sci 6:89–95
Tooze J, Hollinshead M, Frank R, Burke B (1987) An antibody specific for an endoproteolytic cleavage site provides evidence that pro-opiomelanocortin is packaged into secretory granules in AtT20 cells before its cleavage. J Cell Biol 105:155–162
Turk BE, Cantley LC (2003) Peptide libraries: at the crossroads of proteomics and bioinformatics. Curr Opin Chem Biol 7:84–90
Vlot AC, Klessig DF, Park SW (2008) Systemic acquired resistance: the elusive signal(s). Curr Opin Plant Biol 11:436–442
Wheeler JI, Irving HR (2010) Evolutionary advantages of secreted peptide signaling molecules in plants. Funct Plant Biol 37:382–394
Wickner RB (1974) Chromosomal and nonchromosomal mutations affecting the "killer character" of *Saccharomyces cerevisiae*. Genetics 76:423–432
Wu J, Kurten EL, Monshausen G, Hummel GM, Gilroy S, Baldwin IT (2007) NaRALF, a peptide signal essential for the regulation of root hair tip apoplastic pH in Nicotiana attenuata, is required for root hair development and plant growth in native soils. Plant J 52:877–890
Xu H, Shields D (1993) Prohormone processing in the trans-Golgi network: endoproteolytic cleavage of prosomatostatin and formation of nascent secretory vesicles in permeabilized cells. J Cell Biol 122:1169–1184
Zhou A, Martin S, Lipkind G, LaMendola J, Steiner DF (1998) Regulatory roles of the P domain of the subtilisin-like prohormone convertases. J Biol Chem 273:11107–11114

Methods to Isolate and Identify New Plant Signaling Peptides

Sunil Sagar, Chris Gehring, and Kenneth P. Minneman

Abstract Peptides play a key role in plant development and intercellular communication. A variety of novel peptides and peptidic compounds have been isolated over the last few decades. This chapter examines the various methods used to isolate and identify new peptides and peptidic compounds from plants, and provides an overview of these methods. The principles involved in extraction and methods commonly used for separation, isolation, and structural elucidation are summarized. Methods discussed include high-performance liquid chromatography in both normal and reversed phases, ion-exchange chromatography, size-exclusion chromatography, affinity chromatography, hydrophilic interaction liquid chromatography, hydrophobic interaction chromatography, multidimensional liquid chromatography, and various physical identification techniques such as mass spectrometry for determining molecular structures. The application of such techniques to identification of novel plant peptides is stressed.

1 Introduction

The idea that plant development and physiological responses to the environment are governed largely by the five "classical" plant hormones—auxin, gibberellins, cytokinins, abscisic acid, and ethylene (Kende and Zeevaart 1997)—has become outdated. While the "classical" plant hormones include amino acid derivatives, there is an ever-increasing body of evidence that a much larger range of chemically

S. Sagar (✉) • K.P. Minneman
Computational Biosciences Research Center, King Abdullah University of Science and Technology, Thuwal 23955-6900, Saudi Arabia
e-mail: sunil.sagar@kaust.edu.sa

C. Gehring
Division of Chemical and Life Sciences and Engineering, King Abdullah University of Science and Technology, Thuwal 23955-6900, Saudi Arabia

H.R. Irving and C. Gehring (eds.), *Plant Signaling Peptides*,
Signaling and Communication in Plants 16, DOI 10.1007/978-3-642-27603-3_12,

different molecules can also function as hormones and/or molecular signal transducers in plants. These range from nutrients such as nitrate and glucose through jasmonate and salicylate and include larger molecules such as brassinosteroids, nodulation factors, and notably peptides and small proteins. These peptidic molecules include systemin, RALF, POLARIS, the early nodulins (*ENOD*s), phytosulfokines, CLAVATA3, and plant natriuretic peptides (PNPs) (for review, see Lindsey et al. 2002; Germain et al. 2006). It is noteworthy that many of these peptidic molecules are synthesized from precursors and undergo extensive posttranslational processing essential for their biological activity in vivo.

While bioactivity-guided chemical analysis has enabled the discovery of novel peptidic signaling molecules, genome sequencing of different plant species has made possible the prediction of a number of potential signaling molecules and possible receptors for some of these ligands. We foresee that genomic tools will continue to contribute significantly to the discovery of peptidic signaling systems; however, there are a number of important reasons why chemical methods for the discovery of novel plant molecules will continue to play a key role. Firstly, there are still many unannotated genes or genes that are only predicted and may encode unknown peptidic signaling molecules that cannot be identified with genomic and computational methods. Secondly, posttranslational modifications, such as sulfation, that may be essential for downstream signaling and bioactivity [e.g., in the case of the phytosulfokines (Matsubayashi and Sakagami 1996)] do not leave a detectable genomic signature but confers a distinct chemistry that enables chemical identification and analysis. Finally, genomic approaches may not be particularly useful in non-model and/or unsequenced species, many of which may harbor peptidic signaling molecules with novel biological characteristics. For these reasons, chemical and biochemical approaches will continue to have a key role in the discovery and characterization of this class of plant molecules.

Many known methods are currently used to investigate the bioactivity of crude extracts from various organisms, including plants. Various bioassays are used to determine the potential biological activity of compounds from the extracted material. Extracts containing the desired biological activity are then further fractionated using a variety of chemical techniques, such as those described extensively in this chapter, to identify the specific molecules involved. Following purification of such molecules, their structure is determined by chemical techniques, and the desired biological activity of the pure compound is validated. Figure 1 summarizes this process in the schematic format of a flowchart.

2 Overview

Isolation and structural identification of signaling peptides is essential in determining their functional properties and physiological roles in plants. Interest in identification of peptides with novel functional roles has gained momentum in the last decade as new methods for analysis of individual proteomes have emerged. Proteins are generally extracted from plant biomass, resulting in a mixture of

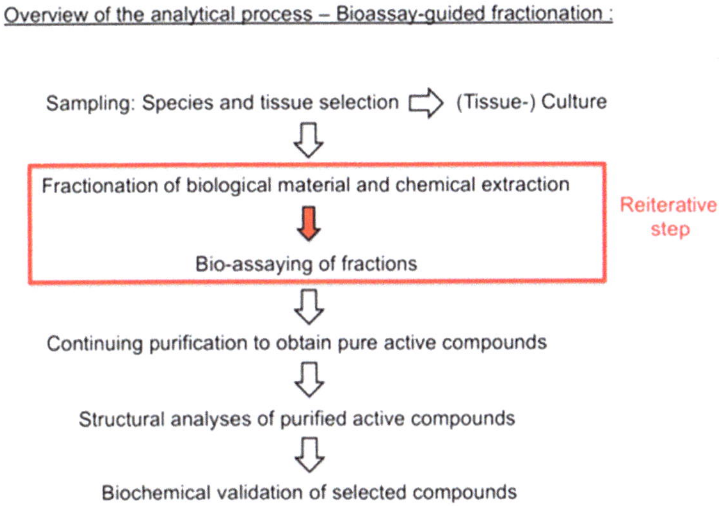

Fig. 1 A flowchart of general approaches for bioassay-guided extraction and screening of compounds

peptides. These peptides are subsequently purified by liquid chromatography and characterized by different types of mass spectrometry. Isolation involves separation of specific peptides from the mixture due to differences in their physicochemical properties. These distinct properties are utilized in chromatographic separation of the peptides using various techniques. However, due to the increasing demand for process efficiency, these techniques are under constant modification for improving the rapidity of processing, purification, and structural identification. Although various separation techniques used in peptide purification have been dealt with in great detail in a number of books, this chapter provides basic information regarding different liquid chromatographic approaches to peptide purification using silica-based and polymeric-based columns. The fundamental concepts and common approaches used for mass spectrometry detection and other structural characterization methods for peptides are also discussed. Purification and identification of peptides often involves multiple discrete steps. In order to avoid the losses associated with each step and for better efficiency, a multidimensional approach where the different techniques are integrated to give better results has also been developed. A brief discussion of this multidimensional approach and integration of the different techniques for peptide purification and identification is also presented.

3 Chemical Properties of Peptides

Peptides are important in a large variety of physiological and biochemical functions of life in all organisms, including plants. Peptides are composed of two or more amino acids where the carboxyl group or C terminus of one amino acid is coupled to the

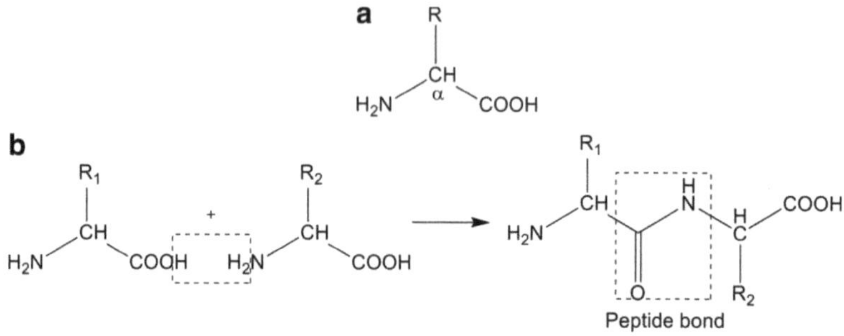

Fig. 2 (**a**) General structure of amino acids and (**b**) condensation of two amino acids leading to peptide bond formation

amino group or N terminus of another amino acid resulting in the formation of peptide or amide bond (Fig. 1). All amino acids (except proline) have an acidic carboxyl group, a basic amino group, and a hydrogen atom bonded to the alpha-carbon atom (Fig. 2). All amino acids have a unique R group, with the R group for glycine being a hydrogen atom. Although there is no universally agreed delineation of the differences between peptides and proteins, molecules with fewer than 50 amino acids are often called peptides whereas longer polypeptides are generally classed as proteins.

Most amino acids exist in two stereoisomeric forms, i.e., D (dextrorotatory) and L (levorotatory). These are the two mirror images of optically active compounds in which the order and bond connections of the constituent atoms remain the same, but their orientation in space differ. In nature, most amino acids exist in the L-conformation. Natural peptides are generally composed of some combination of the 20 standard amino acids. These amino acids differ from each other according to the nature of their R group. Standard amino acids can be divided into five groups depending upon their R substituent, i.e., nonpolar aliphatic (alanine, valine, isoleucine, leucine, methionine, glycine, cysteine, proline), aromatic (phenylalanine, tyrosine, tryptophan), polar uncharged (serine, threonine, asparagine, glutamine), positively charged (arginine, histidine, lysine), and negatively charged (aspartate, glutamate). There may be nonstandard amino acids in peptides not biosynthesized by ribosomes which arise by posttranslational modifications. At neutral pH, the carboxyl group is deprotonated and the amino group is protonated, so that the net charge is zero (i.e., amino acids exist as zwitterion at neutral pH). At alkaline or basic pH, the amino group is deprotonated and the carboxyl group is in the anionic form; however at acidic pH, the carboxyl group is protonated and the amino acid is in cationic form (Fig. 3).

4 Plant Peptides

Plants contain unique biologically active peptides, many of which act primarily in cellular functions such as signaling and defense (Garcia-Olmedo et al. 1998; Marx 1996; Bergey et al. 1996; also see chapters “Plant Antimicrobial Peptides,” “The

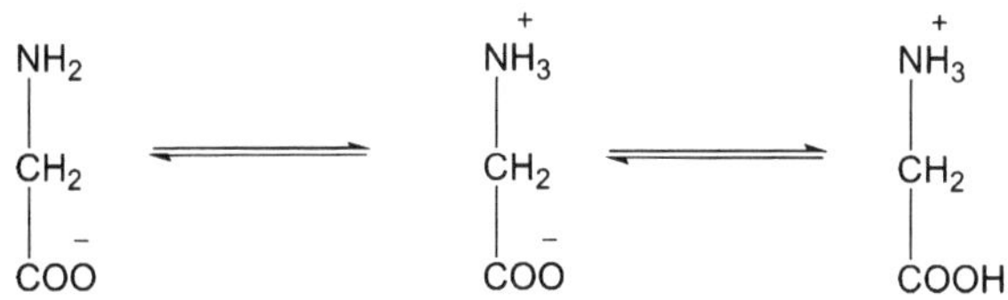

Fig. 3 Ionization state of amino acids is affected by pH

Role of Plant Peptides in Symbiotic Interactions," and "Peptides as Danger Signals: MAMPs and DAMPs"). Interest in the purification and characterization of such peptides has gained momentum in the last decade because of their importance for natural product-based drug discovery programs (Witherup et al. 1994) and proteome analysis. This in turn leads to identification of new molecular targets with therapeutic potential as well as to the development of new biomarkers (Hanash 2003). Only a limited amount of literature has been published for isolation and identification of peptides from plant biomass (Bohlmann and Apel 1991; Gran 1973; Gustafson et al. 1994). The plant biomass contains many components which are not present in animal material, for example, photosynthetic pigments, tannins, unique polysaccharides, and secondary metabolites. Hence, the methods used for isolation of peptides from animals are not always directly applicable to plant material. The general steps involved in isolation and identification of peptides from plant biomass is shown in Fig. 4. Each of the steps is further described below.

4.1 *Plant Material and Extraction*

Isolation and identification of peptides from plants is usually done with freshly collected samples. After collection, the plant sample is usually frozen in liquid nitrogen to prevent molecular degradation. Generally, the sample has to be examined by a botanist for its correct taxonomic classification, and a voucher specimen needs to be deposited at a herbarium for future referencing. The sample is then crushed, homogenized, and extracted with a suitable solvent. Extraction of peptides from plant samples is challenging, as plant tissues are generally low in protein concentration and often high in proteases and oxidative enzymes. Plants also have other unique metabolic characteristics such as pigments, phenolic compounds, lipids, and polysaccharides which generally interfere with further purification and identification (Granier 1988; Tsugita and Kamo 1999; Carpentier et al. 2005).

A number of protocols have been developed to improve protein extraction with minimal apparent contamination. Most of these protocols include the basic strategy of protein precipitation with trichloroacetic acid (TCA) and acetone, followed by resolubilization in buffer containing chaotropes and detergents (Hurkman and Tanaka 1986; Santoni et al. 1994).

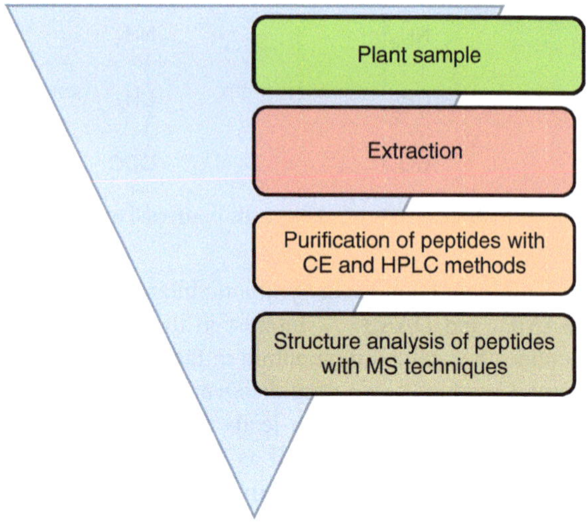

Fig. 4 A flowchart representing the general procedure for peptide purification and identification from plants. *HPLC* high-pressure liquid chromatography, *MS* mass spectrometry

Lyophilization, also called freeze-drying, is a safe method of drying biological materials. Drying sometimes causes some loss of activity or other damage, but this is primarily a problem with larger more complex molecules like proteins. In fact, this technique is also a favored method of drying vaccines, pharmaceuticals, blood fractions, and diagnostics. The choice of an appropriate solvent or solvent system for solid–liquid extractions plays an important role in the quantitative and qualitative composition of the extract.

The properties of the solvent (such as its ability to dissolve the compounds of interest, ease of removal, inertness, toxicity, and flammability) play an important role in selecting the most appropriate extraction solvent. Low-polarity solvents yield more lipophilic compounds, while alcohols extract both nonpolar and polar compounds (Fig. 5). Snyder et al. 1979 calculated the "polarity" and selectivity of more than 80 solvents and classified these solvents into eight groups according to their properties as proton acceptors and donors and their dipole interactions. This study illustrates the importance of careful selection of a suitable solvent system(s) for extraction.

4.2 *Isolation of Peptides*

Isolation of a specific peptide from a mixture is a key step in studying the functions and structures of peptides. A mixture of peptides is commonly resolved by liquid chromatography and then detected and sequenced by tandem mass spectrometry (MS/MS). No single chromatographic technique has the ability to resolve a complex mixture of peptides into all of its individual components. However, high-performance liquid chromatography (HPLC) is the main method used for purification of peptides. HPLC is most commonly used due to its automation,

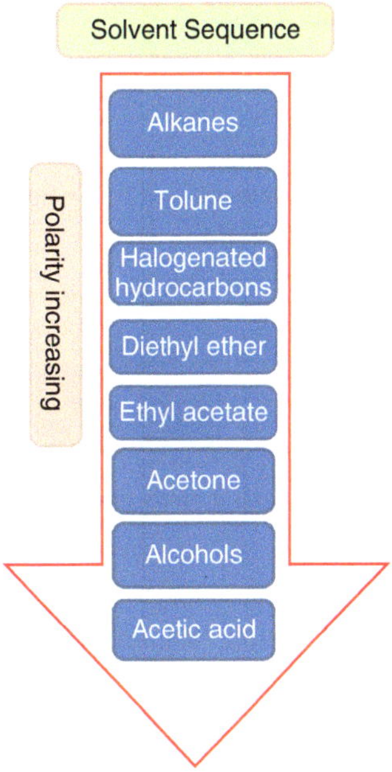

Fig. 5 Solvents used to extract molecules with increasing polarity

reproducibility, high resolution, and the ease with which it can be coupled online to different detection systems (i.e., UV/Vis, conductivity, light scattering, MS, fluorescence, etc.).

HPLC was first used in the mid-1970s for purification of peptides, and soon, it was realized that reversed-phase HPLC (RP-HPLC) with a nonpolar stationary phase had great potential for peptide isolation (Gruber et al. 1976; Frei et al. 1976; Molnar and Horvath 1977). RP-HPLC is now a very commonly used liquid chromatographic method. Different types of reversed-phase columns are commercially available differing in length, particle diameter, internal diameter, pore size, hydrophobicity, pH stability, support material, etc. Usually, gradient conditions are used for peptide separation. Acetonitrile is the main organic solvent used, and the mobile phase typically contains trifluoroacetic acid (TFA), formic acid, etc., as ion-pair reagents (Hancock et al. 1978). Other types of HPLC used for peptide purification are ion-exchange chromatography (IEC), size-exclusion chromatography (SEC), affinity chromatography (AC), hydrophobic interaction chromatography (HIC), and hydrophilic interaction chromatography (HILIC).

Chromatographic separation of peptides is based on the basic principle of interaction with the stationary phase (column packing material) and mobile phase (eluting solvent). The interaction results in the differential adsorption of peptides on the stationary surface. Separation is based on the physicochemical properties of the peptides, i.e., polarity in case of RP-HPLC, HIC, and HILIC; ionic charge in IEC; size in SEC; and shape in AC. Generally, HPLC purification of peptide mixtures is accomplished by using isocratic mobile phases or mobile phase gradients where the components of the mixture are differentially eluted off the column by changing the mobile phase composition with time. However, there are number of experimental

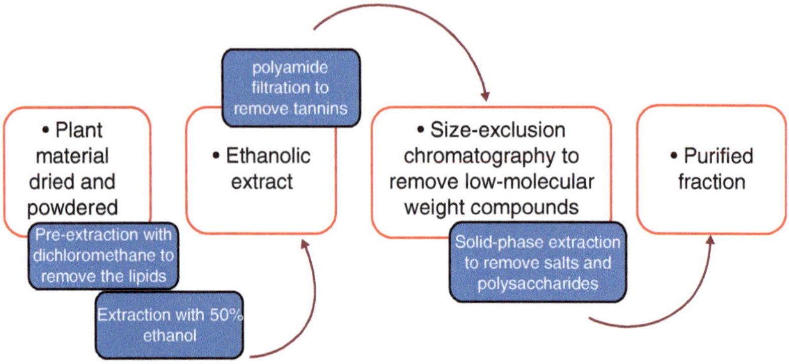

Fig. 6 General protocol for isolation of peptides from plants

parameters which contribute to better separation of the peptides, i.e., peptide composition, solubility and their chemical properties, type of column, packing materials (silica-based and polymeric-based), mobile phase properties and modifiers, etc.

All these HPLC methodologies are normally used in isolation; however, they can be combined to increase the efficiency and resolving power of the purification procedure (Evans and Jorgenson 2004; Issaq et al. 2005; Washburn et al. 2001). Separation of complex peptide mixtures can be done by using two or more orthogonal HPLC separation procedures. Issaq et al. (2005) and Tang et al. (2008) discussed the different multidimensional approaches for purification of peptides. This approach is also discussed briefly here.

Although there is no common method for purification of all of the different types of peptides, Claeson et al. (1998) described a general fractionation protocol for isolation of polypeptides from plants. The utility of the procedure has further been illustrated through the isolation of a 29-residue macrocyclic peptide from plant *Viola arvensis* Murray (Violaceae). The flowchart of the protocol/procedure is shown in Fig. 6. Pre-extraction of the plant material was done by using dichloromethane to remove the lipophilic substances such as lipids, chlorophyll, terpenoids, and phenylpropanoids, and the plant material was then again extracted with 50% aqueous ethanol. To remove the ubiquitous polyphenols (tannins) from the extract, it was then passed through the polyamide column. Peptides were then sequentially eluted with acetic acid (2%) and ethanol (50%)/acetic acid (2%) to elute peptides insoluble in 2% acetic acid. The extract was then passed through a column packed with Sephadex G-10 for SEC, and a mobile phase consisting of ethanol (50%)/acetic acid (2%)/NaCl (0.2 M) was used to elute the various peptides. The high molecular weight fraction (above 700 Da) was eluted at the void volume of the column, while the low molecular weight fraction (<700 Da) was retarded on the column and eluted afterwards. A solid phase extraction (SPE) technique with a C18 reversed-phase column was used to remove the salt and polysaccharides. The lyophilized sample was dissolved in a large volume of

50 mM NH_4HCO_3 and applied to the column. The same buffer was then used to elute the salt and polysaccharides, as these are fairly soluble in this buffer. The peptides were then eluted by using 20, 50, and 80% ethanol, respectively, as highly purified fractions.

5 Specific Isolation Techniques

5.1 Reversed-Phase High-Performance Liquid Chromatography

Reversed-phase high-pressure liquid chromatography (RP-HPLC) is probably the most widely used method for analytical and preparative separation of peptides because of its high resolution. The term "reversed-phase chromatography" signifies the use of a hydrophobic (or nonpolar) stationary phase and a hydrophilic (polar) starting mobile phase. In normal-phase HPLC, the stationary phase is polar and the starting mobile phase is nonpolar. With RP-HPLC, peptides are adsorbed onto the hydrophobic surface of the column, and the most hydrophobic peptide elutes last with the highest concentration of the organic solvent, while the least hydrophobic peptide elutes first. The elution order is related to the hydrophobicity of the solute, since the more hydrophilic the solute, the faster it will be eluted. In RP-HPLC gradient elution, water is used as the starting solvent and acetonitrile (the organic solvent) is added gradually to generate a gradient. The most commonly used organic solvents are acetonitrile, isopropanol, methanol, and ethanol. Acetonitrile is the solvent of choice in RP-HPLC because it is less viscous than methanol, thus causing fewer fluctuations in pressure. Acetonitrile also has better selectivity for peptides and proteins.

Modern RP-HPLC utilizes a wide selection of packing material to separate proteins and peptides. The most popular RP-HPLC column packing material is based on spherical silica. To ensure retention of substrate by the silica-based particles, their surface is modified by attaching alkyl chains varying in length from C4 to C18. The C18 bound stationary phase is the most popular, as it offers retention and selectivity for a wide range of compounds containing different polar and nonpolar groups on their surface. C4 and C8 phases are used preferentially for separation of proteins, while C18 phase is used preferentially for separation of peptides. There are also silica-based packing materials available for many different highly hydrophobic interactions, for example, acidic, basic, halogenated, or phenolic (Henry 2002), as well as ionic with minimal hydrophobic character (Wagner et al. 2002).

The separation efficiency of the silica gel stationary phase depends on its particle size, pore size, surface area, and covalent modifications. Typically, silica gel of particle sizes of 3 and 5 μm is used for analytical separation of peptides. Efficiency of separation increases by 30–40% with the reduction of the particle size from 5 to 3 μm. Unfortunately, smaller particle size also elevates the backpressure beyond the pressure limit of most commercially available HPLC systems. The efficiency of the packing material can be further increased by increasing the available surface area with increased porosity of the particles. In a multidimensional approach for the separation of a complex peptide mixture, the last dimension is usually RP-HPLC because of its high resolving power and the compatibility of its solvents with mass spectrometry (Janini et al. 2004). Since the initial application of RP-HPLC for peptide separation in 1976 (Gruber et al. 1976), several studies have been published regarding the use of this method in plant peptide purification (Kim et al. 2006, 2009; Rivillas-Acevedo and Soriano-Garcia 2007; Lee et al. 2006; Huang et al. 1999; Chen et al. 1998).

5.2 Ion-Exchange Chromatography

IEC is based on the principle that peptides are charged molecules that at specific pH, values exist in anionic, cationic, or zwitterionic (zero net charge) forms (see above). Since peptides can have net positive or negative charges, two types of ion-exchange resins are used. These include anion-exchange (bearing positively charged functional groups) and cation-exchange (bearing negatively charged functional groups) resins. IEC separation has two main steps. The first involves the differential binding of the peptides to the charged resin, and the second is the elution of peptides from the resin. The initial step involves washing the resin with a buffer containing 1 M salt (generally NaCl or KCl) to ensure the presence of counter ions. Usually, a sulfonic acid group [i.e., $–SO_3H$ (strong acid) or carboxylic group, i.e., –COOH (weak acid)] is added to the resin used in cation-exchange chromatography. Quaternary ammonium [$–CH_2N(CH_3)_3$, strong base] or polyalkyl amine group [$–NH(R)_2$, weak base] is the functional group added to the resin in anion-exchange chromatography. Negatively or positively charged peptide molecules bind to the oppositely charged functional groups of the resins. During elution, counter ions in the eluting buffer bind in place of peptides and hence release them (Fig. 7). Separation is based on the relative degree of binding strength of the peptides with the ion-exchange resin. Unbound and weakly bound peptides will be eluted first, while strongly bound peptides will be eluted later.

Choice of buffer is an important factor in IEC. The buffer used must be appropriate for the pH and the resin used. Generally, there are two types of buffers: cationic buffers with anion exchangers (e.g., L-histidine, bis-tris propane, triethanolamine, diethanolamine) and anionic buffers with cation exchangers [e.g., formate, acetate, phosphate, HEPES (4-(2-hydroxyethyl)-1-piperazineethane-sulfonic acid)].

The choice of resin for IEC depends on the stability of peptides at their respective pH and isoelectric point (PI). Isoelectric point (PI) is the pH at which the net charge for the peptide is zero. If peptides are stable when pH > PI, anion-exchange resins are used. However, when peptides are stabile at pH < PI, cation-exchange resins are used. The pH chosen for chromatography should be optimum,

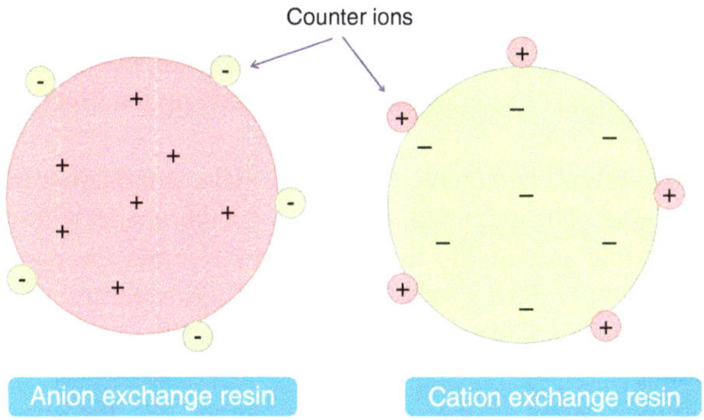

Fig. 7 Diagram showing ion-exchange resins and the counter ions present in the eluting buffer used to replace the bound peptides

i.e., one pH unit above or below the PI is used for particular types of ion-exchange resin. Generally, peptides are stable to conditions such as high concentrations of organic solvents and extreme pH values, as long as they are within the stability range of the chromatography medium. If the pH is high, binding of peptides to the ion-exchange resin would be strong, making them difficult to elute. The ionic strength of the eluting buffer also plays an important role in separation of peptides. Normally, buffers with low ionic strength are used initially. Two different types of elution are used in IEC: isocratic elution and gradient elution. In isocratic elution, the ionic strength of the buffer is manipulated to elute the peptides. In gradient elution, increasing concentrations of salt or ionic strength is used. This is further followed by elution with high salt buffer solution to remove the tightly bound peptides and to regenerate the resin for reuse.

Ion-exchange columns have been used extensively for separation of plant peptides and proteins (Das et al. 2011; Ma et al. 2009; Bhowmick et al. 2008; Lin et al. 2007; Wang and Ng 2005). In a two-dimensional approach, strong cation exchange (SCX) is normally the first dimension for the fractionation of peptides while RP-HPLC is used in the second dimension (Wolters et al. 2001).

5.3 Size-Exclusion Chromatography

SEC is also a commonly used method to separate peptides, proteins, and other biological molecules according to their sizes (Cheng et al. 2010; Pelegrini et al. 2009; Wang et al. 2007; Ooi et al. 2008; Barth et al. 1994; Marquez-Escalante et al. 2006; Chanana et al. 2004). It is also called gel filtration chromatography (GFC) or gel-permeation chromatography (GPC), depending on the nature of mobile phase used. When an aqueous mobile phase is used to elute the sample, the technique is called GFC, while use of an organic solvent is known as GPC. During elution, the larger peptides are excluded from the pores in the beads of the gel and therefore elute faster than the smaller peptides. Smaller peptides enter the porous beads, which slows their progress through the gel, and therefore, they elute after the larger peptides. The differential elution of peptides depends on the diversity of the size of the molecules.

Size-exclusion column beds have two functional components: the pore volume and the void volume. The pore volume refers to the space within the particles. Peptide molecules that are smaller than the pore size enter the particles, penetrate further, and move slowly during sample elution. Their separation is determined by the pore size distribution, and the volume of mobile phase required for elution is called pore volume.

The void volume refers to the excluded volume, i.e., the space between the particles. Proteins too large to enter the pores are excluded and are eluted earlier than the smaller peptides. The volume of the mobile phase required for their elution is called the void volume.

The mobile phase is usually a moderate buffer or salt solution which helps to maintain the biological activity of the peptides. The column packing is made up of porous particles, i.e., cross-linked polymer beads with pores of the selected size to separate peptides of different sizes. Typically used resins in SEC are made from polyacrylamide, dextran, or agarose.

5.4 Affinity Chromatography

Affinity chromatography was first introduced by Pedro Cuatrecasas in 1968 (Cuatrecasas et al. 1968) and is an important biochemical technique used for peptide/protein purification (Kolodzeiskaia 1983; Rudenskaia 1994; Lin et al. 2009; Wang et al. 2009). In AC, the peptide of interest must have a specific property that can be exploited during the chromatography. The column is designed to take advantage of the specific property of the peptide(s) of interest. AC is based on the principle of a unique interaction between the peptide molecules with the ligand bonded to the matrix. The matrix of affinity chromatography columns, composed of inert substances, is generally composed of agarose or other polysaccharides. Matrix beads in the affinity column are cross-linked to ligands that bind specifically to the target peptide molecules. Amino, hydroxyl, carbonyl, and thiol groups placed within the matrix act as ligand binding sites. The peptide mixture loaded to the equilibrated column sieve through the matrix of affinity beads. Depending upon the affinity of the peptide molecules for the ligand, some peptide molecules bind tightly and others bind loosely to the matrix. The unbound peptides wash off the column with the wash buffer. The peptides retained on the column are then eluted by changing the pH of the salt or organic solvent concentration of the eluent. The specificity of the interaction between peptide molecules and matrix beads depends upon three factors, i.e., matrix, ligand, and interaction of the ligand with the matrix. A chosen ligand should bind strongly to the molecule of interest. The ligands used in affinity chromatography are antibodies for antigens, receptor molecules for hormones, substrate analogs and inhibitors for enzymes, lectins for glycoproteins, sugars for lectins, reactive dyes for dehydrogenases and other proteins, and metal ions for metal-binding proteins.

Immobilized metal ion affinity chromatography (IMAC) was introduced by Jerke Porath in 1975 and is based on coordinate covalent binding between peptides and metal ions. Peptides have wide variety of amino acids with range of different affinities for the metal ions. Peptide separation in IMAC generally depends on the metal ion–peptide bond strength. Hence, choosing the type of immobilized ion is very important for success of peptide separation. Peptides containing a higher number of histidine residues would be able to bind to the column containing divalent metal ions more tightly than those with fewer histidine residues, as histidine can generally act as a ligand towards divalent metal cations. Nickel ion columns are the most widely used metal ion affinity columns. Not many naturally occurring peptides have affinity for metal ions, so known peptides can be engineered to contain a ligand (e.g., poly-histidine tag). Other types of immobilized metal ion columns developed to separate various peptides contain Fe, Co, Cd, Ni, or Zn metal ions as ligands.

Titanium dioxide column (Yu et al. 2007), Fe(III) ions immobilized on magnetic beads, and Ga(III) ions immobilized on a chelate resin (Sykora et al. 2007) have been used for enrichment of phosphopeptides. Organomercurial-agarose (Hg beads) columns, which specifically capture cysteine-containing peptides, have been used for the enrichment of such peptides (Raftery 2008), and boronic acid affinity chromatography was used for enrichment of glycated peptides (Frolov and Hoffmann 2008).

5.5 Hydrophilic Interaction Liquid Chromatography

HILIC was first described by Alpert (1990) for separation of hydrophilic substances such as proteins, peptides, and nucleic acids. HILIC is considered as a variant of normal-phase chromatography. Normal-phase liquid chromatography with silica or alumina as the stationary phase is generally used to separate polar compounds with nonaqueous mobile phases, such as hexane or chloroform. This method has limitations for the separation of hydrophilic compounds such as peptides, which are soluble in mobile phases with relatively high water content. HILIC is based on the principle that an increase in the peptide retention time will occur with an increase in the polarity of the stationary phase or decrease in the concentration of organic solvent in the mobile phase (Yoshida 1997; Strege et al. 2000). Peptides retained on the column are eluted by increasing water content or the polarity of the mobile phase (similar to nonaqueous normal-phase chromatography). The stationary phases used in HILIC are gel particles to which polar functional groups such as 2-hydroxyethyl aspartamide, carbamoyl, glycerol, carboxymethyl, and diethylaminoethyl are bound. The mobile phases used in HILIC are water/acetonitrile, buffer, and salt solution. The most common organic solvent used in HILIC is acetonitrile, as it yields a less viscous mobile phase. Gradient elution in HILIC can be accomplished by increasing the polarity of the mobile phase by using either a decrease in the concentration of organic solvent or an increase in the salt concentration. Several examples of peptide separations using HILIC have been published (Tolstikov and Fiehn 2002; Risley and Strege 2000; Boutin et al. 1992; Dunkel et al. 2007; Vacek et al. 2006). TSKgel Amide-80 column is generally used for isolation of peptides in combination with a volatile mobile phase. By using this combination, the peptides are directly amenable to N-terminal protein sequencing and amino acid analysis after evaporation of the organic phase. Polyhydroxyethyl A column is also commonly used in combination with a triethylamine phosphoric acid (TEAP) buffer and high concentration of organic solvent for peptide separation.

5.6 Hydrophobic Interaction Chromatography

HIC separation of peptides is based on the principle of hydrophobic interactions between immobilized hydrophobic ligands attached to the gel matrix and nonpolar regions on the surface of peptides. During the chromatographic procedure, the soluble peptides are loaded on to the column, and the adsorption between gel matrix and peptides are shielded by water molecules. This is then increased by adding high salt concentration in the mobile phase, and hence, the term "salt-promoted adsorption" is also used for this type of chromatography (Porath 1986). Van Oss and coworkers (1986) proposed that van der Waals forces mainly contribute to the hydrophobic interactions. These are weaker interaction forces than ion exchange, reversed phase (Fausnaugh et al. 1984), or affinity chromatography and thus help in maintaining the minimum structural damage and the biological activity of the peptides. The elution of bound peptides is achieved by using a gradient of decreasing salt concentration of the eluent (Melander and Horvath 1977; Fausnaugh and Regnier 1986).

The type of stationary phase and its degree of substitution and type of mobile phase used are the main factors responsible for the efficiency of the HIC techniques. The various kinds of stationary phases differ in the type of ligand and the ligand chain length. The most widely used ligands for HIC are linear chain alkyl groups (ethyl, propyl, pentyl, etc.) and aromatic groups. Straight chain alkyls show only hydrophobic interactions; however, aromatic groups (like phenyl) show mixed hydrophobic and aromatic (π–π) interactions. Peptide binding capacity also increases with increase in the length of alkyl chain and degree of substitution of ligand (ligand density) due to multipoint attachment of peptide. The most widely used matrix supports are cross-linked agarose, silica, or synthetic copolymer materials. The selectivity will not be exactly the same even by using the same type of ligand, if the support matrix is different.

Retention of peptides also depends on the characteristics of the mobile phase used such as the type and concentration of salt. The use of high salt concentration in the equilibration buffer and sample solution promotes ligand–protein interactions

(Oscarsson and Karsnas 1998). Salts such as sodium, potassium, or ammonium sulfates are the most effective to promote ligand–peptide interactions, due to the higher "salting out" effects. The pH of the mobile phase is also an important factor that contributes to peptide retention. Generally, an increase in pH decreases the hydrophobic interaction (Porath et al. 1973), and a decrease in pH results in an increase in hydrophobic interactions. Details of this method have been published in many articles (Sammer et al. 2009; Azad et al. 2006; Peckham et al. 2006; Desimone et al. 2000; Van Sluyter et al. 2009).

5.7 *Multidimensional Separation Approach*

Efficient purification of a peptide mixture is usually not possible using a single chromatographic or electrophoretic method. A multidimensional method that employs two or more orthogonal HPLC separation procedures significantly improves the resolution of a complex peptide mixture. The first step in a multidimensional separation of a peptide mixture is usually fractionation, where the complex mixture is separated into simpler mixtures by using liquid chromatography (first dimension), and these fractions are further analyzed critically with a second dimension HPLC.

Since most current two-dimensional LC systems are interfaced with mass spectrometers, reversed-phase chromatography is usually the main choice for the second dimension separation, because the samples eluted from the RP column are in the most suitable form for injection into the mass spectrometer.

Normally, for good two-dimensional separations of peptides, one type of column is used in the first dimension and a different chemistry column in the second dimension. The most widely used approach involves IEC in the first dimension and reversed-phase chromatography in the second dimension. Washburn et al. (2002) described the multidimensional approach for peptide separation, where the first dimension is SCX chromatography, while the second dimension is reversed phase, using a single biphasic (two stationary phases) column. A biphasic column is a capillary (100 μm internal diameter) packed with strong SCX resin followed by reversed-phase resin. The peptides are separated in the first dimension by charge (SCX) and in the second dimension by hydrophobicity. Affinity chromatography, size exclusion, gel electrophoresis, etc., are other chromatographic modes which are used as first dimension for sample fractionation (Neverova and Van Eyk 2005; Stasyk and Huber 2004). Two types of mobile gradient, i.e., increasing salt concentration elution alternated with an organic solvent gradient, are used to elute the analytes.

Gilar et al. (2005) evaluated different two-dimensional HPLC approaches including SCX-RP, HILIC-RP, and RP–RP and found that the peak capacity is highest in RP–RP at different pH values. The peak capacity of a column is defined as the number peaks that can be separated from one another up to a predefined capacity ratio (Li et al. 2009). The study also showed that SEC at first dimension gave the poorest resolution of the peptide mixture when compared with other chromatographic methods. Another study reported the multidimensional chromatography approach, combining IMAC, HILIC, and RPLC for phosphopeptide purification and fractionation from the yeast *Saccharomyces cerevisiae* (Albuquerque et al. 2008).

Enrichment of phosphopeptides has also been reported by using two-dimensional approaches, i.e., SCX followed by RP/MS analysis (Lim and Kassel 2006), or by the use of mixed anion and cation-exchange resin in the first dimension and RP in the second dimension (Motoyama et al. 2007). In 1997, Opiteck et al. (1998) reported a two-dimensional liquid chromatography system based on eight serially connected size-exclusion columns coupled to two parallel RP columns. This was used for isolating overexpressed proteins of interest from bacterial cell lysates. In cases where two-dimensional LC does not give satisfactory results, a third orthogonal separation procedure is used which may help in the better resolution of the peptide mixture. Various types of three-dimensional approaches used for peptide separation are affinity/SCX/RP (Gygi et al. 1999), RP/SCX/RP (McDonald et al. 2002), SEC/SCX/RP (Jacobs et al. 2004), IEF/SCX/RP (Chan et al. 2004), etc. Several studies have been reported (Zhang et al. 2010; Sandra et al. 2009; Dugo et al. 2008; Link 2002; Liu et al. 2002) including more detailed descriptions of the various coupled multidimensional liquid chromatography. It is clear from this brief discussion that combining two or more orthogonal separation techniques significantly improves overall resolution and helps in identification of a larger number of peptides.

5.8 *Identification Techniques for Peptides*

Peptides are commonly detected by using mass spectrometry, but may also be detected with high sensitivity through chemical labeling of the amino or thiol groups on the peptide with fluorescent dyes (Santa et al. 2008; Lacroix et al. 2005). Conventional fluorescence uses arc lamps as excitation sources and is more sensitive than UV detection, but LIF detector is most sensitive in conventional techniques and is at least 10–100-fold more sensitive than a fluorescence detector (Issaq et al. 2009). Peptides can be detected by UV absorption of the peptide bond at 185–220 nm; however, UV is not a very sensitive method to detect the peptides and one cannot differentiate by UV spectra alone whether two or more peptides are co-eluting. The detection of peptides containing aromatic residues is more selective and sensitive.

The two main approaches used for peptide profiling are mass assisted laser desorption/ionization time-of-flight mass spectrometry (MALDI-TOF-MS) and electrospray ionization mass spectrometry (ESI-MS) (Mann et al. 2001; Shevchenko et al. 1996). Both are soft ionization techniques which keep the molecules of interest fully intact and are used for both characterization and quantitation of the separated peptides.

ESI-MS is an online HPLC-MS method where the peptide sample is dissolved in polar, volatile buffer and pumped through a stainless steel capillary column at a rate of 10–100 mL/min (<1 mL/min for nanospray) (Alomirah et al. 2000; Ostin et al. 2007; Walcher et al. 2003). Strong voltage (i.e., 3–4 kV) is applied to the liquid sample at the tip of the capillary with a flow of nebulizing gas that converts the sample to droplets carrying the charges. The polarity of the voltage applied determines whether positive or negative ions are formed. These gas phase ions are then directed through a high vacuum until they evaporate to near atomic size and hit the detector. The smaller the size of the droplet, the more rapid is the formation of ions. The detection signal is then converted to data, which is presented in the form of peaks where each peak represents single peptide. Mass of ion (or m/z) is normally shown along the x-axis, and the y-axis corresponds to the number of ions detected. Samples of molecular weight up to 1,200 Da produce singly charged ions; however, larger peptides yield multiply charged ions. Peptides which tend to ionize easily also display many peaks.

MALDI-TOF, invented in the late 1980s (Karas and Hillenkamp 1988), is a highly sensitive technique with a superior mass resolution and accuracy. It has been extensively used for the detection of small amounts of peptides (Xiang et al. 2010; Ding et al. 2009; Moravcova et al. 2009; Manzano et al. 2008; Silva-Sanchez et al. 2008). It is an off-line method, and the peptides are deposited on a target plate for MS analysis. The sample peptide mixture is co-crystallized with a light adsorbing compound (matrix assisted), and a laser pulse is used to convert the mixture components to gas phase and ionize them, often with a single charge ($z = 1$). Ions are then accelerated in a high-voltage electrostatic field through a vacuum. Time of flight (TOF) of the ions is proportional to m/z, i.e., the smaller ions reach the detector before larger ions. Data is in the form of peaks similar to ESI data. Unlike ESI, MALDI-TOF generates spectra of peptides that have a singly charged ion. The list of the masses of peptides, which is also called the peptide's mass fingerprint, is then compared to the peptides mass databases to identify various peptides. Peptides can be detected up to sub-picomolar levels with a mass accuracy of 20–100 ppm (parts per million). MALDI-TOF provides optimal performance for the detection of peptides with a mass range (i.e., m/z) from 1,000 to 5,000, although it has no absolute upper analytical limit (Twerenbold et al. 2001). This technique is also more robust than ESI because of its tolerance to salts and nonvolatile compounds. Both MALDI-TOF MS and its variant surface-enhanced laser desorption/ionization (SELDI)-TOF MS are the techniques that are particularly used for detection of small peptides and proteins relative to large proteins.

The application of mass spectrometry for the detection of peptides has been revolutionized with the development of electrospray ionization (ESI) and matrix-assisted laser desorption/ionization (MALDI) techniques. Both these techniques allow the determination of accurate molecular mass and the detection of chemical modification at high sensitivity (picomole to femtomole) and can be coupled to different types of mass analyzers (Zhou and Veenstra 2008).

6 Conclusion

Bioactivity-guided chemical analyses have already led to the discovery of a considerable number of peptidic signaling molecules that have profoundly changed the way we view plant responses, and it is most likely that the numbers of novel molecules that trigger ligand-specific physiological responses at nanomolar levels will increase. This increase will in part be due to ever-improving analytical instrumentation for isolation and structural determination such as some of those described in this chapter and progress in development of high-throughput bioassay and screening technologies.

References

Albuquerque CP, Smolka MB, Payne SH, Bafna V, Eng J, Zhou H (2008) A multidimensional chromatography technology for in-depth phosiphoproteome analysis. Mol Cell Proteomics 7 (7):1389–1396

Alomirah HF, Alli I, Konishi Y (2000) Applications of mass spectrometry to food proteins and peptides. J Chromatogr A 893(1):1–21

Alpert AJ (1990) Hydrophilic-interaction chromatography for the separation of peptides, nucleic acids and other polar compounds. J Chromatogr 499:177–196

Azad MA, Sawa Y, Ishikawa T, Shibata H (2006) Purification and characterization of protein phosphatase 2A from petals of the tulip Tulipa gesnerina. J Biochem Mol Biol 39:671–676

Barth HG, Boyes BE, Jackson C (1994) Size exclusion chromatography. Anal Chem 66:595R–620R

Bergey DR, Howe GA, Ryan CA (1996) Polypeptide signaling for plant defensive genes exhibits analogies to defense signaling in animals. Proc Natl Acad Sci USA 93:12053–12058

Bhowmick R, Kumari NK, Jagannadham MV, Kayastha AM (2008) Purification and characterization of a novel protease from the latex of Pedilanthus tithymaloids. Protein Pept Lett 15:1009–1016

Bohlmann H, Apel K (1991) Thionins. Annu Rev Plant Physiol Plant Mol Biol 42:227–240

Boutin JA, Ernould AP, Ferry G, Genton A, Alpert AJ (1992) Use of hydrophilic interaction chromatography for the study of tyrosine protein kinase specificity. J Chromatogr 583:137–143

Carpentier SC, Witters E, Laukens K, Deckers P, Swennen R, Panis B (2005) Preparation of protein extracts from recalcitrant plant tissues: an evaluation of different methods for two-dimensional gel electrophoresis analysis. Proteomics 5:2497–2507

Chan KC, Lucas DA, Hise D, Schaefer CF, Xiao Z, Janini GM, Buetow KH, Issaq HJ, Veenstra TD, Conrads TP (2004) Analysis of the human serum proteome. Clin Proteomics 1:101–225

Chanana V, Kaur KJ, Salunke DM (2004) Purification, identification and preliminary crystallographic characterization of a novel seed protein from Vigna unguiculata. Acta Crystallogr D Biol Crystallogr 60:2100–2103

Chen ZK, Fan CX, Ye YH, Yang L, Jiang Q, Xing QY (1998) Isolation and characterization of a group of oligopeptides related to oxidized glutathione from the root of Panax ginseng. J Pept Res 52:137–142

Cheng Y, Chen J, Xiong YL (2010) Chromatographic separation and tandem MS identification of active peptides in potato protein hydrolysate that inhibit autoxidation of soybean oil-in-water emulsions. J Agric Food Chem 58:8825–8832

Claeson P, Goransson U, Johansson S, Luijendijk T, Bohlin L (1998) Fractionation protocol for the isolation of polypeptides from plant biomass. J Nat Prod 61:77–81

Cuatrecasas P, Wilchek M, Anfinsen CB (1968) Selective enzyme purification by affinity chromatography. Proc Natl Acad Sci USA 61:636–643

Das S, Mishra B, Gill K, Ashraf MS, Singh AK, Sinha M, Sharma S, Xess I, Dalal K, Singh TP, Dey S (2011) Isolation and characterization of novel protein with anti-fungal and anti-inflammatory properties from Aloe vera leaf gel. Int J Biol Macromol 48:38–43

Desimone M, Kruger M, Wessel T, Wehofsky M, Hoffmann R, Wagner E (2000) Purification and characterization of an aminopeptidase from the chloroplast stroma of barley leaves by chromatographic and electrophoretic methods. J Chromatogr B Biomed Sci Appl 737:285–293

Ding H, Zhang A, Wang J, Lu R, Zhang H, Zhang J, Jiang M (2009) Identity of an ABA-activated 46 kDa mitogen-activated protein kinase from Zea mays leaves: partial purification, identification and characterization. Planta 230:239–251

Dugo P, Cacciola F, Kumm T, Dugo G, Mondello L (2008) Comprehensive multidimensional liquid chromatography: theory and applications. J Chromatogr A 1184:353–368

Dunkel A, Koster J, Hofmann T (2007) Molecular and sensory characterization of gamma-glutamyl peptides as key contributors to the kokumi taste of edible beans (Phaseolus vulgaris L.). J Agric Food Chem 55:6712–6719

Evans CR, Jorgenson JW (2004) Multidimensional LC-LC and LC-CE for high-resolution separations of biological molecules. Anal Bioanal Chem 378:1952–1961

Fausnaugh JL, Regnier FE (1986) Solute and mobile phase contributions to retention in hydrophobic interaction chromatography of proteins. J Chromatogr 359:131–146

Fausnaugh JL, Kennedy LA, Regnier FE (1984) Comparison of hydrophobic-interaction and reversed-phase chromatography of proteins. J Chromatogr 317:141–155

Frei RW, Michel L, Santi W (1976) Post-column fluorescence derivatization of peptides. Problems and potential in high-performance liquid chromatography. J Chromatogr 126:665–677

Frolov A, Hoffmann R (2008) Analysis of amadori peptides enriched by boronic acid affinity chromatography. Ann N Y Acad Sci 1126:253–256

Garcia-Olmedo F, Molina A, Alamillo JM, Rodriguez-Palenzuela P (1998) Plant defense peptides. Biopolymers 47:479–491

Germain H, Chevalier E, Matton D (2006) Plant bioactive peptides: an expanding class of signaling molecules. Can J Bot 84:1–19

Gilar M, Olivova P, Daly AE, Gebler JC (2005) Orthogonality of separation in two-dimensional liquid chromatography. Anal Chem 77(19):6426–6434

Gran L (1973) Oxytocic principles of Oldenlandia affinis. Lloydia 36(2):174–178

Granier F (1988) Extraction of plant proteins for two-dimensional electrophoresis. Electrophoresis 9:712–718

Gruber KA, Stein S, Brink L, Radhakrishnan A, Udenfriend S (1976) Fluorometric assay of vasopressin and oxytocin: a general approach to the assay of peptides in tissues. Proc Natl Acad Sci USA 73:1314–1318

Gustafson K, Sowder R II, Henderson LE, Parsons IC, Kashman Y, Cardellina JH II, McMahon JB, Buckheit RW Jr, Pannell LK, Boyd MR (1994) Circulins A and B: novel HIV-inhibitory macrocyclic peptides from the tropical tree Chassalia parvifolia. J Am Chem Soc 116:9337–9338

Gygi SP, Rist B, Gerber SA, Turecek F, Gelb MH, Aebersold R (1999) Quantitative analysis of complex protein mixtures using isotope-coded affinity tags. Nat Biotechnol 17:994–999

Hanash S (2003) Disease proteomics. Nature 422:226–232

Hancock WS, Bishop CA, Prestidge RL, Harding DR, Hearn MT (1978) Reversed-phase, high-pressure liquid chromatography of peptides and proteins with ion-pairing reagents. Science 200:1168–1170

Henry RA (2002) Highly selective zirconia-based phases for HPLC applications. Am Lab 34:18–25

Huang B, Ng TB, Fong WP, Wan CC, Yeung HW (1999) Isolation of a trypsin inhibitor with deletion of N-terminal pentapeptide from the seeds of Momordica cochinchinensis, the Chinese drug mubiezhi. Int J Biochem Cell Biol 31:707–715

Hurkman WJ, Tanaka CK (1986) Solubilization of plant membrane proteins for analysis by two-dimensional gel electrophoresis. Plant Physiol 81:802–806

Issaq HJ, Chan KC, Janini GM, Conrads TP, Veenstra TD (2005) Multidimensional separation of peptides for effective proteomic analysis. J Chromatogr B Analyt Technol Biomed Life Sci 817:35–47

Issaq HJ, Chan KC, Blonder J, Ye X, Veenstra TD (2009) Separation, detection and quantitation of peptides by liquid chromatography and capillary electrochromatography. J Chromatogr A 1216:1825–1837

Jacobs JM, Mottaz HM, Yu LR, Anderson DJ, Moore RJ, Chen WN, Auberry KJ, Strittmatter EF, Monroe ME, Thrall BD, Camp DG II, Smith RD (2004) Multidimensional proteome analysis of human mammary epithelial cells. J Proteome Res 3:68–75

Janini GM, Chan KC, Conrads TP, Issaq HJ, Veenstra TD (2004) Two-dimensional liquid chromatography-capillary zone electrophoresis-sheathless electrospray ionization-mass spectrometry: evaluation for peptide analysis and protein identification. Electrophoresis 25:1973–1980

Karas M, Hillenkamp F (1988) Laser desorption ionization of proteins with molecular masses exceeding 10,000 daltons. Anal Chem 60:2299–2301

Kende H, Zeevaart J (1997) The five “classical” plant hormones. Plant Cell 9:1197–1210

Kim MH, Park SC, Kim JY, Lee SY, Lim HT, Cheong H, Hahm KS, Park Y (2006) Purification and characterization of a heat-stable serine protease inhibitor from the tubers of new potato variety “Golden Valley”. Biochem Biophys Res Commun 346:681–686

Kim JY, Park SC, Hwang I, Cheong H, Nah JW, Hahm KS, Park Y (2009) Protease inhibitors from plants with antimicrobial activity. Int J Mol Sci 10:2860–2872

Kolodzeiskaia MV (1983) Affinity chromatography of aminopeptidases. Ukr Biokhim Zh 55:577–591

Lacroix M, Poinsot V, Fournier C, Couderc F (2005) Laser-induced fluorescence detection schemes for the analysis of proteins and peptides using capillary electrophoresis. Electrophoresis 26:2608–2621

Lee JH, Lee DH, Yu HE, Kim JH, Lee JS (2006) Isolation and characterization of a novel glutathione S-transferase-activating peptide from the oriental medicinal plant Phellodendron amurense. Peptides 27:2069–2074

Li X, Stoll DR, Carr PW (2009) Equation for peak capacity estimation in two-dimensional liquid chromatography. Anal Chem 81:845–850

Lim KB, Kassel DB (2006) Phosphopeptides enrichment using on-line two-dimensional strong cation exchange followed by reversed-phase liquid chromatography/mass spectrometry. Anal Biochem 354:213–219

Lin P, Xia L, Ng TB (2007) First isolation of an antifungal lipid transfer peptide from seeds of a Brassica species. Peptides 28:1514–1519

Lin P, Wong JH, Xia L, Ng TB (2009) Campesin, a thermostable antifungal peptide with highly potent antipathogenic activities. J Biosci Bioeng 108:259–265

Lindsey K, Casson S, Chilley P (2002) Peptides: new signaling molecules in plants. Trends Plant Sci 7:78–83

Link AJ (2002) Multidimensional peptide separations in proteomics. Trends Biotechnol 20 (Suppl):S8–S13

Liu H, Lin D, Yates JR III (2002) Multidimensional separations for protein/peptide analysis in the post-genomic era. Biotechniques 32:898, 900, 902 passim

Ma DZ, Wang HX, Ng TB (2009) A peptide with potent antifungal and antiproliferative activities from Nepalese large red beans. Peptides 30:2089–2094

Mann M, Hendrickson RC, Pandey A (2001) Analysis of proteins and proteomes by mass spectrometry. Annu Rev Biochem 70:437–473

Manzano C, Abraham Z, Lopez-Torrejon G, Del Pozo JC (2008) Identification of ubiquitinated proteins in Arabidopsis. Plant Mol Biol 68:145–158

Marquez-Escalante JA, Figueroa-Soto CG, Valenzuela-Soto EM (2006) Isolation and partial characterization of trehalose 6-phosphate synthase aggregates from Selaginella lepidophylla plants. Biochimie 88:1505–1510

Marx J (1996) Plants, like animals, may make use of peptide signals. Science 273:1338–1339

Matsubayashi Y, Sakagami Y (1996) Phytosulfokine, sulfated peptides that induce the proliferation of single mesophyll cells of Asparagus officinalis L. Proc Natl Acad Sci USA 93:7623–7627

McDonald WH, Ohi R, Miyamoto DT, Mitchison TJ, Yates JR (2002) Comparison of three directly coupled HPLC MS/MS strategies for identification of proteins from complex mixtures: single-dimension LC-MS/MS, 2-phase MudPIT, and 3-phase MudPIT. Int J Mass Spectrom 219:245–251

Melander W, Horvath C (1977) Salt effect on hydrophobic interactions in precipitation and chromatography of proteins: an interpretation of the lyotropic series. Arch Biochem Biophys 183:200–215

Molnar I, Horvath C (1977) Separation of amino acids and peptides on non-polar stationary phases by high-performance liquid chromatography. J Chromatogr 142:623–640

Moravcova D, Kahle V, Rehulkova H, Chmelik J, Rehulka P (2009) Short monolithic columns for purification and fractionation of peptide samples for matrix-assisted laser desorption/ionization time-of-flight/time-of-flight mass spectrometry analysis in proteomics. J Chromatogr A 1216:3629–3636

Motoyama A, Xu T, Ruse CI, Wohlschlegel JA, Yates JR III (2007) Anion and cation mixed-bed ion exchange for enhanced multidimensional separations of peptides and phosphopeptides. Anal Chem 79:3623–3634

Neverova I, Van Eyk JE (2005) Role of chromatographic techniques in proteomic analysis. J Chromatogr B Analyt Technol Biomed Life Sci 815:51–63

Ooi LS, Tian L, Su M, Ho WS, Sun SS, Chung HY, Wong HN, Ooi VE (2008) Isolation, characterization, molecular cloning and modeling of a new lipid transfer protein with antiviral and antiproliferative activities from Narcissus tazetta. Peptides 29:2101–2109

Opiteck GJ, Ramirez SM, Jorgenson JW, Moseley MA III (1998) Comprehensive two-dimensional high-performance liquid chromatography for the isolation of overexpressed proteins and proteome mapping. Anal Biochem 258:349–361

Oscarsson S, Karsnas P (1998) Salt-promoted adsorption of proteins onto amphiphilic agarose-based adsorbents. II. Effects of salt and salt concentration. J Chromatogr A 803:83–93

Ostin A, Bergstrom T, Fredriksson SA, Nilsson C (2007) Solvent-assisted trypsin digestion of ricin for forensic identification by LC-ESI MS/MS. Anal Chem 79:6271–6278

Peckham GD, Bugos RC, Su WW (2006) Purification of GFP fusion proteins from transgenic plant cell cultures. Protein Expr Purif 49:183–189

Pelegrini PB, Farias LR, Saude AC, Costa FT, Bloch C Jr, Silva LP, Oliveira AS, Gomes CE, Sales MP, Franco OL (2009) A novel antimicrobial peptide from Crotalaria pallida seeds with activity against human and phytopathogens. Curr Microbiol 59:400–404

Porath J (1986) Salt-promoted adsorption: recent developments. J Chromatogr 376:331–341

Porath J, Sundberg L, Fornstedt N, Olsson I (1973) Salting-out in amphiphilic gels as a new approach to hydrophobic adsorption. Nature 245:465–466

Raftery MJ (2008) Enrichment by organomercurial agarose and identification of cys-containing peptides from yeast cell lysates. Anal Chem 80:3334–3341

Risley DS, Strege MA (2000) Chiral separations of polar compounds by hydrophilic interaction chromatography with evaporative light scattering detection. Anal Chem 72:1736–1739

Rivillas-Acevedo LA, Soriano-Garcia M (2007) Isolation and biochemical characterization of an antifungal peptide from Amaranthus hypochondriacus seeds. J Agric Food Chem 55:10156–10161

Rudenskaia GN (1994) Affinity chromatography of proteinases. Bioorg Khim 20:213–228

Sammer UF, Volksch B, Mollmann U, Schmidtke M, Spiteller P, Spiteller M, Spiteller D (2009) 2-amino-3-(oxirane-2,3-dicarboxamido)-propanoyl-valine, an effective peptide antibiotic from the epiphyte Pantoea agglomerans 48b/90. Appl Environ Microbiol 75:7710–7717

Sandra K, Moshir M, D'Hondt F, Tuytten R, Verleysen K, Kas K, Francois I, Sandra P (2009) Highly efficient peptide separations in proteomics. Part 2: bi- and multidimensional liquid-based separation techniques. J Chromatogr B Analyt Technol Biomed Life Sci 877:1019–1039

Santa T, Fukushima T, Ichibangase T, Imai K (2008) Recent progress in the development of derivatization reagents having a benzofurazan structure. Biomed Chromatogr 22:343–353

Santoni V, Bellini C, Caboche M (1994) Use of two-dimensional protein-pattern analysis for the characterization of Arabidopsis thaliana mutants. Planta 192:557–566

Shevchenko A, Jensen ON, Podtelejnikov AV, Sagliocco F, Wilm M, Vorm O, Mortensen P, Boucherie H, Mann M (1996) Linking genome and proteome by mass spectrometry: large-scale identification of yeast proteins from two dimensional gels. Proc Natl Acad Sci USA 93:14440–14445

Silva-Sanchez C, de la Rosa AP, Leon-Galvan MF, de Lumen BO, de Leon-Rodriguez A, de Mejia EG (2008) Bioactive peptides in amaranth (Amaranthus hypochondriacus) seed. J Agric Food Chem 56:1233–1240

Snyder LR, Dolan JW, Gant JR (1979) Gradient elution in high-performance liquid chromatography: I. Theoretical basis for reversed-phase systems. J Chromatogr A 165:3–30

Stasyk T, Huber LA (2004) Zooming in: fractionation strategies in proteomics. Proteomics 4:3704–3716

Strege MA, Stevenson S, Lawrence SM (2000) Mixed-mode anion-cation exchange/hydrophilic interaction liquid chromatography-electrospray mass spectrometry as an alternative to reversed phase for small molecule drug discovery. Anal Chem 72:4629–4633

Sykora C, Hoffmann R, Hoffmann P (2007) Enrichment of multiphosphorylated peptides by immobilized metal affinity chromatography using Ga(III)- and Fe(III)-complexes. Protein Pept Lett 14:489–496

Tang J, Gao M, Deng C, Zhang X (2008) Recent development of multi-dimensional chromatography strategies in proteome research. J Chromatogr B Analyt Technol Biomed Life Sci 866:123–132

Tolstikov VV, Fiehn O (2002) Analysis of highly polar compounds of plant origin: combination of hydrophilic interaction chromatography and electrospray ion trap mass spectrometry. Anal Biochem 301:298–307

Tsugita A, Kamo M (1999) 2-D electrophoresis of plant proteins. Methods Mol Biol 112:95–97

Twerenbold D, Gerber D, Gritti D, Gonin Y, Netuschill A, Rossel F, Schenker D, Vuilleumier JL (2001) Single molecule detector for mass spectrometry with mass independent detection efficiency. Proteomics 1:66–69

Vacek J, Klejdus B, Petrlova J, Lojkova L, Kuban V (2006) A hydrophilic interaction chromatography coupled to a mass spectrometry for the determination of glutathione in plant somatic embryos. Analyst 131:1167–1174

van Oss CJ, Good RJ, Chaudhury MK (1986) Nature of the antigen-antibody interaction. Primary and secondary bonds: optimal conditions for association and dissociation. J Chromatogr 376:111–119

Van Sluyter SC, Marangon M, Stranks SD, Neilson KA, Hayasaka Y, Haynes PA, Menz RI, Waters EJ (2009) Two-step purification of pathogenesis-related proteins from grape juice and crystallization of thaumatin-like proteins. J Agric Food Chem 57:11376–11382

Wagner K, Miliotis T, Marko-Varga G, Bischoff R, Unger KK (2002) An automated on-line multidimensional HPLC system for protein and peptide mapping with integrated sample preparation. Anal Chem 74:809–820

Walcher W, Timperio AM, Zolla L, Huber CG (2003) Characterization of a variant of the spinach PSII type I light-harvesting protein using kinetically controlled digestion and RP-HPLC-ESI-MS. Anal Chem 75:6775–6780

Wang HX, Ng TB (2005) An antifungal peptide from the coconut. Peptides 26:2392–2396

Wang SZ, Ding K, Lin SQ, Lin ZB (2007) Isolation, purification and structural analysis of GL-PP-3A, an active polysaccharide peptide from Ganoderma lucidum. Yao Xue Xue Bao 42:1058–1061

Wang S, Rao P, Ye X (2009) Isolation and biochemical characterization of a novel leguminous defense peptide with antifungal and antiproliferative potency. Appl Microbiol Biotechnol 82:79–86

Washburn MP, Wolters D, Yates JR III (2001) Large-scale analysis of the yeast proteome by multidimensional protein identification technology. Nat Biotechnol 19:242–247

Washburn MP, Ulaszek R, Deciu C, Schieltz DM, Yates JR III (2002) Analysis of quantitative proteomic data generated via multidimensional protein identification technology. Anal Chem 74:1650–1657

Witherup KM, Bogusky MJ, Anderson PS, Ramjit H, Ransom RW, Wood T, Sardana M (1994) Cyclopsychotride A, a biologically active, 31-residue cyclic peptide isolated from Psychotria longipes. J Nat Prod 57:1619–1625

Wolters DA, Washburn MP, Yates JR III (2001) An automated multidimensional protein identification technology for shotgun proteomics. Anal Chem 73(23):5683–5690

Xiang B, Du GH, Wang XC, Zhang SX, Qin XY, Kong JQ, Cheng KD, Li YJ, Wang W (2010) Elucidating the structure of two cyclotides of Viola tianshanica maxim by MALDI TOF/TOF MS analysis. Yao Xue Xue Bao 45:1402–1409

Yoshida T (1997) Peptide separation in normal phase liquid chromatography. Anal Chem 69:3038–3043

Yu LR, Zhu Z, Chan KC, Issaq HJ, Dimitrov DS, Veenstra TD (2007) Improved titanium dioxide enrichment of phosphopeptides from HeLa cells and high confident phosphopeptide identification by cross-validation of MS/MS and MS/MS/MS spectra. J Proteome Res 6:4150–4162

Zhang X, Fang A, Riley CP, Wang M, Regnier FE, Buck C (2010) Multi-dimensional liquid chromatography in proteomics – a review. Anal Chim Acta 664:101–113

Zhou M, Veenstra T (2008) Mass spectrometry: m/z 1983-2008. Biotechniques 44(667–668):670

Methods to Identify New Partners of Plant Signaling Peptides

Melinka A. Butenko, Markus Albert, and Reidunn B. Aalen

Abstract Over the last decade, there has been an increasing awareness of the importance of cell-to-cell communication in plants by peptide ligands and membrane-bound receptors due to striking aberrant phenotypes caused by synthetic peptides, overexpression of peptides, or mutations in peptide or receptor genes. The number of experimentally confirmed peptide-receptor signaling modules represents only a tiny fraction of the number of genes encoding putative ligands and receptors in *Arabidopsis thaliana*. Thus, a major challenge in plant biology is to identify new partners of plant signaling peptides. Here we present an overview of methods and prerequisites for ligand–receptor matchmaking with examples from current literature.

1 Introduction

More than 400 genes encoding proteins with receptor configuration (i.e., with an extracellular domain, a transmembrane domain, and a cytoplasmic kinase domain) have been identified by bioinformatic analyses in *Arabidopsis thaliana* (Shiu and Bleecker 2001). They can be classified into more than 21 structural classes by their extracellular domains. Eight classes, with more than 200 members in total, have extracellular domains with a variable number of leucine-rich repeats (LRRs) (Shiu and Bleecker 2001). The majority of these receptor-like kinases (RLKs) are orphans—without any identified ligand. Less than a dozen ligands have confirmed partners, and most of them relay their signals through class X and XI RLKs with more than 20 LRR domains (Butenko et al. 2009; Butenko and Aalen 2011).

M.A. Butenko • R.B. Aalen (✉)
Department of Molecular Biosciences, University of Oslo, Oslo, Norway
e-mail: reidunn.aalen@imbv.uio.no

M. Albert
Center for Plant Molecular Biology, University of Tübingen, Tübingen, Germany

H.R. Irving and C. Gehring (eds.), *Plant Signaling Peptides*,
Signaling and Communication in Plants 16, DOI 10.1007/978-3-642-27603-3_13,

The majority of the RLKs are without known function, despite the systematic effort to search for aberrant phenotypes of single, double, and even triple mutant lines with T-DNA insertions. A 194 LRR-RLKs have been under investigation in the *Arabidopsis* 2010 program (http://www4.ncsu.edu/~sclouse/Clouse2010objectives.htm), and a resulting 30 or so have been given a name, indicating that a function, mutant phenotype, or overexpression phenotype has been found. Obviously, a biological meaningful matching of receptors and ligands requires functional information—if a peptide and an RLK are involved in the same process, very similar mutant phenotypes are expected. The search for a missing member in a ligand–receptor pair thus requires a phenotype indicative of a biological function either on the side of the putative ligand or a receptor.

While the *Arabidopsis* RLKs are under systematic study, taking advantage of the publicly available collections of T-DNA insertion lines, the same approach is more difficult for putative peptide ligands for several reasons. Firstly, peptide ligands are small proteins, often <100 amino acids, and the encoding genes may not have been recognized as such, due to the cutoff values used for open reading frames (ORFs) in gene prediction programs. There are good reasons not to recognize all small ORFs as genes, as there are more than 30,000 ORFs in *Arabidopsis* potentially giving rise to polypeptides of 25–250 amino acids (Lease and Walker 2006). However, almost a 1,000 genes encoding putative secreted peptides were identified when all these ORFs were screened for the presence of a predicted N-terminal signal peptide (NSP), paralogs that could comprise a gene family, and homologues in other plant species (Lease and Walker 2006). A second problem is the lack of expression data for the majority of these potential peptide ligand genes. Since they have not been recognized as genes, they are also not present on microarray chips commonly used for genome-wide expression studies. Therefore, we do not know when and where the potential genes are expressed, and hence, we cannot pinpoint them as potential ligand candidates for receptors which function in specific tissues, processes, or developmental stages.

Fortunately, due to their important functions in plant growth and development, bioactive peptides that are potential RLK ligands have been discovered either by biochemical or genetic approaches, despite the lack of previous gene annotations and expression data. The biochemical approach has started out from observations that certain plant extracts have promoting or inhibiting effects on differentiation or cell proliferation in tissue cultures. Thereafter, the active substance has been isolated and purified, and the encoding gene has been identified. Examples are phytosulfokine (PSK), plant peptide containing sulfated tyrosine (PSY1), and tracheary element differentiation inhibitory factor (TDIF) (Ito et al. 2006; Matsubayashi and Sakagami 1996). The genetic approach has started out from mutants with phenotypes of interest and subsequent identification of the mutated gene, for example, *CLAVATA 3* (*CLV3*), *TAPETUM DETERMINANT 1* (*TPD1*), *INFLORESCENCE DEFICIENT IN ABSCISSION* (*IDA*), and *EPIDERMAL PATTERNING FACTORS* (*EPFs*) (Butenko et al. 2003; Clark et al. 1995; Hara et al. 2009; Hunt and Gray 2009; Yang et al. 2003). Having found one putative peptide ligand, additional potential ligands have been identified as paralogs by bioinformatics searches, which furthermore make it possible to analyze expression patterns. For these novel ligand candidates, receptors must be identified. In this chapter, we highlight methods for the identification of putative peptide ligand–receptor pairs and discuss potential pitfalls, as well as genetic and biochemical methods used for confirming ligand–receptor interactions.

2 Identification of Common Phenotypes

Mutation in a receptor and its ligand should give the same phenotype, as in both cases signaling is disrupted (Fig. 1a). Thus, if a scorable mutant phenotype for either the receptor or the putative ligand is present, it should be possible to identify the partner in a genetic screen. For instance, the class X LRR-RLK EXCESS MICROSPOROCYTES 1 (EMS1, also known as EXTRA SPOROGENOUS CELLS, EXS) was postulated to be the receptor of the cysteine-rich tapetum determinant 1 (TPD1) peptide and this on the basis of indistinguishable mutant phenotypes (Jia et al. 2008; Yang et al. 2003a, b). The pollen sacs (locules) contain excess microsporocytes and lack the tapetum layer resulting in complete male sterility. Although meiotic nuclear divisions occur in these microsporocytes, they do not complete normal meiotic divisions, resulting in the degeneration of microsporocytes and the presence of only cell debris.

Care has to be taken when dissecting phenotypes to make sure that they are defined unambiguously. Mutants can, for instance, have long roots for a number of reasons, like excessive cell elongation or excessive cell proliferation. On the other hand, the mutant phenotype of ligand and receptor need not be absolutely identical, if additional factors are involved. This is evident from the studies of the CLAVATA 3 (CLV3) ligand which is crucial for the regulation of the number of stem cells in the shoot apical meristem (SAM) (Fletcher and Meyerowitz 2000; Schoof et al. 2000). Mutation in the LRR-RLK *CLAVATA1* gene results in a phenotype similar to *clv3*, as is also the case for *CLAVATA2* which encodes a LRR receptor protein without a kinase domain (Kayes and Clark 1998). Based on these similarities, CLV3 was suggested to be the ligand of a CLV1/CLV2 heterodimer (Trotochaud et al. 1999).

However, *clv1* null alleles confer weaker phenotypes than both *clv3* and dominant negative mutations of *clv1*. This would be explicable if CLV3 is able to signal through additional RLKs. Consistent with this hypothesis, genetic screens have identified CORYNE (CRN), also known as SUPPRESSOR OF OVEREXPRESSION OF LIGAND-LIKE PROTEIN1 2 (SOL2), as a mediator of CLV3 signaling (Miwa et al. 2008; Muller et al. 2008). But even *clv1 crn* double mutants have slightly smaller inflorescence meristems and fewer carpels than single *clv3* mutants, suggesting the presence of yet an additional receptor complex that enables CLV3 signaling (Muller et al. 2008). Recently, an additional LRR-RLK, RECEPTOR-LIKE PROTEIN KINASE 2 (RPK2), also known as TOADSTOOL 2 (TOAD2), has been identified as a receptor candidate for relaying the CLV3 signal (Kinoshita et al. 2010; also see chapters “Peptides Regulating Apical Meristem Development” and “Peptides Regulating Plant Vascular Development”).

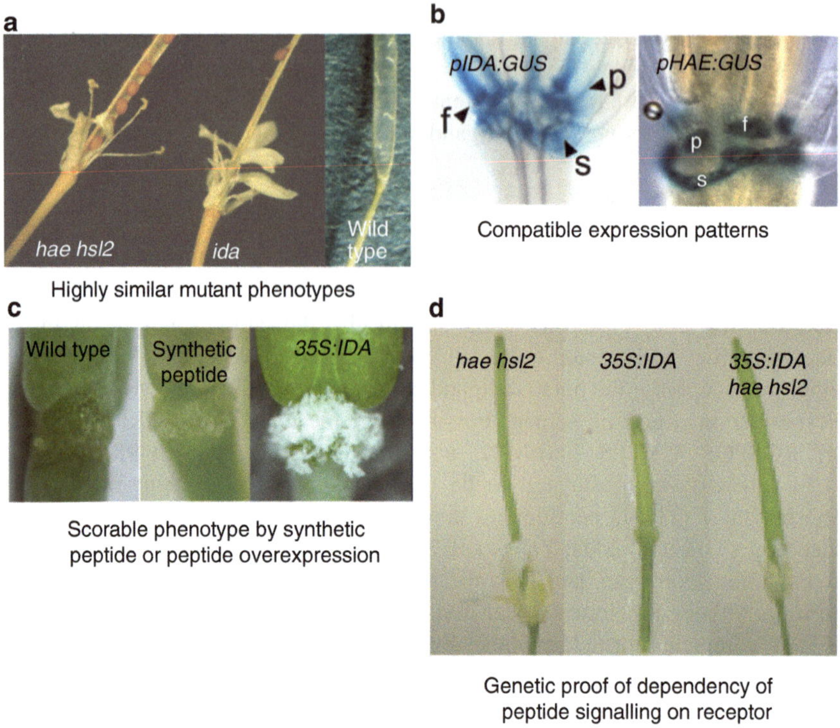

Fig. 1 Crucial steps in ligand–receptor matchmaking. (**a**) Highly similar mutant phenotypes are expected for the ligand and receptor genes, here illustrated with the total deficiency in abscission in *ida* and *hae hsl2* mutants as compared with *Arabidopsis* wild type. (**b**) The ligand and receptor genes encoding a signaling module are expected to have similar expression patterns that facilitate interaction between a secreted peptide and plasma-membrane-bound receptor, here illustrated by *pIDA;GUS* and *pHAE;GUS* in the abscission zone region. Abscission zones are found at the base of sepals (s), petals (p), and filaments (f). (**c**) Identification of a receptor to a ligand is easier if overexpression of the putative ligand has a scorable phenotype, here illustrated by precocious floral organ abscission and enlarged abscission zone cells secreting arabinogalactan protein by using the 35S cauliflower mosaic virus promoter to drive IDA expression (*35S:IDA*) or addition of a synthetic 20 amino acid peptide (10 μM) representing a conserved C-terminal motif in IDA. (**d**) Genetic proof of ligand–receptor interaction, here illustrated with the inability of *35S:IDA* to induce the overexpression phenotype as seen in (**c**) when the two receptor genes *HAE* and *HSL2* are mutated

The converse situation is also conceivable, namely, that the phenotype of the mutant receptor gene confers a stronger phenotype than mutations in the putative ligand gene. This is the case for the double mutants of the LRR-RLK genes *HAESA* (*HAE*) and *HAESA-LIKE2* (*HSL2*) and mutants of the putative ligand gene *IDA* that control the cell separation step during floral organ abscission. Although both *hae hsl2* and *ida* retain their floral organs indefinitely (Fig. 1a), these organs are more strongly attached in the receptor double mutant than the peptide mutant (Butenko et al. 2003; Stenvik et al. 2008). This may suggest that these receptors relay additional signals from peptides similar to *IDA*, for example, encoded by *IDA-LIKE* genes also expressed in the floral organ abscission zone.

3 Expression Patterns Compatible with Ligand–Receptor Interactions

Traditional genetic screen based on random mutagenesis and identification of lines with certain mutant phenotypes, followed by genetic mapping of the mutant gene, is a tedious and time-consuming task. A candidate approach based on compatible expression patterns can therefore be a useful alternative (Fig. 1b). This has been done successfully in the search for the receptor of TDIF which specifically inhibits the differentiation from procambial cells into tracheary elements (TEs) and promotes the procambium proliferation in the vascular system (Ito et al. 2006). T-DNA insertion lines for all of the 56 LRR-RLK genes predicted to be expressed in the procambium were screened for insensitivity to TDIF. This led to the identification of the TDIF receptor (TDR), a LRR-RLK belonging to subclass XI and which is identical to PHLOEM INTERCALATED WITH XYLEM (PXY) (Hirakawa et al. 2008).

The large number of RLKs, and the extensive overlap in their expression patterns, implies that a substantial number of RLKs may be candidate receptors for a single peptide. In addition, it is not trivial to determine what a compatible expressions pattern is for a ligand–receptor pair. From known pairs, several patterns have been found. *TPD1* and *EMS1* are expressed in neighboring cells, *TPD1* in the microsporocytes only, and *EMS1* predominantly in the tapetum (Jia et al. 2008; Yang et al. 2003a, b). Thus, the current model suggests that TPD1 is secreted from microsporocytes or their precursors and then binds to EMS1 localized on the surface of tapetum precursor cells. In the case of floral organ abscission, the expression pattern of *IDA* is broader but seemingly overlapping with that of *HAE* and *HSL2* which are found specifically in the abscission zone cells (Fig. 1b) (Cho et al. 2008; Stenvik et al. 2008). Furthermore, the *CLV3* gene and the genes encoding its receptors are not expressed in the same regions of the shoot apex; *CLV3* is expressed in the central zone (CZ) of the meristem, while *CLV1* for instance is expressed in a small group of cells just beneath the *CLV3* expression domain (Guo et al. 2010; Trotochaud et al. 1999).

Thus, it is not entirely obvious what a compatible expression pattern may be, but it can be useful as an initial approach. As less expression data is available for peptide genes, it becomes more complicated to use this approach to search for putative ligands to known receptors.

4 Detection of Peptide Localization and Mature Peptides

The nonoverlapping mRNA expression patterns between ligand and receptor can still be compatible with an interaction since peptide ligands are secreted and can move in the apoplastic space. Thus, the expression signal from promoter/reporter constructs may not reflect the localization of the peptide ligand itself. Constructs expressing the ligand with a detectable tag can therefore be useful. It has been demonstrated that a CLV3–GFP fusion protein under the control of the endogenous *CLV3* promoter has a broader cellular distribution than its mRNA and spreads outward and downward close to the organizing center (Lenhard and Laux 2003). The fusion construct could almost fully complement the mutant phenotype, and export of CLV3 was shown to be a prerequisite for function (Lenhard and Laux 2003; Sharma et al. 2003).

The use of peptide-reporter fusion proteins is, however, not without pitfalls. In planta, CLV3 is subjected to proteolytic cleavage both N- and C-terminally, resulting in 12-mer or 13-mer CLE peptides (Ito et al. 2006; Kondo et al. 2006; Ohyama et al. 2009). Hence, in theory, a C-terminal tag like green fluorescent protein (GFP) might be digested off the peptide and potentially have a localization that differs from that of the peptide itself. An N-terminal fusion protein is not an option due to the NSP. If the position of C-terminal processing is known, a tag fused directly to the processing site can be an alternative. A large tag on a small peptide may not necessarily be functional, and this must therefore be tested, for instance, by complementation of a mutant phenotype.

An alternative to a tag on the peptide could be detection by antibodies against the ligand itself. However, generation of functional antibodies against peptide ligands like CLV3 has proven difficult. This approach also requires knowledge of the mature peptide, which can only be reliably determined by isolation of the peptide from the plant. Several peptide ligands, PSK, PSY1, and TDIF, were in fact discovered as bioactive compounds isolated from plant tissue culture material (Ito et al. 2006; Matsubayashi and Sakagami 1996). Mature CLV3 peptides have furthermore been isolated from callus overexpressing CLV3 by in situ matrix-assisted laser desorption ionization time-of-flight mass spectrometry (MALDI-TOF MS) analysis (Kondo et al. 2006) and by nano-LC-MS/MS analysis of secreted peptides accumulating in culture medium derived from whole plants submerged in liquid medium (Ohyama et al. 2009). These methods require that the relevant processing enzymes are present and active under culturing conditions, which may not be the case for highly cell or tissue-specific peptides. A number of bioactive peptides have inhibitory effects on growth (Fiers et al. 2005), and sufficient amounts of plant material may be difficult to obtain.

5 Genetic Proofs of Ligand–Receptor Interactions

Basically, a genetic proof of a ligand–receptor relationship should demonstrate that the function of the ligand is dependent on a functional receptor. This requires a specific scorable phenotype for the active putative ligand, most often brought about by applying synthetic peptides to plants or cell cultures and/or by constitutive overexpression of the peptide (Fig. 1c). Together with compatible expression data and similar mutant phenotypes, genetic crosses between overexpression lines and receptor mutants showing loss of the overexpression phenotype or insensitivity of receptor mutants exposed to peptides provide evidence that point to ligand–receptor interactions. Mutations in the gene *IDA* and the genes encoding its putative receptors *HAE* and *HSL2* result in an indefinite retention of floral organs, which normally are abscised in *Arabidopsis* shortly after pollination has taken place (Butenko et al. 2003). Overexpression of *IDA* leads to premature abscission but not in *hae hsl2* background (Fig. 1d) (Cho et al. 2008; Stenvik et al. 2006, 2008), consistent with *IDA* being the ligand of *HAE* and *HSL2* controlling the cell separation step of the abscission process.

Such a genetic test shows that a given phenotype is dependent on a receptor but does not prove that the tested receptor is the native receptor. Overexpression of *IDA-LIKE* (*IDL*) genes, results in a very similar precocious abscission phenotype as for IDA overexpressor lines (Stenvik et al. 2008), and this IDL1, IDL2, and IDL3 overexpression phenotype is lost in the *hae hsl2* mutant background (unpublished results). The expression patterns of the *IDL* genes differ from that of *IDA*, and the severity of the overexpression phenotype is weaker for the IDL proteins which are more distantly related to IDA. Even though *HAE* and *HSL2* have a broader expression pattern than *IDA*, it is questionable whether the IDL proteins normally and/or only signals through these receptors. With high peptide concentrations and potential functional redundancy between related receptors, it is conceivable that the peptide phenotype results from an ectopic interaction not found in wild-type plants.

We have proposed that functional redundancy on the peptide side is mirrored by functional redundancy on the receptor side (Butenko et al. 2009). Such mutual redundancy has been confirmed for some of the CLE peptides related to CLV3. Overexpression of CLE14, CLE19, and CLE20 results in the consumption of the root meristematic cells, and CLE1, CLE6, and CLE7 trigger rapid SAM consumption, all in a *CLV2*-dependent manner (Fiers et al. 2005; Meng et al. 2010; also see chapters "Peptides Regulating Apical Meristem Development" and "Peptides Regulating Root Growth"). But *CLV2* may not necessarily be the native receptor, as CLE1 and CLE19 have been found to bind the related receptors CLV1, barely any meristem (BAM) 1 and (with exception for CLE1) BAM2 (Guo et al. 2010). *CLE1* is expressed in the pericycle and endodermis, as well as the primary root tip, *CLE6* in the shoot apex, *CLE19* in lateral root branch points, and native receptors would be expected to have an expression pattern compatible with the peptide gene.

A genetic proof of a putative interaction requires a rather specific overexpressor phenotype as in the case of *IDA* overexpressed in the abscission zone. For peptides that have a more general effect on growth and development, like dwarfed plants, an alternative is to generate lines with enhanced expression of the ligand using its own promoter. This can, for instance, be achieved with an inducible two-component system (Brand et al. 2006). If such a system generates a molecular or morphological phenotype, this phenotype will not appear in a cross when the right native receptor gene is mutated.

6 Bioassays and In Vivo Interaction Studies for Determining Ligand–Receptor Systems

6.1 Use of Bioassays as Indicators of Ligand–Receptor Interactions

Although the use of genetic screens to find putative interactions between peptide ligands and receptors is a powerful tool, the approach can prove tedious and time-consuming. Furthermore, the success of this procedure requires the availability of a discernable phenotype when both the receptor and ligand encoding genes are mutated, but this is often masked by genetic redundancy. A more direct approach to establish a link between a peptide ligand and its corresponding receptor is the use of bioassays where the biological activity of the active peptide ligand is measured by the effect it triggers in an organism or tissue cultures. Bioassays may be qualitative or quantitative; quantitative bioassays allow for an estimation of the concentration the active substance must have in order to achieve a measurable biological response in a dose-dependent manner.

The first plant peptide that was demonstrated to have signaling activity, systemin (Pearce et al. 1991), was identified by searching for the systemic inducing signal capable of promoting the expression of the proteinase inhibitor I and II genes in tomato leaf extracts (Pearce et al. 1991). In 1972, Green and Ryan showed by immunological techniques that an inhibitor of the animal proteinases chymotrypsin and trypsin accumulated in aerial tissues of tomato and potato plants as a direct consequence of insect damage or mechanical wounding (Green and Ryan 1972). This biological response was later used as an output measurement to identify the active substance causing the proteinase accumulation, an 18 amino acid peptide which, when synthetically synthesized and applied, was as effective as the native polypeptide, and showed activity at nanomolar levels (Pearce et al. 1991).

Systemin and several other pathogen elicitors have been shown to be capable of inducing an increase in 1-aminocyclopropane-1-carboxylate (ACC) synthase activity, the key enzymes of ethylene biosynthesis, and medium alkalinization in suspension-cultured tomato cells (Felix and Boller 1995). During alkalinization, a change of the proton gradient across the membrane is mediated though the inhibition of a plasma membrane H^+-ATPase leading to an influx of protons accompanied by an efflux of potassium ions, which can be measured as a change in pH (Felix and Boller 1995). This, in addition to an oxidative burst causing the transient production of large amounts of reactive oxygen species (ROS) (Doke 1983), has proven to be valuable for bioassays to identify both peptide ligand–receptor pairs involved in pathogen response as well as subdomains relevant for binding and receptor activation. The characterization of a high-affinity receptor binding site and its localization for the pathogen associated molecular pattern (PAMP) EF-Tu was identified by the detection of oxidative bursts and biosynthesis of ethylene in *Arabidopsis* leaves and in *Nicotiana benthamiana* plants transformed with the EF-Tu receptor (EFR) (Albert et al. 2010; Kunze et al. 2004; Zipfel et al. 2006).

The bioassays developed for identifying components of the innate immune system in plants have recently also proven beneficial for the identification of endogenous peptides capable of interacting with receptors involved in immune signaling. It has recently been reported that CLV3p, the mature hydroxyprolinated CLV3 peptide, is capable of interacting with the LRR-RLK FLAGELLIN SENSITIVE 2 (FLS2) to induce the same defense response as the bacterial elicitor flg22 (Lee et al. 2011). However, the CLV3p-mediated SAM arrest and root growth inhibition are not dependent on *FLS2* and are uncoupled from flg22-triggered growth suppression (Lee et al. 2011). This opens the possibility that other secreted plant peptides may bind additional receptors involved in innate immunity.

Although the bioassays developed will make ligand–receptor matching easier, it is not completely unproblematic. When monitoring the bioactivity of peptide ligands in various developmental responses, such as the involvement of CLE41/44 in suppression of xylem development shown in *Zinnia* culture as a bioassay (Ito et al. 2006) or the consumption of the SAM by CLV3 (Kondo et al. 2006) using liquid plant cultures, it became apparent that full activity of the peptides and high-affinity binding to the corresponding receptor(s) requires posttranslational modification, such as glycosylation (Ohyama et al. 2009). The necessity of post-translational modifications for full activity of peptides involved in development may make it difficult to detect responses in bioassays based on immune responses using synthetic peptides, as it is not possible to synthetically introduce these modifications.

In situations where there is genetic evidence for peptide ligand–receptor interaction but no well-developed bioassay to substantiate the results, the option of chimeric receptors can prove useful. By using the kinase domain of a receptor known to give a readable biological output, such as an oxidative burst, to the ectodomain of the putative interacting partner, it is possible to monitor a response by applying the peptide in question. Chimeric receptors also allow the mapping of subdomains relevant for ligand binding and receptor activation (Albert et al. 2010).

6.2 Yeast-Two-Hybrid Interactions to Determine Receptor Binding Sites

Another approach that has proven beneficial to determine ligand–receptor interactions is yeast-two-hybrid (Y-2-H) system experiments followed by pulldown and co-immunoprecipitation. This can be used both to verify an interaction between a putative peptide ligand and its receptor and to determine the site on the ectodomain responsible for ligand binding. As previously mentioned, mutations in the *ems1* receptor gene and the *tpdl* ligand gene generate anthers that lack the tapetum layer but produce more microsporocytes at the expense of tapetal cells (Jia et al. 2008). Genetic studies demonstrated that TPD1 signaling requires functional EMS1 (Jia et al. 2008). To show a direct biochemical interaction between TPD1 and EMS1, a yeast-two-hybrid experiment using the entire EMS1 ectodomain was performed; however, no interaction was detected. There are various reasons why yeast-two-hybrid experiments may fail when trying to infer protein–protein interaction between a putative peptide ligand and its receptor. Firstly, the correct folding of membrane anchored proteins might not occur when localized to the nucleus. Secondly, potential *N*-linked glycosylation present on the LRRs of most RLKs may not be correctly synthesized in yeast and therefore may hamper protein interaction. However, using smaller fragments of the ectodomain as bait in the yeast-two-hybrid system might overcome this problem. Indeed, when using a fragment containing only four typical LRR repeats of EMS1, an interaction with TPD1 was detected (Jia et al. 2008).

A yeast-two-hybrid strategy has also successfully been used for identifying peptide ligands for orphan receptors with known biological functions. Pollen receptor kinases of tomato, LePRK1, 2, and 3, believed to play a role in pollen germination and pollen tube growth, were used as baits in a yeast-two-hybrid screen. This lead to the identification of the cysteine-rich protein LAT52, known to be essential during pollen hydration and pollen tube growth (Tang et al. 2002).

Although yeast-two-hybrid screens and other biochemical methods to show protein–protein interactions have been successfully used to identify interactions between peptide ligands and their receptors, methods employing competitive binding assays with labeled active peptides are far more commonly used and provide more accurate results.

7 Biochemical Proof of Interaction

7.1 Labeled Peptide Ligands for Determining Receptor Binding Properties

After performing genetic screens and obtaining receptor candidates, a direct physical interaction between a ligand and its specific receptor has to be proven. To demonstrate peptide ligand binding to a corresponding receptor, radiolabeled peptides carrying a ^{3}H or ^{125}I isotope on a histidine or tyrosine residue have served as important and helpful tools (Mayers and Klostergaard 1983). The first successful applications of such a binding assay to study plant receptor–peptide ligand interactions were shown for FLS2 and its bacterial flagellin-derived peptide ligand flg22 (Gomez-Gomez et al. 2001). This approach has further been used to prove the interaction of endogenous plant peptides with their receptors biochemically, such as CLV3 with the LRR-RLK CLV1 (Ogawa et al. 2008) or PSK with its receptor from *Daucus carota* DcPSKR1 (Matsubayashi et al. 2002). To determine specific binding, competition by a corresponding unlabeled ligand in a high excess is necessary. When such a competition works, the radioactive labeling associated with the receptor will be significantly lower in samples with competitor compared to those without, and only such differences are diagnostic for specific binding. Subsequent, additional studies with a stepwise increased unlabeled competitor lead to a competition curve, from which the dissociation constant (K_d) for the ligand can be deduced and the half maximal inhibitor concentration (IC_{50}) for antagonists. Both values are important measures of the affinity of a ligand for its receptor and customarily obtained in biochemistry and pharmacology.

Apart from radiolabeled ligands, additional alternative labeling techniques have become available, for example, peptides labeled with an acridinium ester that can be used in a chemiluminescent approach. This type of assay was first established to study cytokine–receptor interactions (Joss and Towbin 1994) and was compared to binding studies using iodinated ligands. Due to its high sensitivity and the ability to detect bound labeled molecules in a range of 1–10 atomol, this assay might serve as a valuable alternative to radioactive methods. However, this assay still needs to be adapted for studying plant receptor–peptide interactions. In addition, fluorescently labeled peptides could also serve in studies of receptor–ligand interactions. Although this labeling technique has not been used to monitor a direct ligand–receptor interaction in vivo, promising experiments have been performed. Recently, Goto et al. (2011) reported the successful labeling of the pollen attracting peptide LURE2 with the Alexa488 fluorescence dye and could demonstrate that this peptide was still active as a pollen tube attractor. Earlier, Alexa488 labeled phytosulfokine (Alexa488-PSK) was used to visualize the binding of this peptide to its receptor DcPSKR1 (Shinohara and Matsubayashi 2007). Following overexpression and extraction of C-terminally Halo-tagged receptors, they were immobilized to a solid phase (HaloLink resin, Promega, www.promega.com). These sepharose beads with covalently bound receptors were incubated with the fluorescently labeled ligand Alexa488-PSK, and it was possible to monitor a specific binding of the ligand to the beads via confocal laser scanning microscopy. The authors further suggested that this method would be suitable for proving more ligand–receptor interactions and for identifying corresponding receptors of known potential peptide ligands. However, due to the large number of candidates (e.g., 600 in *A. thaliana* and 1,200 in rice), this method might only be helpful if interaction partners have already been identified genetically or by other methods (see Sects. 5 and 6) that limit the number of candidates. Otherwise, the need for purified, tagged, and immobilized receptors might be too labor intensive and time consuming. However, as the detection of fluorescence becomes more and more sensitive, Alexa Fluor-labeled peptides, for example, could substitute for radioactive-labeled peptides and turn out to be the tool of choice for future for receptor–ligand studies.

7.2 *Cross-Linking of Labeled Ligands: Tools to Identify Receptor Proteins?*

An alternative or complementary technique to those described above is chemical cross-linking of ligands to their corresponding binding proteins. This method might be useful for finding new receptors of known active peptides or for identifying the potential binding site within a receptor.

So far, cross-linking has been successfully applied to label specific peptide binding sites for the PAMP flagellin (represented by the peptide flg22) (Chinchilla et al. 2006) or to visualize the interaction of the fungal xylanase with its plant receptor EIX1/2 (Hanania and Avni 1997). Furthermore, chemical cross-linking of a ^{125}I labeled 13 amino acid long peptide of *Phytophthora megasperma* f. sp. *glycinea* (Pep-13) was successful, and the authors could identified a 91-kDa parsley protein as a potential receptor (Nürnberger et al. 1995).

Endogenous plant peptides have also been used to study the interaction with corresponding receptor proteins. In a few cases, the cross-linking technique even helped to isolate and to identify the specific binding proteins. For the *A. thaliana* 23 amino acid peptide AtPep1, an ^{125}I-labeled azido-Cys-AtPep1 photoaffinity analog was used (Yamaguchi et al. 2006). Photoaffinity labeling utilizes photoactivatable groups which can be activated by UV light and allow unspecific covalent connections between the ligand and the interacting protein (Dorman and Prestwich 1994). Theoretically, the advantage of this method is that the ligand is cross-linked to the protein only at positions where it is specifically bound, as it is only here that the low distance between the interaction partners facilitates a connection by the cross-linker. The AtPep1 peptide carried an UV-light activatable cross-linker and was utilized to label a 170-kDa membrane-bound protein in *A. thaliana* cell suspension culture. The analysis of the tryptic peptides led to the identification of the LRR receptor kinase AtPEPR1. Further studies confirmed that AtPEPR1 is the receptor of AtPep1. Similarly, the phytosulfokine receptor PSKR1 (Matsubayashi and Sakagami 1996) was isolated from carrot microsomal fractions and identified after ligand-based affinity chromatography (Matsubayashi et al. 2002). The *D. carota* phytosulfokine receptor (DcPSKR1) belongs to the LRR-RLK receptors, has a size of ~120 kDa, and contains 21 LRRs with an island domain of 36 amino acids between LRR 17 and 18. This is so far the single case where cross-linking helped to provide valuable information about the PSK ligand binding domain within the extracellular LRR domain of the receptor. After cross-linking of the photoactivatable PSK analog [^{125}I]-[*N*-(4-azidosalicyl)Lys^5] with DcPSKR1 using UV irradiation, the cross-linked region was narrowed down following chemical and enzymatic fragmentation. The labeled tryptic fragment of DcPSKR1 was isolated via anti-PSK antibodies and further analyzed via mass spectrometry. The deduced sequence showed clear hits for the protein sequence of the 36 amino acid extracellular island domain flanked by LRRs 17 and 18 within the receptor LRR domain (Shinohara and Matsubayashi 2007; Shinohara et al. 2007). The importance of this PSKR1 island region for PSK binding was further proven with a PSKR1-ΔID receptor which lacks the island domain and was not able to bind the PSK ligand.

These results look promising and imply that this method is an easy and powerful tool for finding and identifying receptors with photoactivatable and labeled ligands. On the other hand, many more receptors should have been identified for known ligands if cross-linking is indeed easy to handle and reliably leads to clear results. Unfortunately, it is often quite difficult to identify the cross-linked residues by mass spectrometry analysis due to the low efficiency of cross-linking reactions and the low amounts of receptor proteins present in cells. The number of flg22 binding sites on *A. thaliana* cells for instance was estimated at 1.6 pmol per g of tissue (Bauer et al. 2001). It is also problematic as apparently almost all receptor proteins are posttranslationally modified. In particular, glycosylations of unknown size and number and the composition of unknown sugar monomers drastically complicate the mass spectrometry analysis. Therefore, this method might not be applicable to in all cases of receptor–ligand studies and it might also not always lead to conclusive results as described for the DcPSKR1–PSK interaction.

8 Future Perspectives

It is to be expected that the strategies for matching peptide ligands with receptors discussed here will reveal that fundamental processes in plants are regulated by peptide signals. It will, however, take a substantial amount of effort before novel receptors for the predicted orphan plant peptide ligands have been found and their interactions characterized. The findings that closely related peptide ligands are likely to interact with families of closely related RLKs (Butenko et al. 2009) might make the task easier, as it may facilitate the discovery of candidate receptors for members of peptide families. On the other hand, recent evidence indicates there may not necessarily be a one-to-one relationship between ligands and receptors since some ligands may bind several receptors and some receptors more than one ligand (Butenko and Aalen 2011; De Smet et al. 2009; Lee et al. 2011). Therefore, the development of new high-throughput methodology will be welcome, for instance, based on ligand fishing by the ectodomain of RLKs from complex natural samples or synthetic peptides. A starting point for such a technology could possibly be Halo-tagged extracellular domains immobilized on microbeads that has been used for showing direct interaction with peptide ligands (Shinohara and Matsubayashi 2007). Interactions in such a system will need confirming in planta as well as genetic verification.

The few peptide ligand–receptor pairs identified to date have given exciting new insights into the molecular mechanisms underlying plant processes. We anticipate further truly rewarding revelations in this field in the years ahead.

References

Albert M, Jehle AK, Lipschis M, Mueller K, Zeng Y, Felix G (2010) Regulation of cell behaviour by plant receptor kinases: pattern recognition receptors as prototypical models. Eur J Cell Biol 89:200–207

Bauer Z, Gomez-Gomez L, Boller T, Felix G (2001) Sensitivity of different ecotypes and mutants of *Arabidopsis thaliana* toward the bacterial elicitor flagellin correlates with the presence of receptor-binding sites. J Biol Chem 276:45669–45676

Brand L, Horler M, Nuesch E, Vassalli S, Barrell P, Yang W, Jefferson RA, Grossniklaus U, Curtis MD (2006) A versatile and reliable two-component system for tissue-specific gene induction in *Arabidopsis*. Plant Physiol 141:1194–1204

Butenko MA, Aalen RB (2011) Receptor ligands in development. In: Kemmerling B, Tax F (eds) Receptor-like kinases in plants: from development to defense. Springer, Berlin

Butenko MA, Patterson SE, Grini PE, Stenvik GE, Amundsen SS, Mandal A, Aalen RB (2003) *INFLORESCENCE DEFICIENT IN ABSCISSION* controls floral organ abscission in *Arabidopsis* and identifies a novel family of putative ligands in plants. Plant Cell 15:2296–2307

Butenko MA, Vie AK, Brembu T, Aalen RB, Bones AM (2009) Plant peptides in signalling: looking for new partners. Trends Plant Sci 14:255–263

Chinchilla D, Bauer Z, Regenass M, Boller T, Felix G (2006) The *Arabidopsis* receptor kinase FLS2 binds flg22 and determines the specificity of flagellin perception. Plant Cell 18:465–476

Cho SK, Larue CT, Chevalier D, Wang H, Jinn TL, Zhang S, Walker JC (2008) Regulation of floral organ abscission in *Arabidopsis thaliana*. Proc Natl Acad Sci USA 105:15629–15634

Clark SE, Running MP, Meyerowitz EM (1995) *CLAVATA3* is a specific regulator of shoot and floral meristem development affecting the same processes as *CLAVATA1*. Development 121:2057–2067

De Smet I, Voss U, Jurgens G, Beeckman T (2009) Receptor-like kinases shape the plant. Nat Cell Biol 11:1166–1173

Doke N (1983) Generation of superoxide anion by potato tuber protoplasts during the hypersensitive response to hyphal wall components of *Phytophthora infestans* and specific inhibition of the reaction by suppressors of hypersensitivity. Physiol Plant Pathol 23:359–367

Dorman G, Prestwich GD (1994) Benzophenone photophores in biochemistry. Biochemistry 33:5661–5673

Felix G, Boller T (1995) Systemin induces rapid ion fluxes and ethylene biosynthesis in *Lycopersicon peruvianum* cells. Plant J 7:381–389

Fiers M, Golemiec E, Xu J, van der Geest L, Heidstra R, Stiekema W, Liu CM (2005) The 14-amino acid CLV3, CLE19, and CLE40 peptides trigger consumption of the root meristem in *Arabidopsis* through a *CLAVATA2*-dependent pathway. Plant Cell 17:2542–2553

Fletcher JC, Meyerowitz EM (2000) Cell signaling within the shoot meristem. Curr Opin Plant Biol 3:23–30

Gomez-Gomez L, Bauer Z, Boller T (2001) Both the extracellular leucine-rich repeat domain and the kinase activity of FSL2 are required for flagellin binding and signaling in *Arabidopsis*. Plant Cell 13:1155–1163

Goto H, Okuda S, Mizukami A, Mori H, Sasaki N, Kurihara D, Higashiyama T (2011) Chemical visualization of an attractant peptide, LURE. Plant Cell Physiol 52:49–58

Green TR, Ryan CA (1972) Wound-induced proteinase inhibitor in plant leaves: a possible defense mechanism against insects. Science 175:776–777

Guo YF, Han LQ, Hymes M, Denver R, Clark SE (2010) CLAVATA2 forms a distinct CLE-binding receptor complex regulating *Arabidopsis* stem cell specification. Plant J 63:889–900

Hanania U, Avni A (1997) High-affinity binding site for ethylene-inducing xylanase elicitor on *Nicotiana tabacum* membranes. Plant J 12:113–120

Hara K, Yokoo T, Kajita R, Onishi T, Yahata S, Peterson KM, Torii KU, Kakimoto T (2009) Epidermal cell density is autoregulated via a secretory peptide, EPIDERMAL PATTERNING FACTOR 2 in *Arabidopsis* leaves. Plant Cell Physiol 50:1019–1031

Hirakawa Y, Shinohara H, Kondo Y, Inoue A, Nakanomyo I, Ogawa M, Sawa S, Ohashi-Ito K, Matsubayashi Y, Fukuda H (2008) Non-cell-autonomous control of vascular stem cell fate by a CLE peptide/receptor system. Proc Natl Acad Sci USA 105:15208–15213

Hunt L, Gray JE (2009) The signaling peptide EPF2 controls asymmetric cell divisions during stomatal development. Curr Biol 19:864–869

Ito Y, Nakanomyo I, Motose H, Iwamoto K, Sawa S, Dohmae N, Fukuda H (2006) Dodeca-CLE peptides as suppressors of plant stem cell differentiation. Science 313:842–845

Jia G, Liu X, Owen HA, Zhao D (2008) Signaling of cell fate determination by the TPD1 small protein and EMS1 receptor kinase. Proc Natl Acad Sci USA 105:2220–2225

Joss UR, Towbin H (1994) Acridinium ester labelled cytokines: receptor binding studies with human interleukin-1 alpha, interleukin-1 beta and interferon-gamma. J Biolumin Chemilumin 9:21–28

Kayes JM, Clark SE (1998) *CLAVATA2*, a regulator of meristem and organ development in *Arabidopsis*. Development 125:3843–3851

Kinoshita A, Betsuyaku S, Osakabe Y, Mizuno S, Nagawa S, Stahl Y, Simon R, Yamaguchi-Shinozaki K, Fukuda H, Sawa S (2010) RPK2 is an essential receptor-like kinase that transmits the CLV3 signal in *Arabidopsis*. Development 137:3911–3920

Kondo T, Sawa S, Kinoshita A, Mizuno S, Kakimoto T, Fukuda H, Sakagami Y (2006) A plant peptide encoded by CLV3 identified by in situ MALDI-TOF MS analysis. Science 313:845–848

Kunze G, Zipfel C, Robatzek S, Niehaus K, Boller T, Felix G (2004) The N terminus of bacterial elongation factor Tu elicits innate immunity in Arabidopsis plants. Plant Cell 16:3496–3507

Lee H, Chah OK, Sheen J (2011). Stem-cell-triggered immunity through CLV3p-FLS2 signalling. Nature 473:376–379.

Lease KA, Walker JC (2006) The *Arabidopsis* unannotated secreted peptide database, a resource for plant peptidomics. Plant Physiol 142:831–838

Lenhard M, Laux T (2003) Stem cell homeostasis in the *Arabidopsis* shoot meristem is regulated by intercellular movement of CLAVATA3 and its sequestration by CLAVATA1. Development 130:3163–3173

Matsubayashi Y, Sakagami Y (1996) Phytosulfokine, sulfated peptides that induce the proliferation of single mesophyll cells of *Asparagus officinalis L.* Proc Natl Acad Sci USA 93:7623–7627

Matsubayashi Y, Ogawa M, Morita A, Sakagami Y (2002) An LRR receptor kinase involved in perception of a peptide plant hormone, phytosulfokine. Science 296:1470–1472

Mayers GL, Klostergaard J (1983) The use of protein A in solid-phase binding assays: a comparison of four radioiodination techniques. J Immunol Methods 57:235–246

Meng L, Ruth KC, Fletcher JC, Feldman L (2010) The roles of different CLE domains in *Arabidopsis* CLE polypeptide activity and functional specificity. Mol Plant 3:760–772

Miwa H, Betsuyaku S, Iwamoto K, Kinoshita A, Fukuda H, Sawa S (2008) The receptor-like kinase SOL2 mediates CLE signaling in *Arabidopsis*. Plant Cell Physiol 49:1752–1757

Muller R, Bleckmann A, Simon R (2008) The receptor kinase CORYNE of *Arabidopsis* transmits the stem cell-limiting signal CLAVATA3 independently of CLAVATA1. Plant Cell 20:934–946

Nürnberger T, Nennstiel D, Hahlbrock K, Scheel D (1995) Covalent cross-linking of the Phytophthora megasperma oligopeptide elicitor to its receptor in parsley membranes. Proc Natl Acad Sci USA 92:2338–2342

Ogawa M, Shinohara H, Sakagami Y, Matsubayashi Y (2008) *Arabidopsis* CLV3 peptide directly binds CLV1 ectodomain. Science 319:294

Ohyama K, Shinohara H, Ogawa-Ohnishi M, Matsubayashi Y (2009) A glycopeptide regulating stem cell fate in *Arabidopsis thaliana*. Nat Chem Biol 5:578–580

Pearce G, Strydom D, Johnson S, Ryan CA (1991) A polypeptide from tomato leaves induces wound-inducible proteinase inhibitor proteins. Science 253:895–897

Schoof H, Lenhard M, Haecker A, Mayer KFX, Jürgens G, Laux T (2000) The stem cell population of *Arabidopsis* shoot meristems is maintained by a regulatory loop between the CLAVATA and WUSCHEL genes. Cell 100:635–644

Sharma VK, Ramirez J, Fletcher JC (2003) The *Arabidopsis* CLV3-like (CLE) genes are expressed in diverse tissues and encode secreted proteins. Plant Mol Biol 51:415–425

Shinohara H, Matsubayashi Y (2007) Functional immobilization of plant receptor-like kinase onto microbeads towards receptor array construction and receptor-based ligand fishing. Plant J 52:175–184

Shinohara H, Ogawa M, Sakagami Y, Matsubayashi Y (2007) Identification of ligand binding site of phytosulfokine receptor by on-column photoaffinity labeling. J Biol Chem 282:124–131

Shiu SH, Bleecker AB (2001) Receptor-like kinases from *Arabidopsis* form a monophyletic gene family related to animal receptor kinases. Proc Natl Acad Sci USA 98:10763–10768

Stenvik G-E, Butenko MA, Urbanowicz BR, Rose JK, Aalen RB (2006) Overexpression of INFLORESCENCE DEFICIENT IN ABSCISSION activates cell separation in vestigial abscission zones in *Arabidopsis*. Plant Cell 18:1467–1476

Stenvik G-E, Tandstad NM, Guo Y, Shi C-L, Kristiansen W, Holmgren A, Clark SE, Aalen RB, Butenko MA (2008) The EPIP peptide of INFLORESCENCE DEFICIENT IN ABSCISSION is sufficient to induce abscission in *Arabidopsis* through the receptor-like kinases HAESA and HAESA-LIKE2. Plant Cell 20:1805–1817

Tang W, Ezcurra I, Muschietti J, McCormick S (2002) A cysteine-rich extracellular protein, LAT52, interacts with the extracellular domain of the pollen receptor kinase LePRK2. Plant Cell 14:2277–2287

Trotochaud AE, Hao T, Wu G, Yang Z, Clark SE (1999) The CLAVATA1 receptor-like kinase requires CLAVATA3 for its assembly into a signaling complex that includes KAPP and a Rho-related protein. Plant Cell 11:393–406

Yamaguchi Y, Pearce G, Ryan CA (2006) The cell surface leucine-rich repeat receptor for AtPep1, an endogenous peptide elicitor in *Arabidopsis*, is functional in transgenic tobacco cells. Proc Natl Acad Sci USA 103:10104–10109

Yang S-L, Xie L-F, Mao H-Z, Puah CS, Yang W-C, Jiang L, Sundaresan V, Ye D (2003) TAPETUM DETERMINANT1 is required for cell specialization in the *Arabidopsis* anther. Plant Cell 15:2792–2804

Zipfel C, Kunze G, Chinchilla D, Caniard A, Jones JD, Boller T, Felix G (2006) Perception of the bacterial PAMP EF-Tu by the receptor EFR restricts Agrobacterium-mediated transformation. Cell 125:749–760

Computational-Based Analysis to Associate the Function of Plant Signaling Peptides with Distinct Biological Processes

Stuart Meier and Lara Donaldson

Abstract Over 1,000 putative secreted peptides have been identified in the *Arabidopsis* genome, but little is known about their functional roles. Here we demonstrate, using PROPEP2 in a worked example, how a computational-based co-expression analysis can be used to associate the function of a specific plant peptide with a specific biological process. The analysis reveals that PROPEP2 is, highly co-expressed with pathogen defense genes including, the PROPEP3 paralog and the extracellular Pep1 and Pep2 receptors. It additionally reveals that the co-expressed genes are induced in response to a broad range of pathogens and their associated elicitors. The results are completely supportive of previously known functional roles of this peptide and thus clearly demonstrate that the computational-based analysis presented can be used to associate the function of a specific plant peptide with a specific biological process and can also accurately identify functionally related genes.

1 Introduction

It has become increasingly accepted that plant peptides can function as endogenous phytohormones that have important signaling roles in numerous biological processes including reproduction, growth, development, and defense responses (Matsubayashi and Sakagami 2006; Ryan and Pearce 2010). A high proportion of plant peptides are secreted out of the cell and mediate both local and long distance intercellular communication (Lindsey et al. 2002).

Secreted plant peptides are initially synthesized as precursor pre-propeptides that are processed by the removal of the N-terminal signal sequence that functions

S. Meier (✉) • L. Donaldson
Division of Chemistry, Life Science and Engineering, King Abdullah University of Science and Technology, Thuwal 23955-6900, Kingdom of Saudi Arabia
e-mail: stuart.meier@kaust.edu.sa

H.R. Irving and C. Gehring (eds.), *Plant Signaling Peptides*,
Signaling and Communication in Plants 16, DOI 10.1007/978-3-642-27603-3_14,

to direct their extracellular secretion, resulting in the synthesis of the prohormone (Wheeler and Irving 2010; also see chapters “Plant Peptide Signaling: An Evolutionary Adaptation,” “Processing of Peptides,” and “Methods to Identify New Partners of Plant Signaling Peptides”). The prohormones are then packaged into membrane-bound secretary vesicles and can be exocytosed into the apoplast in response to specific stimuli. Some propeptides require further processing before being secreted in order to yield the mature active peptide hormone (Matsubayashi 2011). Secreted peptides have been shown to bind to specific extracellular localized receptors of the leucine-rich repeat receptor-like kinase (LRR-RLK) class, the largest class of receptors in plants (Butenko et al. 2009).

In the majority of cases, peptide hormones are encoded by single exon genes with small open reading frames (<200 amino acids) and are characterized by the presence of an N-terminal signal sequence that direct their extracellular secretion (Matsubayashi and Sakagami 2006; Wheeler and Irving 2010). In *Arabidopsis*, over 1,000 genes have been predicted to encode secreted peptides (Lease and Walker 2006); however, few of these mature peptides have been biochemically characterized hampering the identification of their target receptors (Matsubayashi and Sakagami 2006). In addition, the biological processes in which many of the predicted plant peptides function are unknown.

2 Co-expression to Infer Gene Function

Gene expression analysis has proved useful in associating the function of a gene to a specific cellular response (also see chapters “Plant Peptide Signaling: An Evolutionary Adaptation” and “Methods to Identify New Partners of Plant Signaling Peptides”). A gene that is determined to be differentially expressed in response to a specific stimuli (e.g: stress) can be functionally linked to a specific biological response. With the advent of full genome transcriptomic technologies, such as microarray, the entire transcriptome of an organism can be studied in a single experiment providing valuable experimentally derived transcriptional data for thousands of genes.

A range of computational-based tools have been developed to analyze and identify expression patterns across microarray experiments. Genome-wide expression correlation analysis has been shown to be very useful in identifying networks of co-expressed genes that function coherently to achieve a common biological response. In eukaryotes, it is well established that cellular responses to changing environmental conditions are rarely mediated by the regulation and action of a single gene or protein. Rather, they require the coordinated participation of a network of often structurally diverse proteins, including receptors, signaling molecules, transporters, and biosynthesis genes (Sweetlove and Fernie 2005). This is supported by observations of changes in expression of hundreds or thousands of genes in response a specific stimuli in transcriptomics studies (Eisen et al. 1998; Tamayo et al. 1999; Tavazoie et al. 1999).

A number of studies have shown that genes that have been confirmed to be co-expressed in response to a large number of diverse conditions have correlated functional relationships, including physical interactions between their proteins (Allocco et al. 2004; Stuart et al. 2003; Jansen et al. 2002). Further, genes encoding metabolic enzymes that function within the same biosynthesis pathway have been shown to form co-expression modules (Wei et al. 2006; Ihmels et al. 2004; Meier et al. 2011). Co-expression analysis performed on a genome-wide scale can therefore be used to identify networks of genes that function in common biological processes. If a gene of interest (GOI) has an unknown function and is found to be highly co-expressed with a gene(s) that has a well-defined role in a specific biological process, the function of the unknown gene can be associated with this process. Alternatively, if a GOI that functions in a well-defined cellular process is used, a co-expression analysis can be used to identify unknown genes that share close functional relationships.

It is relevant to note here that changes in gene transcription do not necessarily translate to changes in protein abundance and cellular activity due to posttranscriptional regulatory mechanisms. However, as these mechanisms rely on a gene being transcribed in the first place, gene transcription can be considered the primary level of regulation of protein synthesis. Further, while cells can alter protein/enzyme activity to fine-tune cellular responses, the protein must be present at appropriate quantities for this to occur. Thus, changes in gene transcription in response to specific stimuli can be considered a primary regulatory response that reflects a change in requirement for a specific protein(s) at a specific point in time. Furthermore, when a group of functionally related genes are confirmed to be co- expressed, it is a strong indicator that their expression is in fact co-regulated.

Here, we demonstrate how the use of a number of openly available Web-based tools can be used to associate the function of an unknown protein/gene or, specifically in this case, a plant peptide encoding gene, with a specific biological process. In addition, the methodology can identify factors that are important in regulating the genes' transcription. We perform the analysis using the *ELICITOR PEPTIDE 2 PRECURSOR* (*PROPEP2*) gene (AT5G64890) that encodes a precursor peptide that belongs to a paralog gene family that has established roles in plant pathogen defense signaling. We demonstrate the use of this methodology on a gene with a known function so the results obtained can be compared to experimental-based literature to verify the accuracy of the analysis. We have previously used this methodology to relate the function of the *Arabidopsis* plant natriuretic peptide (Meier et al. 2008; also see chapter "Peptides and the Regulation of Plant Homeostasis") and wall-associated kinase-like 10 (Meier et al. 2010) to specific biological processes.

3 PROPEPs and Defense Signaling

The PROPEP family of peptides have been shown to function as endogenous amplifiers of innate immune defense responses through a microbe-associated molecular patterns (MAMP) initiated feedback loop which amplifies signaling through both the jasmonic acid/ethylene and salicylic acid signaling pathways and induces the transcription of defense-related genes (Huffaker and Ryan 2007; Huffaker et al. 2006; Yamaguchi et al. 2010). These endogenous elicitors are also referred to as damage-associated molecular patterns (DAMPs).

The Peps are synthesized from the C-terminal portion of their precursor proteins (Huffaker et al. 2006), and following processing, the Pep proteins are secreted into the apoplast where they specifically interact with the cell-surface-localized LRR-RLK, PEP RECEPTOR (PEPR)-1 (At1G73080) (Huffaker and Ryan 2007), and PEPR2 (At1G17750) (Yamaguchi *et al.* 2010) and amplify defense signaling. A recent study using a candidate-based yeast two-hybrid screen revealed that PEPR1 and PEPR2 both interact with BRASSINOSTEROID-INSENSITIVE 1-ASSOCIATED RECEPTOR KINASE (BAK1) (Postel et al. 2010) and, like MAMP receptors, this interaction is essential for full activation of defense responses against pathogens (Krol et al. 2010; also see chapter "Peptides as Danger Signals: MAMPs and DAMPs").

Studies have revealed that the *AtPROPEP* paralog genes are differentially expressed with *PROPEP2* and *PROPEP3* being the most highly expressed family members in response to pathogens and MAMPs and are the only ones induced by salicylic acid (Huffaker et al. 2006). In contrast, *PROPEP1* expression is not induced in response to any of these treatments but has been shown to be activated in response to wounding and jasmonic acid, suggesting that it might function as a defense signal similar to that of the systemins.

4 Analysis Methodology

In summary, the analysis starts by performing an expression correlation analysis with a specific GOI. This identifies genes that are most highly co-expressed with the GOI, and from this, an expression correlated gene group (ECGG) is established. This group is then subjected to a functional annotation analysis in order to identify any specific enrichment in functional terms associated with the group. Conditions are then identified that induce large collective changes in expression of the genes in order to confirm their co-expression in response to specific stimuli and additionally associate the genes with specific biological responses. Further, transcriptional responses in signaling-related mutants or in response to hormone and chemical treatments can help elucidate factors that regulate transcription of the genes. Finally, promoters of the co-expressed genes can be assessed for enrichments in known transcription factor binding sites (TFBS) that may mediate their co-regulation by binding specific transcription factors. The Web-based tools used in this analysis are all open access, do not require any programming knowledge, and are relatively easy to use. The output files are easily manageable and easy to understand, making them ideal for laboratory-based scientists who can use them to plan future experimental studies or complement experimental results.

4.1 Expression Correlation Analysis

The first step in the analysis is to perform a co-expression analysis in order to identify a group of genes in the genome with which a specific GOI is most highly co-expressed. There are a number of methodologies available that can identify groups of co-expressed genes. We choose to use an expression correlation analysis since it allows a defined GOI to be used as the driving or central gene in the analysis, and the co-expression of all other genes in the genome is measured relative to the GOI. In comparison, clustering algorithms group sets of genes according to similarities in their expression patterns; however, the grouping is not based around a single GOI.

There are a number of co-expression databases that are available that cover a range of different organisms including *Arabidopsis* (Manfield et al. 2006; Obayashi and Kinoshita 2011), rice (Obayashi and Kinoshita 2011), mammals (Obayashi et al. 2008) *Escherichia coli*, and *Saccharomyces cerevisiae* (Steinhauser et al. 2004). For this analysis, we use the *Arabidopsis* co-expression tool (ACT) (http://www.arabidopsis.leeds.ac.uk/act/coexpanalyser) (Manfield et al. 2006). Like many similar databases, this tool uses hybridization signal intensities from microarray experiments to calculate a Pearson correlation coefficient (*r*-value), which is a measure of expression similarity that reflects the strength and direction of the linear relationship between the reference gene (GOI) and other genes represented on the selected *Arabidopsis* microarray chip. This tool uses NASC/GARNet microarray data that has been labeled, hybridized, and analyzed using a standardized procedure, thus providing a homogeneous and readily comparable data set. The microarray probe ID must be used as the input; the tool conveniently has an ID exchanger to convert *Arabidopsis* genome initiative (AGI) codes to probe IDs.

There are a number of options available within the ACT. The correlation analysis can be performed using the AtGenome1 (8K) or the ATH1 full genome (22K) chip. In addition to providing greater genome coverage, the analysis using the ATH1 chip is performed over a larger number of arrays (322 compared to 121 for AtGenome1) which increases the accuracy of the analysis. The arrays used in this analysis encompass a range of diverse experimental conditions that include specific tissue types and developmental stages, abiotic and biotic stress treatments, and selected mutants. The use of such a large number of diverse experimental conditions has been shown to be critical in accurately identifying co-expressed genes using expression data. A study in yeast revealed that the use of a minimum of 100 microarray experiments performed over a broad range of experimental conditions was necessary to accurately identify genes that bind common transcription factors which presumably coordinate their co-expression (Allocco et al. 2004). The predictive power was found to increase with the use of up to 100 microarrays with minimal improvement with numbers >100 (Allocco et al. 2004).

Using Analysis Option 1, entitled "Co-expression Analysis in the Specified Experiment," one can select specific experiments over which the co-expression analysis is performed. This can be useful to identify genes that are co-expressed in specific tissue types or in response to select conditions such as pathogen challenges. Indeed, a different *Arabidopsis* expression correlation tool named CressExpress (http://www.cressexpress.org) is specifically designed to perform gene expression analyses in specific tissue types (Srinivasasainagendra et al. 2008). It is important to note, however, that if the co-expression analysis is performed over a small number of arrays, the accuracy of the results will be compromised. The main aim of this type of analysis is to identify networks of co-expressed genes that are likely to be co-regulated and consistently function in common biological responses. When using a small number of arrays to identify co-expressed genes, the identified genes may be regulated by distinct regulatory mechanisms that are commonly activated in response to a specific condition. This would be similar to inferring the co-expression of a set of genes based on their collective induction in response to a single stimulus. In contrast, the expression of genes determined to be co-expressed over a large number of diverse experimental conditions would be similarly induced, repressed, or unaltered in response to a range of specific conditions. Indeed, conditions that do not induce a change in the expression of a gene provide an important baseline for other conditions that induce differential expression. Thus, in order to accurately identify co-expressed genes that are likely to be co-regulated and consistently function in specific cellular responses, it is important to perform the co-expression analysis over a large number of diverse experimental conditions.

The co-expression analysis for *PROPEP2* was performed using Analysis Option 2 in the ACT suite, named "Co-expression analysis over available array experiments." Options within this tool include selecting the number of most positively or negatively correlated genes that are returned in the analysis. In our analysis, the list limit is left blank, so we obtain the correlation values for all genes, ranging from the most positively ($r = +1$) to the most negatively($r = -1$) expression correlated genes.

The result output page can be copied and pasted as Unicode text into an Excel spreadsheet. The driving GOI, in this case, *PROPEP2*, is listed at the top of the file. The file will contain around 22,000 rows that represent each probe on the array. The results are listed in decreasing order of positive correlation. For each probe, there is an additional five columns of data that include the r-value, p-value, e-value; the corresponding gene ID; and the annotation/description of the gene. The statistical significance of the calculated r-value is indicated by the p-value and the e-value which accounts for multiple hypothesis testing.

A group of the most highly expression correlated genes is next selected from the list; this group forms the ECGG for the specific GOI and is used for subsequent analysis. The size of the ECGG can be selected in a number of ways. It can be based on the strength of the expression correlation using the r-value as a cutoff. Correlation r-values above 0.9 are considered very high and likely to be of biological significance (Manfield et al. 2006); however, we have found that r-values around 0.7 contain enrichments in functionally associated terms (Meier et al. 2008) which also supports a biologically significant relationship. A study in yeast revealed that r-values above 0.8 can be useful in accurately identifying genes that bind common transcription factors (Allocco et al. 2004). If a GOI is not found to be highly co-expressed with any other genes ($r < 0.6$) in the genome, it suggests that the gene is under complex regulatory control. This may be mediated by the presence of multiple TFBS in the promoter of the gene that are differentially regulated by different transcription factors in response to different stimuli. Regardless of low r-values for a specific GOI, the identified genes still represent the most highly expression correlated genes in the genome. Generally speaking, genes that share a higher level of correlation are more likely to contain similar TFBS in their promoters, be co-regulated, and have closely associated functions.

In our analysis, we select the top 100 genes that are co-expressed with *PROPEP2* for subsequent analysis, herein referred to as the *PROPEP2-ECGG* (Table 1). This number of genes was chosen so as to have a good representative sample size for future analysis. The r-value range of the *PROPEP2-ECGG* was 0.89–0.72.

4.2 Functional Enrichment Analysis

The *PROPEP2-ECGG* was next subjected to a functional enrichment analysis in order to test for any overrepresented functional terms associated with the groups. There are many tools available that can reveal functional annotation information for individual genes (Al Shahrour et al. 2007; Maere et al. 2005; Zhou and Su 2007; Carbon et al. 2009; Harris et al. 2004). For the purpose of this analysis, however, it is important that the annotation tool identifies functional terms that are statistically enriched in the ECGG. Enrichment in a particular term is usually determined by comparing the frequency of occurrence of the term in a specific gene group to that in the entire genome. For example, if only 3% of genes in the entire genome are annotated to function in osmotic stress responses but 80% of the genes in a particular group are found to have this annotation, then the group would likely be enriched in this functional annotation category.

We performed this analysis for the *PROPEP2-ECGG* using the FatiGo functional analysis tool (version 4.2), which is part of the Babelomics suite (http://www.babelomics.org/) (Al Shahrour et al. 2007). The analysis is run selecting to compare an ID list (*PROPEP2-ECGG*) to the rest of the genome. Under options, the Fisher exact test to identify "Over-represented terms in list 1 (genome comparison)" and "Never" remove duplicates are selected. All available *Arabidopsis* databases are selected; these include INTERPRO and KEGG pathways, and for Gene Ontology, these include biological process, molecular function, and cellular component.

The analysis revealed that the *PROPEP2-ECGG* was enriched in a number of biological process terms that were associated with pathogen defense (Table 1), including "defense response (20 genes)" and "response to biotic stimulus (13 genes)." It is particularly interesting to note that the *PROPEP2* paralog, *PROPEP3*, was present in the *PROPEP2-ECGG*. There was also an enrichment in genes annotated to have "receptor activity" (15 genes) and "signal transducer activity" (18 genes) and contain "serine/threonine-protein kinase" domains (18 genes).

Table 1 List of genes co-expressed with PROPEP2 (AT5G64890)

No.	Gene ID	*r*-Value	Primary gene symbol	Functional annotation[a]
1	AT3G53600	0.89	C2H2-type zinc finger family protein	RC
2	AT5G65600	0.89	Concanavalin A-like lectin protein kinase (PK)	STA, RA, STPK
3	AT4G11170	0.88	Disease resistance protein (TIR-NBS-LRR class)	DR, CD, STA, RA
4	AT4G21390	0.88	S-locus lectin PK family protein	STPK
5	AT3G21230	0.88	4-COUMARATE:COA LIGASE 5 (4CL5)	
6	AT1G57650	0.86	Disease resistance protein (NBS-LRR class), apoptosis	DR
7	AT5G47730	0.85	Sec14p-like phosphatidylinositol transfer family protein	
8	AT1G61560	0.85	MILDEW RESISTANCE LOCUS O 6 (MLO6)	DR, RBS, CD
9	AT1G79680	0.84	WALL-ASSOCIATED KINASE-LIKE (WAKL) 10	STPK
10	AT5G67340	0.84	RM repeat protein; ubiquitin-protein ligase	
11	AT2G02310	0.84	PHLOEM PROTEIN 2-B6 (AtPP2-B6)	
12	AT1G26380	0.83	FAD-binding Berberine family protein	
13	AT5G44990	0.82	Glutathione S-transferase family protein	
14	AT1G76070	0.82	Expressed protein (Expr Prot.)	
15	AT4G11370	0.82	Putative RING-H2 finger protein RHA1a	
16	AT1G15520	0.81	PLEIOTROPIC DRUG RESISTANCE 12 (PDR12)	DR, RBS
17	AT2G16900	0.81	Phospholipase-like protein (PEARLI 4) family	
18	AT1G77500	0.81	Expr Prot., N-terminal prot. Myristoylation (Myris)	
19	AT1G66090	0.81	Disease resistance protein (TIR-NBS class)	DR, CD, STA, RA
20	AT5G12340	0.81	Expr Prot.	
21	AT5G67080	0.81	MAP KINASE KINASE KINASE 19 (MAPKKK19)	STPK
23	AT2G35980	0.8	YELLOW-LEAF-SPECIFIC GENE 9 (YLS9)	DR, RBS
24	AT3G63380	0.8	ATPase E1-E2 type family protein	STA, RA
26	AT1G70130	0.79	Concanavalin A-like lectin PK family protein	STPK
27	AT5G02490	0.79	Heat shock protein 70 (Hsp 70) family protein	RBS, RtB
28	AT3G09010	0.79	PK superfamily protein	STA, RA, STPK
33	AT1G51890	0.78	Leucine-rich repeat (LRR) PK family protein	STPK
37	AT4G33430	0.77	BAK1	DR, RBS, RtB, CD, STA, RA, STPK
40	AT1G66550	0.77	WRKY67	
41	AT2G32140	0.77	Transmembrane receptor activity	DR, STA, RA

42	AT2G39380	0.77	EXOCYST SUBUNIT EXO70 FAMILY PROTEIN H2	CSR, STA
47	AT1G21010	0.76	Expr Prot., N-terminal protein Myris	CSR
48	AT2G32240	0.76	Expr Prot., viral A-type inclusion protein repeat	STA
49	AT4G39030	0.76	ENHANCED DISEASE SUSCEPTIBILITY 5	DR, RBS
50	AT3G11840	0.76	PLANT U-BOX 24, Neg. regul. of PAMP trig. immun	DR, RC, RB
54	AT1G29690	0.76	CONSTIT ACTIV. CELL DEATH, Neg SA regul	CD
55	AT3G25780	0.75	ALLENE OXIDE CYCLASE 3 (AOC3) JA biosyn	RBS
57	AT5G64905	0.75	PROPEP3	DR
58	AT4G34390	0.75	EXTRA-LARGE GTP-BINDING PROTEIN 2 (XLG2)	CSR, STA
59	AT3G26830	0.75	PHYTOALEXIN DEFICIENT 3 (PAD3)	RBS, RtB
62	AT4G17500	0.75	ETHYLENE RESP. ELEM. BIND FACTOR 1 (ERF-1)	DR, RC
63	AT1G51800	0.75	LRR PK family protein	STA, RA, STPK
64	AT4G11470	0.74	CYST-RICH RECEPTOR-LIKE PRO. KINASE 31	STA, RA, STPK
66	AT5G13080	0.74	WRKY75	
69	AT5G48400	0.74	Glutamate receptor family protein (GLR1.2)	STA, RA
70	AT2G38490	0.74	CBL-INTERACTING PK 22 (CIPK22)	STPK
72	AT2G35930	0.74	PLANT U-BOX 23, Neg. regul. of PAMP trig. immun	DR, RC, RB
78	AT3G52400	0.74	SYNTAXIN OF PLANTS 122 (SYP122)	DR, RBS, CD
79	AT2G19190	0.74	FLG22-IND. RECEPTOR-LIKE KINASE 1 (Flg22RK1)	DR, RBS, RtB, STA, RA, STPK
81	AT1G74360	0.73	LRR PK family protein	CSR, STA, RA, STPK
82	AT1G61360	0.73	S-locus lectin PK family protein	STA, RA, STPK
83	AT1G19180	0.73	JA-ZIM-DOMAIN PROTEIN 1 (JAZ1), JA signaling	DR, RBS, RtB
86	AT5G66890	0.73	LRR family protein (NBS-LRR)	DR
87	AT5G61560	0.73	U-box domain-containing PK family protein	STPK
88	AT1G73080	0.73	PEP1 RECEPTOR 1 (PEPR1)	DR, CSR, STA, RA, STPK
93	AT1G17750	0.73	PEP1 RECEPTOR 2 (PEPR2)	CSR, STA, RA, STPK
95	AT4G31550	0.72	WRKY11	DR, RBS, RC, RtB
99	AT1G18890	0.72	CALCIUM-DEPENDENT PK 1	STPK
5,207	AT5G64900	0.04	PROPEP1	

Note: To save space, only the top 20 genes and genes that were found to be enriched in a functional category are presented

[a] Functional enrichment abbreviations. *Biological process*: *DR* defense response, *RBS* response to biotic stimulus, *CD* cell death, *RC* response to chitin, *CSR* cell surface receptor linked signal transduction, *RtB* response to bacterium, *RB* respiratory burst; *Molecular function*: *STA* signal transducer activity, *RA* receptor activity; *INTERPRO*: *STPK* serine/threonine-protein kinase.

Genes present in these groups include *PEPR1* (AT1G73080, $r = 0.73$) and *PEPR2* (AT1G17750, $r = 0.73$), which are known receptors for the Peps (Yamaguchi et al. 2010), and *BAK1* (AT4G33430, $r = 0.77$), which is the central regulator of innate immunity in plants and is known to specifically interact with the PEPRs and be an essential component of Pep-mediated signaling (Krol et al. 2010). In addition, *PROPEP2* was found to be highly correlated with genes that encode other membrane-localized receptors implicated in the detection of MAMPs including *FLG22-INDUCED RECEPTOR-LIKE KINASE 1* (*Flg22RK1*, AT2G19190, $r = 0.74$), which is implicated in bacterial flagella detection, and *CHITIN ELICITOR RECEPTOR KINASE 1* (*CERK1*, AT3G21630, $r = 0.71$, rank 110, data not shown), which plays a critical role in fungal MAMP perception in plants (Miya et al. 2007).

A number of the correlated genes are known to function in the biosynthesis and regulation of major pathogen defense signaling molecules; these genes include *ENHANCED DISEASE SUSCEPTIBILITY 5* (*EDS5*, At1G29690, $r = 0.76$) (Nawrath et al. 2002), *ISOCHORISMATE SYNTHASE 1* (*ICS1*) (AT1G74710, $r = 0.68$, data not shown) (Wildermuth et al. 2001), and *CONSTITUTIVELY ACTIVE CELL DEATH* (*CAD*)*-1* (At1G29690, $r = 0.76$) (Tsutsui et al. 2006; Morita-Yamamuro et al. 2005), which function in the regulation and biosynthesis of salicylic acid; *ALLENE OXIDE CYCLASE 3* (AOC3, AT3G25780, $r = 0.76$) (Truman et al. 2007) and *JASMONATE-ZIM-DOMAIN PROTEIN* (*JAZ1*) (AT1G19180, $r = 0.73$) (Chini et al. 2007), which function in jasmonic acid biosynthesis and signaling, respectively; and *PHYTOALEXIN DEFICIENT 3* (*PAD3*) (AT3G26830, $r = 0.75$), which encodes the enzyme that catalyzes the final step in the synthesis of the broad range antimicrobial phytoalexin, camalexin (Schuhegger et al. 2006), synthesized early and locally in an salicylic acid- and jasmonic acid-dependent manner (Schuhegger et al. 2007).

It is interesting to note that a number of the genes annotated to have, "receptor activity", "signal transducer activity", "serine/threonine-protein kinase domains", and "cell surface receptor linked signal transduction" as well as 15 genes encoding expressed proteins have not been linked to any specific biological process. The high expression correlation of these genes with that of *PROPEP2* and many well-characterized pathogen defense encoding genes makes them interesting candidates that may also have functionally related roles.

In contrast to *PROPEP3*, the expression of *PROPEP1*, *-4*, *-5* and *-6* was not correlated with that of *PROPEP2*. This may suggest that these paralog PROPEPs function in distinct biological responses.

4.3 Co-correlation

In order to obtain more information regarding the co-expression of these genes, Analysis Option 3, the "Co-correlation scatter plot," was used in the ACT suite. This tool can be used to illustrate the level of correlation that all genes share with two defined query genes. In this tool, the option is available to highlight the co-correlation that specific genes share with the query genes. This analysis is best performed using query genes that are highly co-expressed and have closely associated functions. Genes that are found to have a high level of co-expression with both query genes could be considered to be more likely to have closely associated functions.

In our analysis, we perform the co-correlation using *PROPEP2* and *PEPR1* as the query genes and select other Pep signaling components to be highlighted, including *PROPEP1*, *-3, -4, -5, -6*, and *PEPR2*. We additionally select a number of well-characterized pathogen defense-related genes including the MAMP receptors, namely, *CERK1* for fungi; *FLAGELLIN-SENSITIVE 2* (*FLS2*), *EF-TU RECEPTOR* (*EFR*), and *Flg22RK1* for bacteria; and the functionally related *BAK1*, and genes involved in the synthesis of defense-related compounds, including *EDS5* and *ICS1* for SA biosynthesis; *AOC3* and *JAZ1* which function in jasmonic acid biosynthesis and signaling, respectively; and *PAD3* which catalyzes the final step in camalexin biosynthesis. The pathogen-inducible *PATHOGENESIS-RELATED GENES* (*PR*), *PR1*, *PR2*, and *PR5*, were also included (Fig. 1).

It is clear that the expression of both *PROPEP2* and *PEPR1* is highly correlated with *PROPEP3* and *PEPR2*, suggesting that these genes function in closely associated processes. In contrast, the expression of *PROPEP1*, *-4*, *-5*, and *-6* show little correlation with that of either query gene which further suggests that these genes may function in distinct biological responses. This was also the case for the *PR* genes. A co-expression analysis performed using *PROPEP1* as a GOI additionally revealed that *PROPEP1* is not correlated with *PROPEP*3, *PEPR1*, or *PEPR2* and there is no enrichment in pathogen defense-related genes in the *PROPEP1-ECGG* (data not shown). This illustrates that *PROPEP1* has a distinct expression profile to that of *PROPEP2* and *PROPEP3* and thus may function in distinct biological processes. For the sake of comparison, *PROPEP1* was added to the list for the stimuli-specific expression analysis.

The high correlation with the salicylic acid and jasmonic acid synthesis and signaling genes and the *PAD3* camalexin biosynthesis genes suggests that the function of Pep2 and PEPR1 is closely associated with the synthesis of these pathogen defense-related compounds. The high correlation with *CERK1* indicates that the function of these genes is linked to fungal perception, while the lower co-expression with *EFR* and *FLS2* indicates that expression of the genes is less related to bacterial perception, although *PROPEP2* is highly correlated with the bacterial flagellin-inducible gene *Flg22RK1*.

In general, the high co-expression of *PROPEP2* with genes enriched in functional terms associated with pathogen defense responses is fully consistent with the established role of PROPEP2 in biotic defense signaling. Further, the high co-expression of *PROPEP2* with genes that are known to have closely associated functional relationships including the *PROPEP3* paralog, the *PEPR1* and *PEPR2* receptors that bind the PEPs, and *BAK1* that is known to interact with and be an essential component of PEPR signaling illustrates that the co-expression analysis

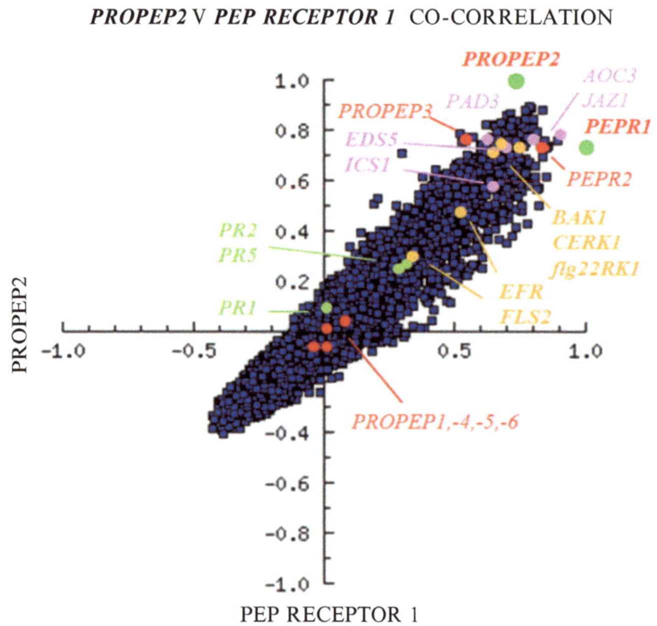

Fig. 1 Co-correlation scatter plot illustrating the level of correlation that all genes share with *PROPEP2* and *PEPR1*. Color-coded genes include *green*, proPep2 and PEPR1; *red*, proPep paralog genes and receptors; *pink*, pathogen defense signaling and biosynthesis genes; *orange*, MAMP-related receptors; and *green*, PR genes

does indeed identify genes that have very closely associated functional relationships with a GOI.

4.4 Stimulus-Specific Expression Profile

The next step in the analysis is to examine the expression of the individual genes in response to specific experimental conditions. The expression correlation analysis provides a generalized measure of how a specific GOI is co-expressed with other genes in the genome across broad range of experimental conditions and can thus serve to identify the most highly co-expressed genes. In order to further relate the function of the co-expressed genes to specific biological responses, conditions need to be identified that induce differential expression of the genes. This analysis can also provide some evidence to support the accuracy of the co-expression analysis since it reveals if the genes are coordinately expressed in response to specific experimental conditions. In addition, transcription profiling in response to chemical treatments and signaling-related mutants can help identify factors that regulate expression of the genes.

The Genevestigator tool (https://www.genevestigator.com/gv/index.jsp) (Zimmermann et al. 2005) can be used to screen for specific experiments that induce differential expression of the genes. In addition to *Arabidopsis*, this tool contains microarray data for rice and other non-plant organisms including *Homo sapiens*. There is currently data for over 5,000 *Arabidopsis* and 300 rice full genome experiments available. For our analysis, in step 1, we select all available 22K high-quality arrays. In step 2, we enter *PROPEP2* and the top 5 expression correlated genes to screen their transcriptional responses across the experiments. In the main window, there are a number of tabs that allow the user to select the way the data is viewed. The "conditions" tab contains the ratio fold change for the selected genes over all nonmutant experimental conditions. The results are displayed in descending order with the condition that induced the largest fold change increase in expression of the selected gene at the top. When scrolling over the experiment name, a box pops up that contains a link to the repository where the complete normalized data for the specific experiment is stored. The "genotype" tab contains experimental expression data for different mutant types presented in a similar manner to the "conditions" tab. The "anatomy" and "development" tabs can also be used to identify the relative expression level of genes in different anatomical structures and different developmental stages, respectively. By viewing the "conditions" and "mutant" experiments, conditions that induce the largest change (either up or down) in expression of the GOI can be easily identified.

Following the screening in Genevestigator, it is important to obtain complete microarray data for interesting experiments from relevant repositories for the final analysis. The results displayed in Genevestigator are in some cases an average response over multiple time points and thus do not provide complete information. In addition, it is not practical to obtain detailed data for up to 100 genes across multiple experiments using Genevestigator. As mentioned, links to microarray repositories are present in Genevestigator. Microarray data can also be obtained by searching repository sites. Normalized microarray data for most *Arabidopsis* experiments can be obtained from the NCBI Gene Expression Omnibus (GEO, http://www.ncbi.nlm.nih.gov/geo/), NASCArrays (http://affymetrix.arabidopsis.info/narrays/experimentbrowse.pl), or ArrayExpress (http://www.ebi.ac.uk/arrayexpress/) Web sites.

For our analysis, it was apparent in Genevestigator that *PROPEP2* expression was generally most strongly induced in response to pathogens and their associated MAMPs elicitors. The related experiments were downloaded from the relevant Web sites, and the fold change ratios (log2) were calculated using Microsoft Excel. The ratios for genes in the *PROPEP2-ECGG* were extracted, and the results are presented as heat maps in Fig. 2. The heat maps were generated using the Multiple Array Viewer program from the MultiExperiment Viewer (MeV) software package (version 4.2.01) created by The Institute for Genomic Research (TIGR) (Saeed et al. 2003).

The results show that expression of the genes was strongly induced by a range of MAMP-related elicitors, including the fungal-derived chitooctaose and the bacterial-derived, flg22, elf26 and hypersensitive reaction and pathogenicity (Hrp)-Z (Lee et al. 2001). The chitooctaose- and flg22-induced expression appears to be

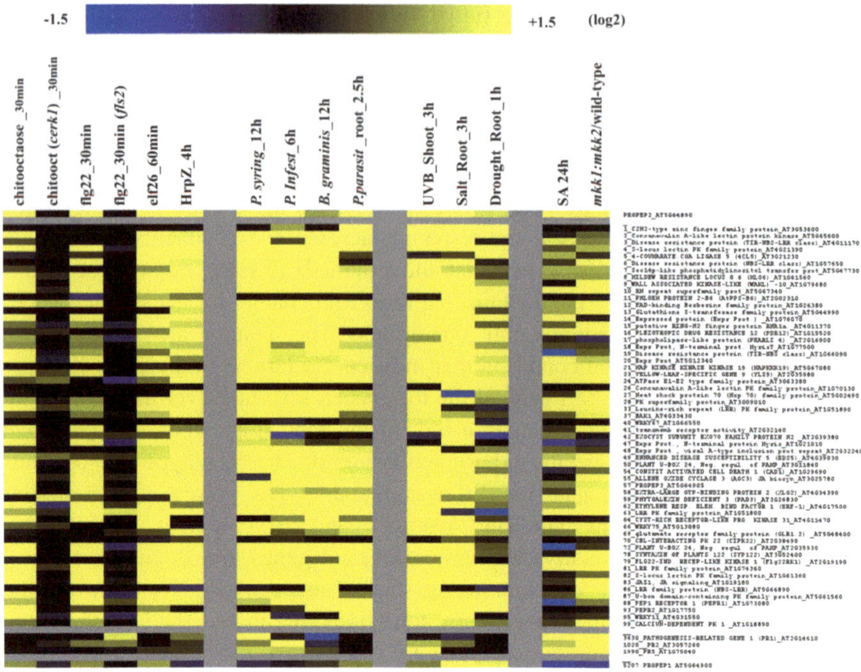

Fig. 2 Heat map illustrating the fold change (log2) in expression of individual genes PROPEP2-ECGG in response to various experimental conditions

mediated through the CERK1 and FLS2 receptors, respectively, since the induction was largely suppressed in the respective mutants. The expression of the genes was also strongly induced by a number of pathogens including *Pseudomonas syringae pv. syringae*, *Phytophthora infestans*, and *Blumeria graminis* and in root by the fungal-like pathogen *Phytophthora parasitica*. The expression of the genes was also induced in response to a number of abiotic stresses including UVB, salt, and drought. The genes were induced in response to salicylic acid, and their expression was also elevated in the untreated *mitogen-activated protein kinase kinase* (*mkk*)*-1* and *mkk2* double mutant (*mkk1:mkk2*) that has elevated levels of salicylic acid (Qiu et al. 2008), suggesting that their expression is positively regulated by salicylic acid.

In general, the expression results support that expression of the genes is induced in response to a broad range of pathogens and their associated MAMP elicitors which is completely consistent with previous studies that have shown that these peptides function in pathogen defense responses. The results also reveal that activation of the genes in response to the chitin and flg22 MAMP elicitors is dependent on the CERK1 and FLS2 receptors respectively.

4.5 Promoter Enrichment Analysis

The final step in the analysis involves examining the promoters of the ECGG for the presence of enriched putative regulatory motifs (*cis* elements) that may be causative for their co-expression. The rationale behind this analysis is that the regulation of eukaryotic gene expression is critically dictated by the presence of specific *cis*-regulatory motifs in the promoter region of a gene; these motifs in turn bind specific transcription factors that modulate expression of the gene (Allocco et al. 2004; Brazma et al. 1998). It is thought that the presence of specific TFBS in the promoter regions of co-expressed genes at a frequency that is different to that of other genes in the genome is responsible for mediating their distinct expression profiles. As most TFBS are present in the promoters of many genes, and often in multiple copies, enrichment is rarely defined by the simple presence or absence of a specific motif in a promoter (Werner 2001). Rather, enrichment of a specific motif in a co-expressed gene set most often results from the motif being present at a higher average copy number per promoter than that present in a background group which is commonly the rest of the promoters in the genome.

There are many computational-based tools that can be used to identify putative motif enrichment in groups of genes. These methods can be broadly divided into two distinct approaches. In the first, promoters are searched for an enrichment or overrepresentation of motif sequences that correspond to experimentally determined TFBS. In the second, a defined group of promoters is searched for the presence of any computationally derived motif sequences that are overrepresented.

Both approaches have their inherent advantages and disadvantages. The identification of motifs that have been experimentally verified provides increased confidence that they are of biological relevance. However, due to tedious experimental procedures required to verify TFBS, very few have been well characterized (Tompa et al. 2005). Thus, this type of promoter analysis is restricted to the identification of a limited and incomplete set of regulatory elements. The biological significance of this limitation is exemplified by the fact that the transcriptional regulation of individual genes in eukaryotic organisms is determined by the combinational presence of multiple *cis*-regulatory sequences in their promoters (Tuch et al. 2008). The use of computationally derived motifs avoids the limitations imposed by experimentally derived motifs; however, the biological significance of identified motifs is unknown. We choose to use a combination of the two approaches to maximize the chances of accurately identifying biologically relevant motifs.

4.5.1 Experimentally Derived Motifs

The visualization tool in the *Arabidopsis thaliana* expression network analysis (ATHENA; http://www.bioinformatics2.wsu.edu/cgi-bin/Athena/cgi/visualize_select.pl) suite allows the rapid identification of known plant TFBS that are enriched in defined groups of *Arabidopsis* promoters (O'Connor et al. 2005). The analysis is run by providing *Arabidopsis* accession IDs (e.g., AT3G57260) for the genes to be analyzed. The main options include selecting the upstream promoter length to be analyzed and whether to cut off at adjacent genes. In our analysis, we input all genes in the *PROPEP2-ECGG*, select a promoter length of 1,000 bp upstream, and do not cut off at adjacent genes.

The result output reveals enrichments in a number of putative motif sequences that correspond to known plant TFBS (Table 2). Of note, one of these motifs is the stringent W-box [TTGAC(C/T)] which corresponds to the binding site of WRKY transcription factors that are known to regulate the expression of abiotic and biotic defense-related genes in plants (Eulgem et al. 1999; Nimchuk et al. 2003). The W-boxes were present in 91 (90%) of our genes compared to them being present in 67% of promoters in the genome. They are present in a total of 241 times (2.4 copies/promoter). The promoter of *PROPEP2* was determined to contain five copies of the W-box. It is important to ensure that enriched motifs are present in a high percentage of the promoters in the group since we are attempting to identify motifs that are responsible for coordinating expression of the gene set as a whole. It is interesting to note that three annotated WRKY transcription factors (WRKY67, WRKY75, and WRKY11) are present in the *PROPEP2-ECGG*, and these represent interesting candidate transcription factors that may regulate transcription of the expression correlated genes.

4.5.2 Computationally Derived Motifs

There are many programs that can be used to identify enriched computationally derived motifs; the accuracy of some of these has been reviewed (Tompa et al. 2005). For our analysis, the POXO (http://ekhidna.biocenter.helsinki.fi/poxo) (Kankainen et al. 2006) promoter analysis suite was used. The POXO suite contains an array of different tools for performing promoter analysis. The sequence download tool in POXO can be used to obtain promoter sequences of defined genes for the analysis. In our analysis, we obtained upstream promoter sequences of 1,000 nucleotides in length for all genes in the *PROPEP2-ECGG*. These sequences were uploaded into the POCO tool that was used to identify overrepresented motifs. Options selected in POCO included the following: to search for motifs 6 nucleotide (max) in length, report the top 20 patterns found, and that the motif must occur in 80 promoters (~80% of genes). The analysis was run against *Arabidopsis* background group.

The POCO results confirm the enrichment of the W-boxes found in Athena with the top three enriched motifs being variations of W-boxes (Table 2). The enriched motifs identified can be analyzed for similarities to known plant TFBS using the pattern screener tool in POXO; this compares identified motifs to sequences stored in PLACE, JASPAR, or TRANSFAC *cis* element public databases. The absence of any matches with known TFBS (indicated in Table 2 by?), however, should not be used to dismiss the biological significance of the enriched putative motif identified since, as mentioned, only a small percentage of TFBS have been characterized. Indeed, such motifs can be considered unique putative regulatory sequences that could be targets for future experimental work.

Table 2 Promoter analysis of *PROPEP2-ECGG*

Motif	Name	Copies/prom			Frequency (%)		
		PROPEP2	ECGG	BG	ECGG	BG	*P*-Value
ATHENA							
TTGAC(T/C)	W-box	5	2.4			67	<10^{-11}
AMCWAMC	MYB4 BS	4	4			85	<10^{-3}
POCO							
TTGAC(T/C)	W-box	5	2.7	1.2	90	66.7	5.36^{-09}
TTGAC	W-box	7	4.3	2.2	100	87.8	4.81^{-08}
TTTGAC	W-box	2	1.8	0.9	90	86.1	8.52^{-05}
TNGAC	?	8	7.9	5.7	100	99.3	2.61^{-04}
AAAGNC	?	3	4.0	2.4	95	89.2	7.61^{-04}
GTCNAA	?	2	3.2	2.0	100	84.8	1.33^{-03}
GNCAA	?	11	9.6	7.3	100	99.8	1.56^{-03}
GACTT	?	3	3.2	1.9	95	83.4	1.93^{-03}
AAANTC	?	3	6.7	4.9	100	98.7	2.10^{-03}
GNCAAA	?	4	4.3	2.9	100	93.2	2.64^{-03}

Confidence in the biological significance of computationally derived motifs can be improved by performing phylogenetic foot printing which entails comparing orthologous regulatory regions of a single gene across several species (Tagle et al. 1988). The rationale behind this approach is that selection pressure restricts the evolution of functional elements, and consequently, they retain greater conservation than nonfunctional sequences (Brazma et al. 1998; Blanchette and Tompa 2002). Therefore, a putative motif identified in the promoter of a specific gene in *Arabidopsis* could be tested for its presence in an orthologous gene promoter in other plants species. This type of comparative analysis has been used to identify conserved regulatory elements between humans and mouse (Jareborg et al. 1999; Wasserman et al. 2000).

5 Conclusions

In summary, the co-expression analysis performed here using *PROPEP2* as an example has demonstrated how the use of computational Web-based tools can be used to accurately link a specific GOI to a specific biological process. The expression correlation analysis revealed that the *PROPEP2* gene is co-expressed with genes that function in biotic stress responses. In addition, *PROPEP2* is highly co-expressed with the *PROPEP3* paralog gene and the *PEPRs* that are known to bind the PEPs. This clearly illustrates that *PROPEP2* is co-expressed with genes that have close functional relationships. The stimuli-specific expression analysis revealed that the expression of the *PROPEP2-ECGG* was largely universally induced in response to a range of pathogens and their associated MAMP elicitors. The universal induction confirms that the genes are indeed co-expressed and strongly indicates that they have a functional role in biotic stress responses.

The identification of an enrichment in W-boxes in the promoters of the *PROPEP2-ECGG* that bind WRKY transcription factors which are known to regulate the expression of defense-related genes indicates that WRKY transcription factors may function to mediate the coordinated expression of these genes. Collectively, these results are entirely consistent with the known role of *PROPEP2* in pathogen defense response signaling in plants and thus support the accuracy of this type of analysis in identifying functionally related genes.

References

Al Shahrour F, Minguez P, Tarraga J, Medina I, Alloza E, Montaner D, Dopazo J (2007) FatiGO +: a functional profiling tool for genomic data. Integration of functional annotation, regulatory motifs and interaction data with microarray experiments. Nucleic Acids Res 35:W91–W96

Allocco DJ, Kohane IS, Butte AJ (2004) Quantifying the relationship between co-expression, co-regulation and gene function. BMC Bioinforma 5:18

Blanchette M, Tompa M (2002) Discovery of regulatory elements by a computational method for phylogenetic footprinting. Genome Res 12:739–748

Brazma A, Jonassen I, Vilo J, Ukkonen E (1998) Predicting gene regulatory elements in silico on a genomic scale. Genome Res 8:1202–1215

Butenko MA, Vie AK, Brembu T, Aalen RB, Bones AM (2009) Plant peptides in signalling: looking for new partners. Trends Plant Sci 14:255–263

Carbon S, Ireland A, Mungall CJ, Shu S, Marshall B, Lewis S (2009) AmiGO: online access to ontology and annotation data. Bioinformatics 25:288–289

Chini A, Fonseca S, Fernandez G, Adie B, Chico JM, Lorenzo O, Garcia-Casado G, Lopez-Vidriero I, Lozano FM, Ponce MR, Micol JL, Solano R (2007) The JAZ family of repressors is the missing link in jasmonate signalling. Nature 448:666–671

Eisen MB, Spellman PT, Brown PO, Botstein D (1998) Cluster analysis and display of genome-wide expression patterns. Proc Natl Acad Sci USA 95:14863–14868

Eulgem T, Rushton PJ, Schmelzer E, Hahlbrock K, Somssich IE (1999) Early nuclear events in plant defense signalling: rapid gene activation by WRKY transcription factors. EMBO J 18:4689–4699

Harris MA, Clark J, Ireland A, Lomax J, Ashburner M, Foulger R, Eilbeck K, Lewis S, Marshall B, Mungall C, Richter J, Rubin GM, Blake JA, Bult C, Dolan M, Drabkin H, Eppig JT, Hill DP, Ni L, Ringwald M, Balakrishnan R, Cherry JM, Christie KR, Costanzo MC, Dwight SS, Engel S, Fisk DG, Hirschman JE, Hong EL, Nash RS, Sethuraman A, Theesfeld CL, Botstein D, Dolinski K, Feierbach B, Berardini T, Mundodi S, Rhee SY, Apweiler R, Barrell D, Camon E, Dimmer E, Lee V, Chisholm R, Gaudet P, Kibbe W, Kishore R, Schwarz EM, Sternberg P, Gwinn M, Hannick L, Wortman J, Berriman M, Wood V, de la Cruz N, Tonellato P, Jaiswal P, Seigfried T, White R (2004) The gene ontology (GO) database and informatics resource. Nucleic Acids Res 32:D258–D261

Huffaker A, Ryan CA (2007) Endogenous peptide defense signals in Arabidopsis differentially amplify signaling for the innate immune response. Proc Natl Acad Sci USA 104:10732–10736

Huffaker A, Pearce G, Ryan CA (2006) An endogenous peptide signal in Arabidopsis activates components of the innate immune response. Proc Natl Acad Sci USA 103:10098–10103

Ihmels J, Bergmann S, Barkai N (2004) Defining transcription modules using large-scale gene expression data. Bioinformatics 20:1993–2003

Jansen R, Greenbaum D, Gerstein M (2002) Relating whole-genome expression data with protein-protein interactions. Genome Res 12:37–46

Jareborg N, Birney E, Durbin R (1999) Comparative analysis of noncoding regions of 77 orthologous mouse and human gene pairs. Genome Res 9:815–824

Kankainen M, Pehkonen P, Rosenstom P, Toronen P, Wong G, Holm L (2006) POXO: a web-enabled tool series to discover transcription factor binding sites. Nucleic Acids Res 34: W534–W540

Krol E, Mentzel T, Chinchilla D, Boller T, Felix G, Kemmerling B, Postel S, Arents M, Jeworutzki E, Al Rasheid KA, Becker D, Hedrich R (2010) Perception of the Arabidopsis danger signal peptide 1 involves the pattern recognition receptor AtPEPR1 and its close homologue AtPEPR2. J Biol Chem 285:13471–13479

Lease KA, Walker JC (2006) The Arabidopsis unannotated secreted peptide database, a resource for plant peptidomics. Plant Physiol 142:831–838

Lee J, Klusener B, Tsiamis G, Stevens C, Neyt C, Tampakaki AP, Panopoulos NJ, Noller J, Weiler EW, Cornelis GR, Mansfield JW, Nurnberger T (2001) HrpZ(Psph) from the plant pathogen Pseudomonas syringae pv. phaseolicola binds to lipid bilayers and forms an ion-conducting pore in vitro. Proc Natl Acad Sci USA 98:289–294

Lindsey K, Casson S, Chilley P (2002) Peptides: new signalling molecules in plants. Trends Plant Sci 7:78–83

Maere S, Heymans K, Kuiper M (2005) BiNGO: a Cytoscape plugin to assess overrepresentation of gene ontology categories in biological networks. Bioinformatics 21:3448–3449

Manfield IW, Jen CH, Pinney JW, Michalopoulos I, Bradford JR, Gilmartin PM, Westhead DR (2006) Arabidopsis co-expression tool (ACT): web server tools for microarray-based gene expression analysis. Nucleic Acids Res 34:W504–W509

Matsubayashi Y (2011) Post-translational modifications in secreted Peptide hormones in plants. Plant Cell Physiol 52:5–13

Matsubayashi Y, Sakagami Y (2006) Peptide hormones in plants. Annu Rev Plant Biol 57:649–674

Meier S, Bastian R, Donaldson L, Murray S, Bajic V, Gehring C (2008) Co-expression and promoter content analyses assign a role in biotic and abiotic stress responses to plant natriuretic peptides. BMC Plant Biol 8:24

Meier S, Ruzvidzo O, Morse M, Donaldson L, Kwezi L, Gehring C (2010) The Arabidopsis wall associated kinase-like 10 gene encodes a functional guanylyl cyclase and is co-expressed with pathogen defense related genes. PLoS One 5:e8904

Meier S, Tzfadia O, Vallabhaneni R, Gehring C, Wurtzel ET (2011) A transcriptional analysis of carotenoid, chlorophyll and plastidial isoprenoid biosynthesis genes during development and osmotic stress responses in Arabidopsis thaliana. BMC Syst Biol 5:77

Miya A, Albert P, Shinya T, Desaki Y, Ichimura K, Shirasu K, Narusaka Y, Kawakami N, Kaku H, Shibuya N (2007) CERK1, a LysM receptor kinase, is essential for chitin elicitor signaling in Arabidopsis. Proc Natl Acad Sci USA 104:19613–19618

Morita-Yamamuro C, Tsutsui T, Sato M, Yoshioka H, Tamaoki M, Ogawa D, Matsuura H, Yoshihara T, Ikeda A, Uyeda I, Yamaguchi J (2005) The Arabidopsis gene CAD1 controls programmed cell death in the plant immune system and encodes a protein containing a MACPF domain. Plant Cell Physiol 46:902–912

Nawrath C, Heck S, Parinthawong N, Metraux JP (2002) EDS5, an essential component of salicylic acid-dependent signaling for disease resistance in Arabidopsis, is a member of the MATE transporter family. Plant Cell 14:275–286

Nimchuk Z, Eulgem T, Holt BF III, Dangl JL (2003) Recognition and response in the plant immune system. Annu Rev Genet 37:579–609

O'Connor TR, Dyreson C, Wyrick JJ (2005) Athena: a resource for rapid visualization and systematic analysis of Arabidopsis promoter sequences. Bioinformatics 21:4411–4413

Obayashi T, Kinoshita K (2011) COXPRESdb: a database to compare gene coexpression in seven model animals. Nucleic Acids Res 39:D1016–D1022

Obayashi T, Hayashi S, Shibaoka M, Saeki M, Ohta H, Kinoshita K (2008) COXPRESdb: a database of coexpressed gene networks in mammals. Nucl Acids Res 36:D77–D82

Postel S, Kufner I, Beuter C, Mazzotta S, Schwedt A, Borlotti A, Halter T, Kemmerling B, Nurnberger T (2010) The multifunctional leucine-rich repeat receptor kinase BAK1 is implicated in Arabidopsis development and immunity. Eur J Cell Biol 89:169–174

Qiu JL, Zhou L, Yun BW, Nielsen HB, Fiil BK, Petersen K, Mackinlay J, Loake GJ, Mundy J, Morris PC (2008) Arabidopsis mitogen-activated protein kinase kinases MKK1 and MKK2 have overlapping functions in defense signaling mediated by MEKK1, MPK4, and MKS1. Plant Physiol 148:212–222

Ryan CA, Pearce G (2010) Peptide hormones for defense, growth, development and reproduction. In: Davies PJ (ed) Plant hormones. Springer, Netherlands, pp 700–716

Saeed AI, Sharov V, White J, Li J, Liang W, Bhagabati N, Braisted J, Klapa M, Currier T, Thiagarajan M, Sturn A, Snuffin M, Rezantsev A, Popov D, Ryltsov A, Kostukovich E, Borisovsky I, Liu Z, Vinsavich A, Trush V, Quackenbush J (2003) TM4: a free, open-source system for microarray data management and analysis. Biotechniques 34:374–378

Schuhegger R, Nafisi M, Mansourova M, Petersen BL, Olsen CE, Svatos A, Halkier BA, Glawischnig E (2006) CYP71B15 (PAD3) catalyzes the final step in camalexin biosynthesis. Plant Physiol 141:1248–1254

Schuhegger R, Rauhut T, Glawischnig E (2007) Regulatory variability of camalexin biosynthesis. J Plant Physiol 164:636–644

Srinivasasainagendra V, Page GP, Mehta T, Coulibaly I, Loraine AE (2008) CressExpress: a tool for large-scale mining of expression data from Arabidopsis. Plant Physiol 147:1004–1016

Steinhauser D, Usadel B, Luedemann A, Thimm O, Kopka J (2004) CSB.DB: a comprehensive systems-biology database. Bioinformatics 20:3647–3651

Stuart JM, Segal E, Koller D, Kim SK (2003) A gene-coexpression network for global discovery of conserved genetic modules. Science 302:249–255

Sweetlove LJ, Fernie AR (2005) Regulation of metabolic networks: understanding metabolic complexity in the systems biology era. New Phytol 168:9–24

Tagle DA, Koop BF, Goodman M, Slightom JL, Hess DL, Jones RT (1988) Embryonic epsilon and gamma globin genes of a prosimian primate (Galago crassicaudatus). Nucleotide and amino acid sequences, developmental regulation and phylogenetic footprints. J Mol Biol 203:439–455

Tamayo P, Slonim D, Mesirov J, Zhu Q, Kitareewan S, Dmitrovsky E, Lander ES, Golub TR (1999) Interpreting patterns of gene expression with self-organizing maps: methods and application to hematopoietic differentiation. Proc Natl Acad Sci USA 96:2907–2912

Tavazoie S, Hughes JD, Campbell MJ, Cho RJ, Church GM (1999) Systematic determination of genetic network architecture. Nat Genet 22:281–285

Tompa M, Li N, Bailey TL, Church GM, De Moor B, Eskin E, Favorov AV, Frith MC, Fu Y, Kent WJ, Makeev VJ, Mironov AA, Noble WS, Pavesi G, Pesole G, Regnier M, Simonis N, Sinha S, Thijs G, van Helden J, Vandenbogaert M, Weng Z, Workman C, Ye C, Zhu Z (2005) Assessing computational tools for the discovery of transcription factor binding sites. Nat Biotechnol 23:137–144

Truman W, Bennett MH, Kubigsteltig I, Turnbull C, Grant M (2007) Arabidopsis systemic immunity uses conserved defense signaling pathways and is mediated by jasmonates. Proc Natl Acad Sci USA 104:1075–1080

Tsutsui T, Morita-Yamamuro C, Asada Y, Minami E, Shibuya N, Ikeda A, Yamaguchi J (2006) Salicylic acid and a chitin elicitor both control expression of the CAD1 gene involved in the plant immunity of Arabidopsis. Biosci Biotechnol Biochem 70:2042–2048

Tuch BB, Li H, Johnson AD (2008) Evolution of eukaryotic transcription circuits. Science 319:1797–1799

Wasserman WW, Palumbo M, Thompson W, Fickett JW, Lawrence CE (2000) Human-mouse genome comparisons to locate regulatory sites. Nat Genet 26:225–228

Wei H, Persson S, Mehta T, Srinivasasainagendra V, Chen L, Page GP, Somerville C, Loraine A (2006) Transcriptional coordination of the metabolic network in Arabidopsis. Plant Physiol 142:762–774

Werner T (2001) The promoter connection. Nat Genet 29:105–106
Wheeler JI, Irving HR (2010) Evolutionary advantages of secreted peptide signalling molecules in plants. Funct Plant Biol 37:382–394
Wildermuth MC, Dewdney J, Wu G, Ausubel FM (2001) Isochorismate synthase is required to synthesize salicylic acid for plant defence. Nature 414:562–565
Yamaguchi Y, Huffaker A, Bryan AC, Tax FE, Ryan CA (2010) PEPR2 is a second receptor for the Pep1 and Pep2 peptides and contributes to defense responses in Arabidopsis. Plant Cell 22:508–522
Zhou X, Su Z (2007) EasyGO: gene ontology-based annotation and functional enrichment analysis tool for agronomical species. BMC Genomics 8:246
Zimmermann P, Hennig L, Gruissem W (2005) Gene-expression analysis and network discovery using Genevestigator. Trends Plant Sci 10:407–409

Index

H.R. Irving and C. Gehring (eds.), *Plant Signaling Peptides*,
Signaling and Communication in Plants 16, DOI 10.1007/978-3-642-27603-3,

P